DESIGNING WITH STRUCTURAL CERAMICS

Europhysics Industrial Workshop

organised by:

- Commission of the European Communities
  Institute for Advanced Materials
  JRC Petten, The Netherlands
- Netherlands Energy Research Foundation (ECN)
  Petten, The Netherlands

held at:

Petten, The Netherlands, 3–6 April 1990

Organising Committee

G. Thomas	European Physical Society, Geneva, Switzerland
M. H. Van de Voorde	CEC, JRC Petten, The Netherlands
H. J. Veringa	ECN Petten, The Netherlands

DESIGNING WITH STRUCTURAL CERAMICS

Edited by

R. W. DAVIDGE

Consultant to the Commission of the European Communities,
2 Locks Lane, Wantage, Oxon OX12 9DB, UK

and

M. H. VAN DE VOORDE

CEC Joint Research Centre, Institute for Advanced Materials, Petten,
The Netherlands

ELSEVIER APPLIED SCIENCE
LONDON and NEW YORK

ELSEVIER SCIENCE PUBLISHERS LTD
Crown House, Linton Road, Barking, Essex IG11 8JU, England

Sole Distributor in the USA and Canada
ELSEVIER SCIENCE PUBLISHING CO., INC.
655 Avenue of the Americas, New York, NY 10010, USA

WITH 35 TABLES AND 227 ILLUSTRATIONS

British Library Cataloguing in Publication Data

Designing with structural ceramics.
I. Davidge, R. W. II. Van de Voorde, M. H.
691.5

ISBN 1-85166-740-7

Library of Congress CIP data applied for

Publication arrangements by Commission of the European Communities, Directorate-General Telecommunications, Information Industries and Innovation, Scientific and Technical Communication Unit, Luxembourg

EUR 14057EN

PREFACE

The last 30 years have seen a steady development in the range of ceramic materials with potential for high temperature engineering applications: in the 60s, self-bonded silicon carbide and reaction-bonded silicon nitride; in the 70s, improved aluminas, sintered silicon carbide and silicon nitrides (including sialons); in the 80s, various toughened ZrO_2 materials, ceramic matrix composites reinforced with silicon carbide continuous fibres or whiskers. Design methodologies were evolved in the 70s, incorporating the principles of fracture mechanics and the statistical variation and time dependence of strength. These have been used successfully to predict the engineering behaviour of ceramics in the lower range of temperature.

In spite of the above, and the underlying thermodynamic arguments for operations at higher temperatures, there has been a disappointing uptake of these materials in industry for high temperature use. Most of the successful applications are for low to moderate temperatures such as seals and bearings, and metal cutting and shaping. The reasons have been very well documented and include:

- Poor predictability and reliability at high temperature.
- High costs relative to competing materials.
- Variable reproducibility of manufacturing processes.
- Lack of sufficiently sensitive non-destructive techniques.

With this as background, a Europhysics Industrial Workshop sponsored by the European Physical Society (EPS) was organised by the Netherlands Energy Research Foundation (ECN) and the Institute for Advanced Materials of the Joint Research Centre (JRC) of the EC, at Petten, North Holland, in April 1990 to consider the status of thermomechanical applications of engineering ceramics. About 50 experts from industry and research institutes participated. These proceedings present in written form the lectures given by various EC experts, and have been arranged under four headings: Engineering Properties; Ceramic Matrix Composites; Technological Aspects; and Industrial Applications. A final section presents some current research topics based on a poster session.

It is for the readers to consider the implications of these expert views relevant to their personal interests and affiliations, but some of the major conclusions are clear:

- An increasing range of materials is available with potential for high temperature use. However, there remain needs for improved materials to operate in stringent environments.
- The application of fracture mechanics has made considerable advances in

understanding of properties and performance, but there are severe gaps in knowledge for high temperature conditions (creep regime) for monolithic ceramics, and most topics concerning composites.

- Enabling technologies such as machining, joining, NDT and fabrication procedures have made significant recent progress.
- Engineering applications have been established for the high-tech aerospace/military industry, where material costs are not paramount, but for a broadening of use, e.g. automobiles and energy conversion, there is still market resistance and a medium-term view is essential.
- A ceramic materials data base is lacking so that much design is on an ad hoc basis and there is a hindrance to the expansion of markets.

Grateful thanks are due to Professor J. A. Goedkoop, who initiated the idea of the EPS Symposium at Petten; to the members of ECN, JRC and EPS for provision of services, which led to the success of the Symposium, particularly Mr M. Merz (JRC) and Mrs C. A. L. Ruitenburg (ECN); to all the authors for collaboration in preparing their papers; and finally to Anita Harvey for her careful preparation of the camera-ready manuscript.

R. W. DAVIDGE
M. H. VAN DE VOORDE

CONTENTS

Technological Aspects

Industrial Applications

Current Research Activities

PERSPECTIVES OF STRUCTURAL CERAMICS AND PRESENT R&D EFFORTS

M.H. VAN DE VOORDE
Commission of the European Communities,
Institute for Advanced Materials, POB2 1755ZG Petten,
The Netherlands.

ABSTRACT

As an introduction to this workshop, general information is presented concerning: potential structural ceramics; the materials data base for mechanical and engineering properties, and chemical and thermophysical behaviour; and test methodology. Comment is made on growth rates for structural ceramics, and on various international programmes.

INTRODUCTION

In modern technologies metallic materials often operate at the limit of their capabilities particularly in high temperature applications. For the next generation of high temperature materials metals were not considered as candidates. For example in aerospace technology, metals were unsuitable for the hot structure of the Hermes vehicle due to their poor behaviour at high temperatures, their high density and problems caused by their high coefficient of thermal expansion. A real breakthrough in the search for high temperature materials is the development of ceramic composites, such as silicon carbide fibres in a silicon carbide matrix or carbon fibres in a silicon carbide matrix. The inherent high temperature limits of metals are also handicapping other major technologies as for example modern energy plants where efficiency is determined by the availability of new high temperature materials for gas and steam turbines, heat exchanges, etc., (Figure 1). New materials are also needed in the automotive, petrochemical and aeronautical industries (Figure 2).

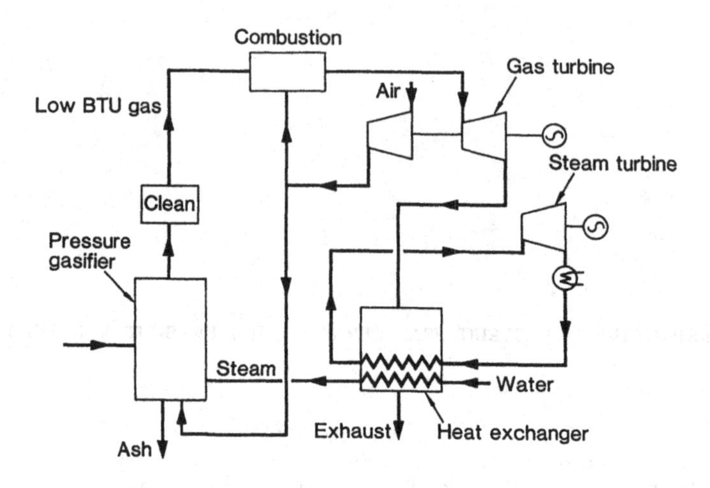

Figure 1. Total gasification cycle for power generation.

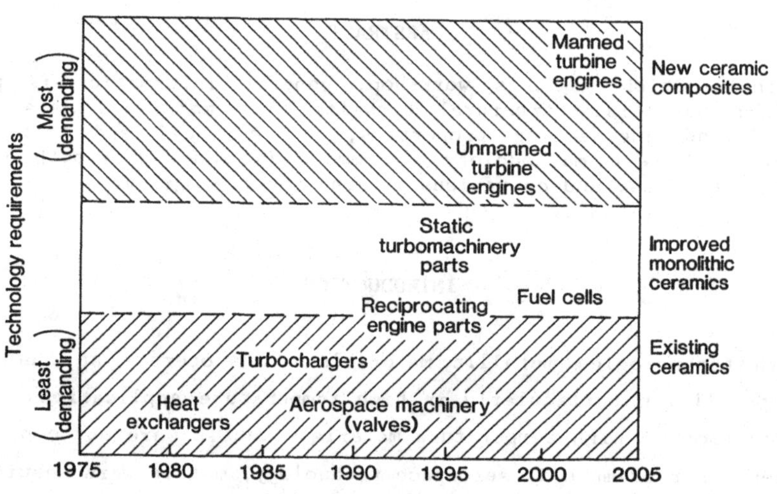

Figure 2. Advanced ceramics are required to meet many new performance goals.

Ceramics are among mankind's oldest technological materials, as revealed by the many archaeological finds of knapped flints and pottery shards. They are, however, also among mankind's newest technological materials that expand their uses from traditional applications such as high temperature containment of molten metals and thermal or electrical insulators, to become crucial structural materials in advanced engineering projects.

The main applications of mass-produced advanced ceramics are currently in electronic and electro-engineering industries (Figure 3). Structural ceramics which are used as components in heat engines, in chemical equipment, in aerospace/defence related applications, etc., only represent about 20% of the entire field of advanced ceramics.

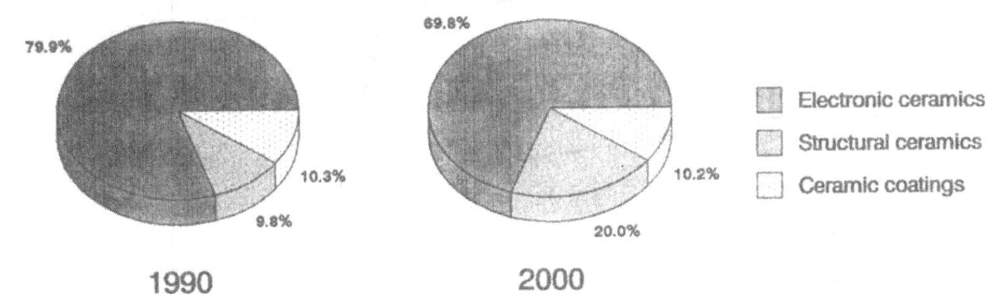

Figure 3. The market for the US advanced ceramics industry by percentage, 1990 and 2000. (Business Communications Co. Inc. (BCC)).

During the last decade, important progress has been made to improve ceramic processing technology and new 'high quality' structural ceramics have been developed. The ceramic materials in modern plants have to operate under stringent conditions of mechanical, thermal and corrosion constraints. These demands on material performance together with the limited available knowledge in designing with brittle ceramics, are most probably the reasons for the slow introduction of structural ceramics in advanced technology.

The purpose of this workshop is to publicise the available ceramics database and to familiarize the European industrial community with modern design principles for structural ceramics.

POTENTIAL STRUCTURAL CERAMICS

Monolithic ceramics, either oxides or non-oxides, are commonly used in engineering designs. The oxides are mainly based on Al_2O_3 and ZrO_2 while the non-oxides are SiC and Si_3N_4 (Figure 4 and Table 1). These materials should be considered alloys and the additives are needed for processing or property optimization. Their chemical composition and microstructure strongly determine the mechanical properties (Figure 5). Developments during the last decade have resulted in the processing of 'high quality' ceramic composites:

- Transformation toughened ZrO_2;
- (SiC) Whisker-(platelet) reinforced ceramics (Al_2O_3);
- Fibre reinforced materials, particularly C-C, C-SiC and SiC-SiC.

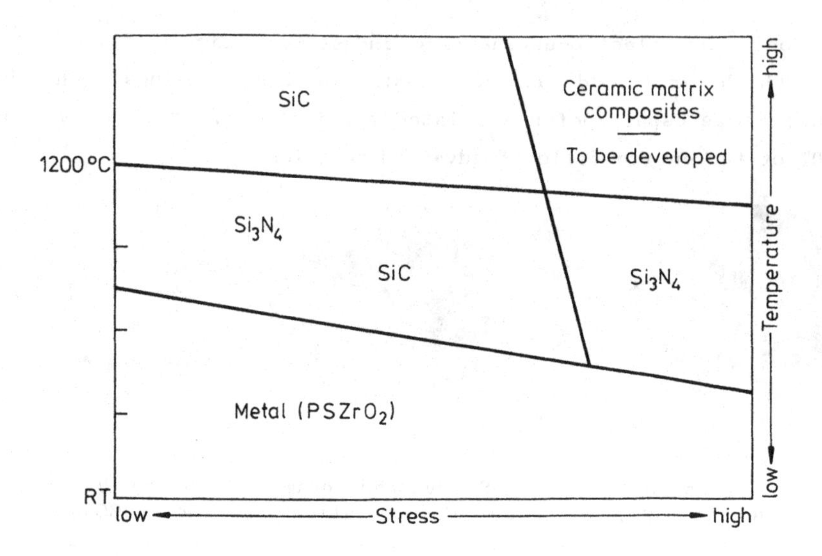

Figure 4. Application map for use of materials.

TABLE 1
Mechanical properties of various engineering ceramics

Main Composition	Al_2O_3	ZrO_2	Si_3N_4 SSN	Si_3N_4 RBSN	SiC
Bending strength (MPa)					
R.T.	440	1020	880	296	500
1000°C	340	450 (800°C)	510	300	475
Fracture toughness K_{1c} (MPa√m)	4.5	8.5	7.0	3.6	2.4
Thermal shock resistance (into H_2O ΔT°C)	200	350	900	600	370

These composites are characterized by a large decrease in strength and fracture toughness compared with monolithics. The continuous fibre reinforced ceramics have the highest potential because of their increased toughness (≈ 15 MPa√m for SiC-SiC), high temperature capability and the non-catastrophic fracture behaviour (Figures 6 and 7).

MATERIALS DATABASE

In modern installations, structural ceramics are exposed to high mechanical and high thermal loading, under corrosive and erosive environments. The

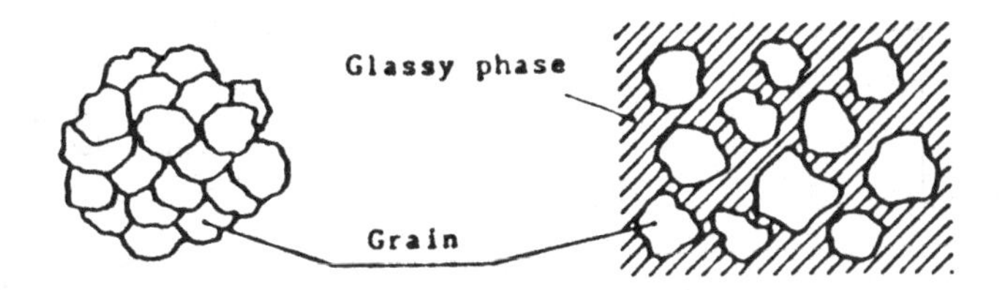

Almost no glassy phase at grain boundaries
(SiC with B and C additives
reaction sintered SiC in Si_3N_4
Si_3N_4 with $BeAl_2O_4$ additive).
* Little degradation of strength at high temperature.
* Low fracture toughness.

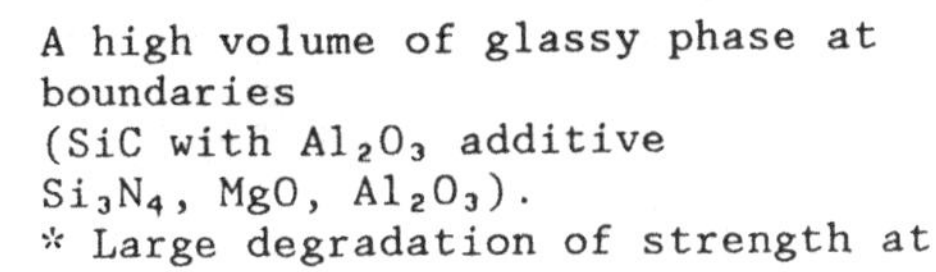

A high volume of glassy phase at boundaries
(SiC with Al_2O_3 additive
Si_3N_4, MgO, Al_2O_3).
* Large degradation of strength at high temperature.
* High fracture toughness.

Small aspect ratio of grain.
* Large deformation at high temperature.
  → Superplasticity.

Large aspect ratio of grain.
* High resistance to creep.
* High fracture toughness.
  → Whisker reinforced ceramics?

Figure 5. Effect of microstructure on mechanical properties.

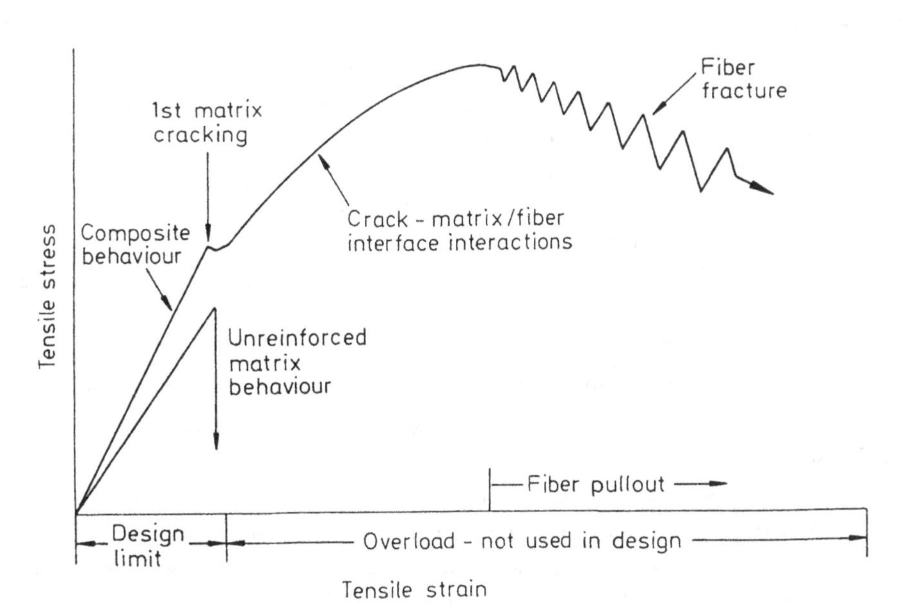

Figure 6. A typical stress-strain curve for a fibre-reinforced composite.

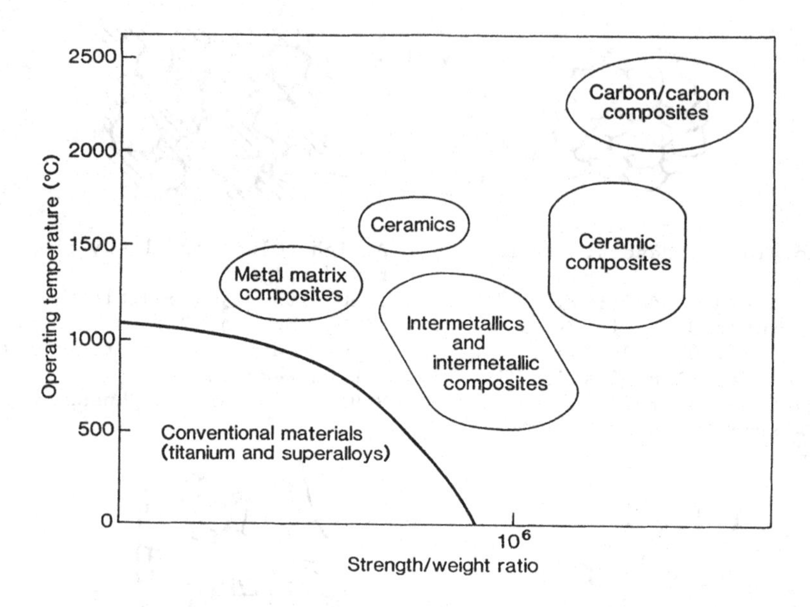

Figure 7. Strength of structural materials.

mechanical stresses often complex, are dynamic in nature; therefore they change with time and temperature. To assure reliability and for reasons of lifetime considerations the following data for materials are required:

Mechanical Properties

- Short term behaviour for failure probability purposes:
  - Strength distribution; bending and tensile tests.
  - Fracture toughness (K_{IC}).
  - Thermal shock resistance.
- Long term degradation for lifetime prediction:
  - Creep behaviour.
  - Fatigue and sub-critical crack growth (static, cyclic and thermal).
- Erosion behaviour.

The designer needs very complex property data, demanding unique test facilities for which there is a shortage in the international community. Typical examples are:

- Highly advanced closed-loop testing machines and biaxial tension-torsion machines, equipped to operate under controlled (C-O-S) bearing gases to study the simultaneous interaction between mechanical loading and corrosive exposure at high temperatures (up to 1600°C), simulating industrial operating conditions (Figure 8).

Figure 8. View of half of the split furnace in cyclic fatigue testing of ceramics.

- For the study of the damaging and life-limiting mechanisms, a computer vision system is necessary for in-situ monitoring of surface damage development on specimens subjected to thermo-mechanical loading. This equipment (Figure 9) can yield important information on the initiation and growth of damage, in addition to the results from post-mortem fractographic investigations.
- Figure 10 shows the lifetime for a hot pressed silicon nitride at 1200°C under uniaxial push-pull cyclic fatigue condition. The cyclic fatigue life is insensitive to frequency for frequencies exceeding 10 Hz. It is also apparent that the cyclic life at a given stress level far exceeds the static (creep) life, indicating a beneficial effect due to cyclic loading.
- The loading under service conditions is seldom of a pure bending nature and consequently the conventional four-point bending tests are often of limited value for the designer. Thus correlation of bending to tensile data and tensile tests are necessary at high temperatures. In addition multi-axial test data are often a better representation of the real stress situation in components and simple test data give too optimistic results.

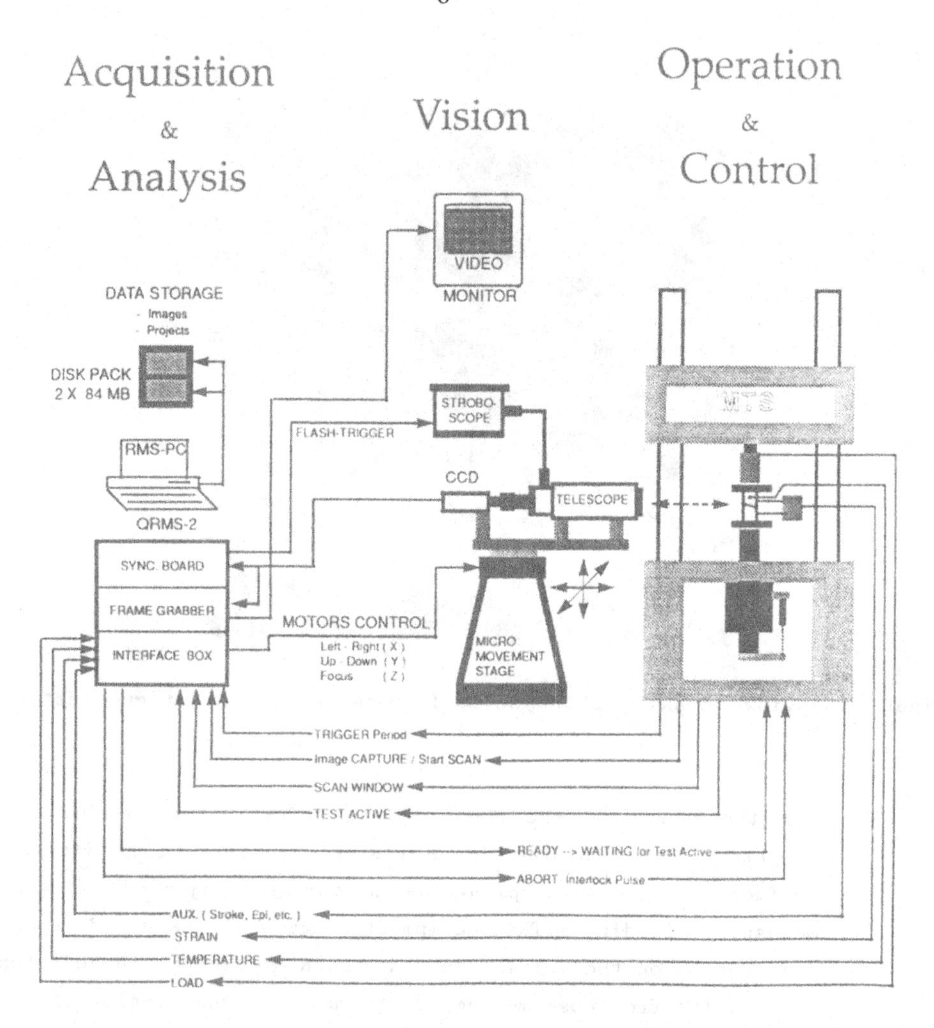

Figure 9. Configuration of computer vision/system and mechanical testing machine.

Chemical Properties

- Gaseous corrosion: oxidation, carburization, sulfidation, etc.
- Hot corrosion: molten salts.

Figure 11 shows a unique high temperature (< 1600°C) facility for testing corrosion in simulated industrial environments. Figure 12 summarizes test data on gaseous corrosion of silicon nitrides, up to 1400°C and pinpoints corrosion effects. Experiments have confirmed the poor oxidation behaviour of C-C, C-SiC ceramic composites at temperatures above 1200°C in air.

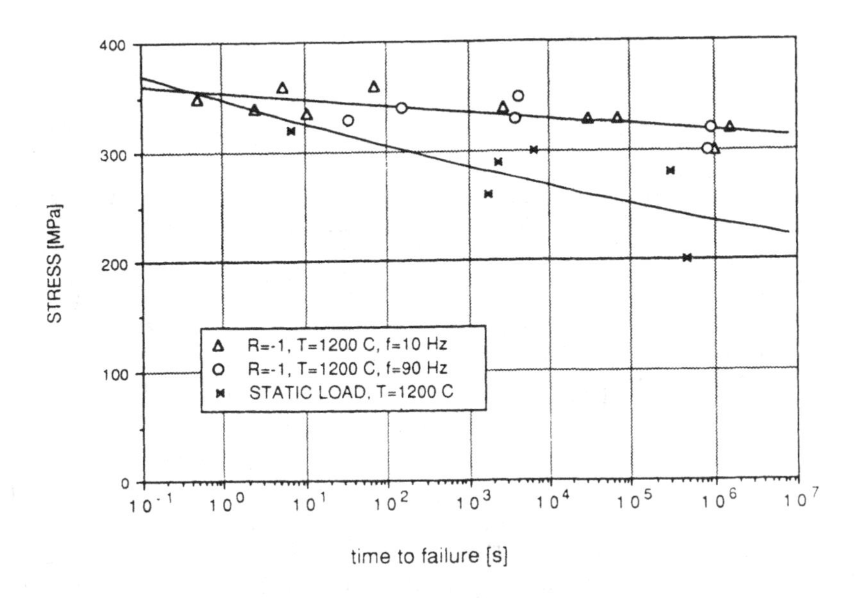

Figure 10. Lifetime for hot pressed silicon nitride at 1200°C under push-pull cyclic fatigue conditions.

Thermo Physical Properties

- Thermal conductivity, of capital importance for heat exchanger design and thermal barrier coating developments.
- Thermal expansion plays a vital role in ceramic/metal joining techniques.

ENGINEERING PROPERTIES

- Ceramic Joining.

Fabrication processes must be able to satisfy user requirements which are generally defined in terms of maintainable strengths and sometimes hermeticity, Table 2. Ceramics can be bonded using a wide spectrum of techniques ranging from fusion welding to mechanical attachment (Figure 13), each of which has advantages and disadvantages, Tables 3 and 4. The most applicable techniques in practice for high temperature ceramics such as silicon nitride are glazing, brazing and diffusion bonding (Figure 14).

- Residual Stresses.

Residual stresses are produced during the manufacturing processes. In ceramic composites the difference in the thermal expansion of the whiskers or fibres and the matrix gives rise to internal stresses. Residual stresses

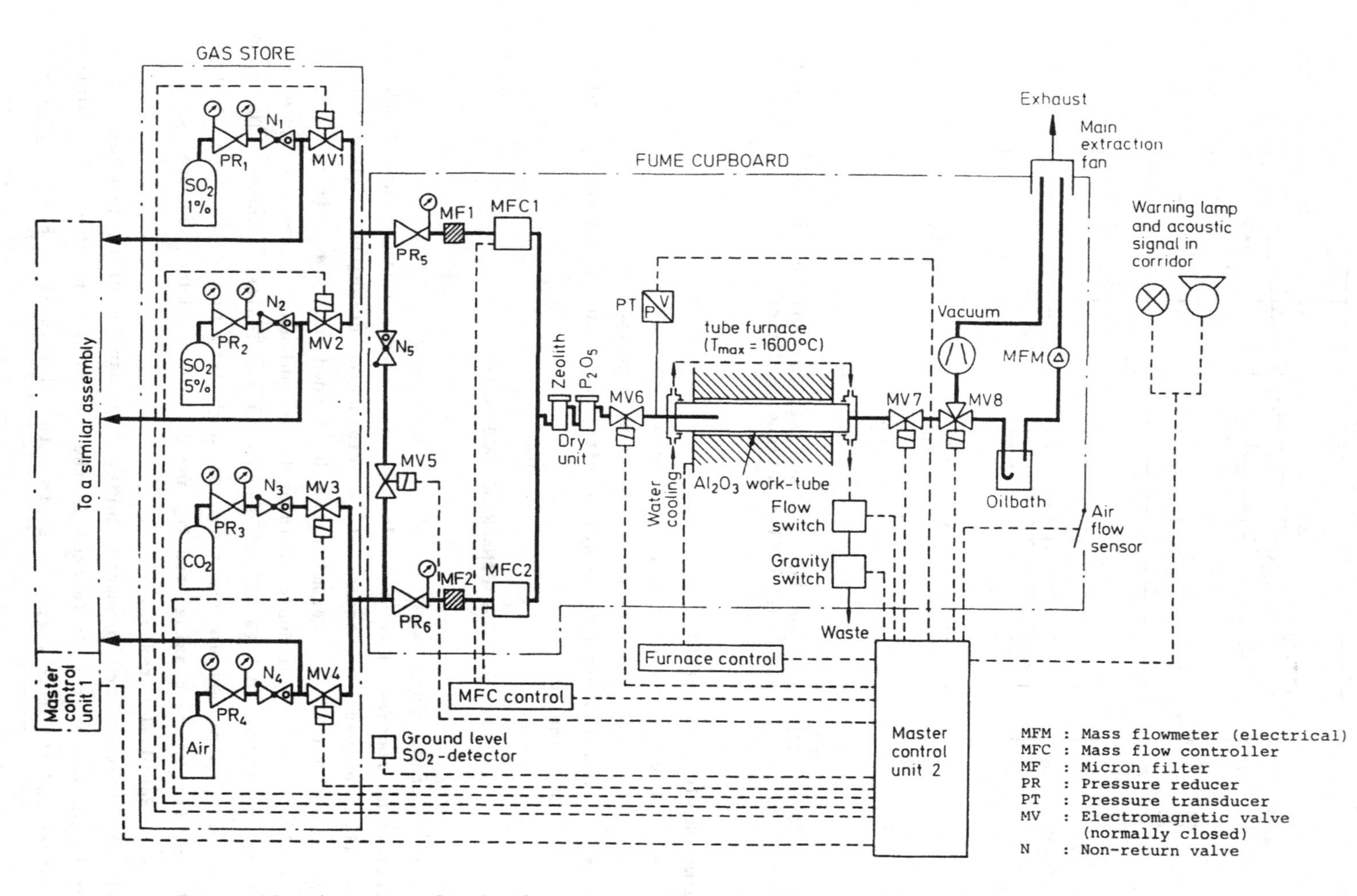

MFM : Mass flowmeter (electrical)
MFC : Mass flow controller
MF : Micron filter
PR : Pressure reducer
PT : Pressure transducer
MV : Electromagnetic valve (normally closed)
N : Non-return valve

Figure 11. Apparatus for isothermal corrosion of silicon nitride in SO_2/air atmosphere.

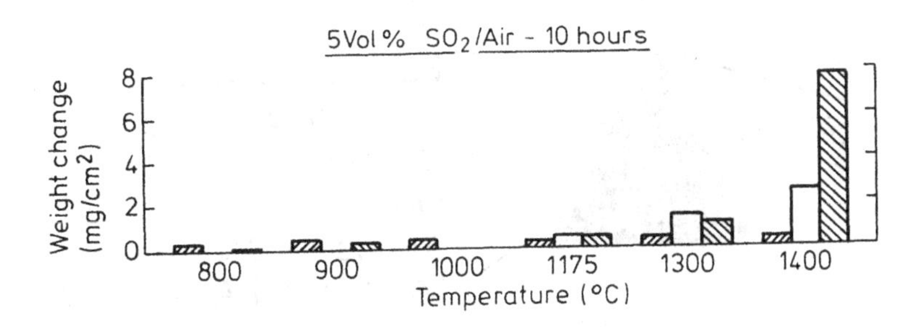

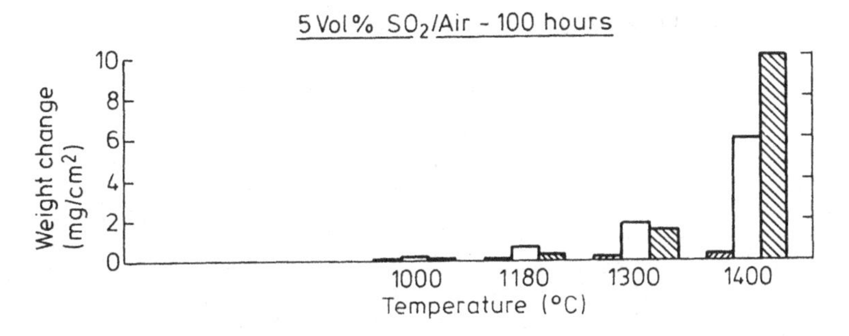

SN1 (Hot-Pressed: Y_2O_3)

SN2 (Hot-Pressed: MgO)

SN3 (Pressureless-Sintered: Y_2O_3, MgO)

Figure 12. Gaseous corrosion of three different silicon nitrides.

TABLE 2
Common requirements for ceramic-metal joints

High and reproducible	- room and high temperature strength
	- fracture resistance
	- thermal shock resistance
	- fatigue resistance
Hermeticity	
High electrical resistance	
High or low thermal conductivity	
High corrosion resistance	
Microstructural stability	

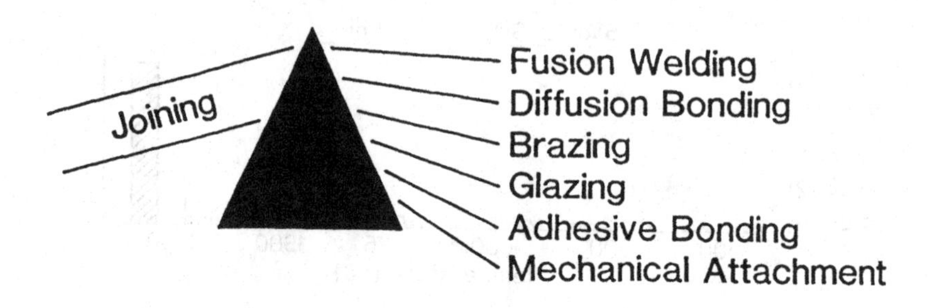

Figure 13. The spectrum of joining processes.

TABLE 3
Direct joining techniques: some advantages and disadvantages

Technique	Advantages	Disadvantages
Fusion welding	Flexible process. Refractory joints. Welding widely accepted.	Melting point must be similar. Expansion coefficients must be similar. Very high localised fabrication temperature. Laboratory demonstration only.
Diffusion bonding	Refractory joints. Minimised corrosion problems.	Special equipment needed. Expansion coefficients should match. Long fabrication times or high fabrication temperatures or both. Limited application so far.

may also result from machining and joining operations.

These stresses can reach high values and could affect the strength and lifetime of the component subjected to external stresses. X-ray analysis techniques allow an accurate description of induced stresses even allowing determination of stress profiles in sub-surface damaged zones. The techniques are as yet in their infancy and need considerable modelling and verification testing.

- Non Destructive Testing.

Quality control of engineering ceramics is an essential element of the design philosophy. Table 5 gives an overview of available techniques.

TABLE 4
Indirect joining techniques: some advantages and disadvantages

Technique	Advantages	Disadvantages
Diffusion bonding using metal interlayers	Joints have some compliance. Does not require workpiece deformation. Modest fabrication temperature. Has found significant application.	Primarily butt joints. Use temperature limited by interlayers. Special equipment needed.
Brazing metallised ceramics	Uses conventional ductile brazes. Uses widely available fabrication equipment. Butt and sleeve joints. Compliant joints.	Metallisation is costly extra step. Fully commercialised only for Al_2O_3.
Brazing using active metal alloys	Does not require metallisation. Butt and sleeve joints. Applicable to non-oxide and oxide ceramics. Being applied in advanced projects.	Requires vacuum equipment. Alloys are brittle. Limited available alloy range.
Glazing	Can fabricate in air. Butt and sleeve joints. Very widely used.	Joints are brittle. Expansion coefficients must match.
Adhesive bonding	Fabricate in air at low temperature. Expansion coefficients relatively unimportant. Butt and sleeve joints. No special equipment needed.	Temperature capability poor. Relatively weak joints. Joints can degrade in moist air.

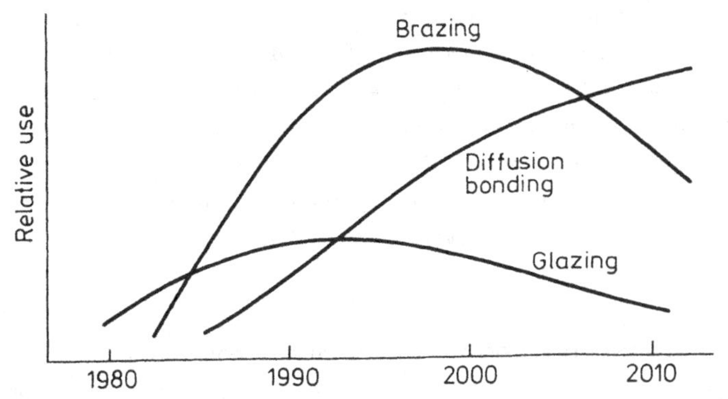

Figure 14. A scenario for the development of joining technologies for advanced engineering ceramics.

TABLE 5
Non-destructive techniques for ceramics

Technique	Efficiency		Shape of Testing Sample
Ultrasonic Test	100 μm	Internal flaw	Simple
Scanning Acoustic Microscopy (SAM)	4 μm	Surface flaw	Simple
Scanning Laser Acoustic Microscopy (SLAM)	10 μm	Internal flaw	Simple
Ultrasonic Wave Computer Tomography	>100 μm	Internal flaw	Complex
Acoustic Emission	Small	Moving crack	Complex
X-ray Photography	1000 μm	Internal flaw	Simple
X-ray Computer Tomography	100 μm	Internal flaw	Complex
Photoacoustic Spectroscopy (PAS)	50 μm	Surface flaw	Simple

TEST METHODOLOGY

Mechanical Properties

At present, the European Normalization Organization is conducting a pan European exercise to develop standards for testing techniques. Aspects requiring quantification include:

- The effect of geometry (shape, volume) of specimen on strength, lifetime, etc.
- The relationship between measured parameters, such as tensile and bending strength.
- Factors which affect the strength such as slow crack growth.

There is a need to collect more data and to have a better insight into fracture mechanisms before we can specify a standard test method for the strength of ceramics. However, standards for test techniques are urgently required to allow comparison of data for different materials and between different laboratories. An analysis of the measuring techniques worldwide has been made and the results summarized in Tables 6-10.

Hot Corrosion

Most aggressive industrial atmospheres contain contaminant vapours of salts and oxides which cause accelerated attack known as 'hot corrosion'. Current and foreseeable applications for which hot corrosion is critical are shown in Table 11.

TABLE 6
Comparison of strength test methods

	Bending Test	Tensile Test
Preparation of Specimen	Easy, simple shape	Difficult, complex shape
Testing	Easy	Difficult alignment
Stress	Not uniform	Uniform
Effective Volume	Small	Large

The bending test is easy, but its effective volume is small. Therefore, this method is suitable for checking the sintering process, but it is difficult to predict the strength of a component in which the effective volume may be much larger than the effective volume of the bending specimen. The tensile test is difficult but its effective volume is large. Therefore, the distribution of tensile strength obtained is useful for competent design.

TABLE 7
Measurement of fracture toughness

Technique	Advantages	Disadvantages
Indentation Fracture (IF) (Indentation Microfracture (IM)	Easy to measure. Small specimen.	Many different theoretical formulae. Difficult to measure crack length accurately.
Indentation Strength (IS) (Indentation Strength-in-Bending (ISB)	Easy to measure. Measurement of crack length unnecessary.	Empirical formula.
Controlled Surface Flaw (CSF) Controlled Microfracture (CM) Indentation-Induced Flaw (IIF)	Easy to introduce sharp crack.	Difficult to measure crack length and shape. Effect of residual stress by indentation.
Single Edge Notched Beam (SENB)	Easy to introduce the notch.	Notch width dependency.
Single Edge Pre-Cracked Beam (SEPB)	Sharp crack.	Difficult to measure crack length. Effect of residual stress if initial crack is introduced by indentation.
Chevron Notched Beam (CNB)	Sharp crack.	Difficult to introduce chevron notch.

TABLE 8
Cyclic fatigue test methods

Testing Method	Stress	Preparation of Specimen	Testing
Bending 3- or 4-point	Not uniform. Tension-tension.	Easy.	Easy.
Cantilever	Not uniform. Tension-compression or tension-tension.	Easy.	Easy.
Direct push-pull (Tensile)	Uniform. Tension-compression or tension-tension.	Difficult (usually cylindrical specimen).	Difficult alignment.
Rotational Bending	Not uniform. Tension-compression.	Difficult (cylindrical specimen).	

TABLE 9
Measurement of K_I-V diagram

Method	Advantages	Disadvantages
Double Cantilever Beam (DCB)	$V = 10^{-10}$ to 10^{-4} m/sec.	Difficult to measure crack length.
Indentation - Induced Flaw (IIF)	$V = 10^{-10}$ to 10^{-4} m/sec.	Difficult to measure crack length.
Double Torsion (DT)	Measurement of crack length unnecessary.	Difficult to discern crack tip (usually V obtained is larger than that from DCB).
Chevron Notched Beam (CNB)	Easy (bending test). Easy to get stable fracture. Measurement of crack length unnecessary.	$V \geq 10^{-6}$ m/sec.
Dynamic Fatigue	Easy, e.g. bending test. Measurement of crack length unnecessary.	Need to assume that $V = C.K_I^n$.

TABLE 10
Hardness measurement methods for ceramics

Measuring Method	Advantages	Disadvantages
Vickers	K_{IC} value can be measured.	The geometry of indentation has to be measured which causes a source of error.
Micro Vickers Hardness	Useful for the measurement of change of mechanical properties in small region.	Load dependence of hardness which causes a source of error.
Knoop Hardness		Load dependence of hardness which causes a source of error.
Rockwell Hardness	Unnecessary to measure the geometry of indentation.	Physical meaning of this value is different from that of Vickers hardness. Low accuracy.

TABLE 11
Current and future applications of advanced ceramics where hot corrosion is critical

Current	-	Burner nozzles and furnace fixtures.
│	-	Diesel engine components.
│	-	Heat exchangers for industrial furnaces. (especially recuperators for glass and aluminium melters).
│	-	Gas turbine components.
↓	-	Coal gasifiers.
Future	-	Waste burners.

Test for hot corrosion include:

- Crucible tests (immersion in liquid salt) give a poor simulation of real conditions since the test gas is largely excluded.
- Burner rig tests give the most reliable and most realistic simulation of real conditions but are mostly dedicated to one particular application, and different burner rigs tend to produce widely differing results.
- Salt evaporation tests have the advantage that contaminant is supplied continuously, but it is difficult to control the rate of deposition.

- In salt coating tests a thin layer of salt is applied to the specimen before exposure to the test at high temperature. This is the simplest and most reproducible test and therefore the obvious choice for a simple ranking test for hot corrosion. The salt applied can be chosen to match the contaminant in a particular application, but sodium is the most common contaminant, and is recommended for pre-standard research.

GROWTH PREDICTION FOR STRUCTURAL CERAMICS

The advanced ceramics industry, comprised of structural ceramics, electronic ceramics and ceramic coatings, seems to progress dynamically. In terms of growth, structural ceramics will lead the advanced ceramic components industry in the US with 18% average annual growth from 1990 to 2000 (Figure 3). Representing a 250 MEcu market in 1990, structural ceramics are predicted to account for 1400 MEcu by 2000. The growth rates foreseen in Europe and Japan in the next decade are much lower; 5% is an optimistic figure for Europe.

Major market sectors include high temperature components, burner nozzles, heat exchangers, wear parts, bearings, cutting tools, engine and gas turbine parts.

INTERNATIONAL R&D EFFORT

Japan is most aggressive in the development of structural ceramics (Figures 15 and 16). The R&D activities are mainly sponsored by MITI and executed under the auspices of JFCA (Japan Fine Ceramics Association). The work is jointly carried out by Industry and Research Laboratories. The JFCC (Japan Fine Ceramics Centre) at Nagoya co-ordinates the ceramics standardization activities.

The structural ceramics R&D in the US was sponsored by the Departments of Defense and Energy and NASA. Much experience has been gathered and industries are motivated to introduce ceramics in high technology applications. The producers and users are prepared to take risks; their technology is strongly supported by the research executed in recognized universities and materials research centres.

The situation in Europe is less flourishing notwithstanding the fact that all EC-nations have now a ceramic materials programme. The German

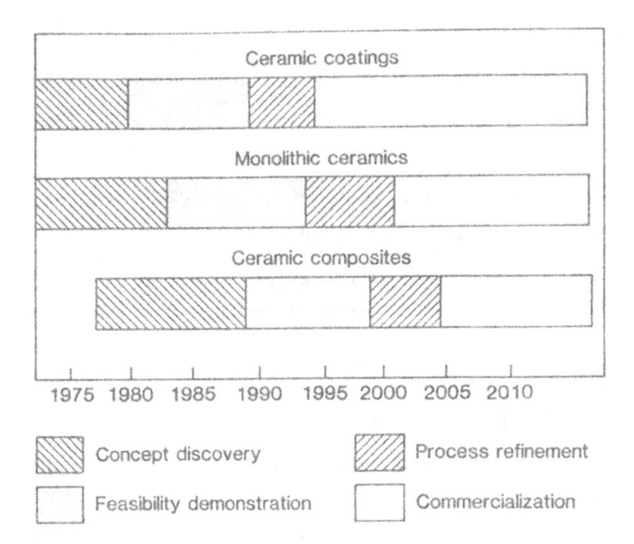

Figure 15. Advanced ceramics for engines - commercial status.

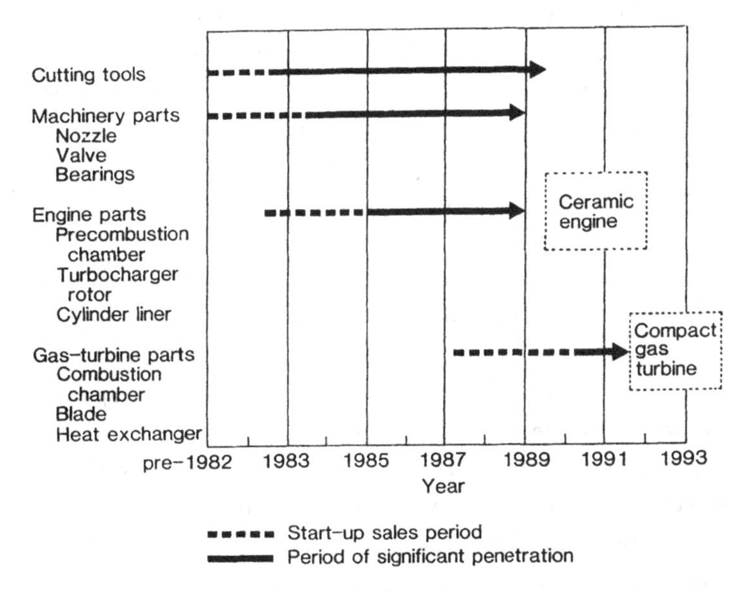

Figure 16. Estimated time of arrival for structural ceramic components.

Government is still devoting significant effort and many industries and research institutes/universities are participating actively. France is focusing attention in the development of ceramic composites. The training in ceramics engineering in Europe has been improved; many metallurgy departments at universities in Europe have been expanded into materials science departments including engineering ceramics. The chemical industry in Europe is developing high quality powders, the aeronautics and aero-engine companies are gradually introducing ceramic components while the automotive industry takes a more passive position.

The European activities are focused around CEC Programmes BRITE, EURAM and JOULE. EUREKA has a few projects on developing ceramics for aero-engine and industrial applications. The Joint Research Committee - Automotive Industry - has also sponsored ceramic projects for diesel engines and for exhaust systems.

VAMAS organises ceramics projects in the field of ceramics standardization. Similar standardization work is carried out in the framework of the IEA (International Energy Agency) and the European Normalization Organization (CEN).

The Institute for Advanced Materials of the European Communities studies the high temperature mechanical and corrosion properties of engineering ceramics as well as ceramic joining, machining and nondestructive testing. The R&D is directed towards engineering applications: aero-engines, aerospace, advanced energy applications, petrochemistry, automobile industry, etc. The Institute promotes European actions on engineering ceramics standardization.

CONCLUSIONS

- High quality engineering ceramics are increasingly becoming available and designers and industrial users have obtained confidence in the use of structural ceramics.
- Designing with engineering ceramics requires preparation of a comprehensive data base and a new design philosophy.
- The progress in high temperature technology is strongly determined by the availability of modern monolithic ceramics and ceramic composites.
- Europe experiences a shortage of high temperature test capabilities; facilities for studying the behaviour of materials in simulated industrial environments are practically non-existent.
- Difficulties arise in comparing test data from various laboratories. Internationally adopted test methodologies are needed.
- Structural ceramics are the materials for the next engineering generation.

MECHANICAL PROPERTIES OF CERAMICS

G. De PORTU and G.N. BABINI
CNR - Research Institute for Ceramics Technology,
via Granarolo 64, 48018 Faenza,
Italy.

ABSTRACT

To evaluate the interest of the scientific community in assessing the mechanical properties of ceramics, a statistical review of the literature has been prepared. Starting from the analysis of the results of such an investigation, the fundamental ideas which contribute to define the mechanical behaviour of ceramics are discussed.

The influence of different flaws, microstructure, environment and time on the failure behaviour are briefly presented. The statistical analysis and the influence of the fabrication processes on the properties and reliability are mentioned. Several mechanisms for toughening ceramic materials are discussed.

INTRODUCTION

In 1981 at the Science of Ceramics Meeting 10 Davidge (1) presented a lecture in which he examined historically the background to current understanding in mechanical properties of ceramics and tried to indicate the possible development in the near future. Ten years later we try to verify how realistic these perspectives were, evaluating the progress made and what problems are still unresolved. We considered first the publications reviewed by the American Ceramic Society in the last decade and related these to the main areas of development of structural ceramics. After that the basic ideas and the parameters which contribute to define the mechanical properties of ceramics were analyzed.

STATISTICAL ANALYSIS OF LITERATURE

To determine the incidence of literature papers on mechanical properties of structural ceramics in the last decade we used as information source the Ceramic Abstracts compiled by the American Ceramic Society. Three sections were analyzed which were considered the most representative for this topic: Instruments and Test Methods, Engineering Materials, Chemistry and Physics.

First we assessed the percentage of papers containing some typical key words for mechanical properties (e.g. strength, fracture toughness, fracture mechanics, etc.) in the total number of articles included in the three sections.

Among the papers on mechanical properties different topics were selected (e.g. zirconia based materials, composites, reliability, proof testing) to verify the interest of the scientific community in these subjects.

The results of this investigation are summarized in Figure 1. From the data it can be deduced that the attention of scientists and engineers on the mechanical behaviour of ceramics has been reasonably high and almost constant in the period considered, with some evidence of a recent increase of interest. Within the papers that directly include mechanical properties, the activity on new tough material became dominant especially from 1985. It may be that interest in mechanical properties and some specific topics is greater than that evidenced in Figure 1 as demonstrated by the numerous conferences and books devoted to these subjects. However the data seem sufficiently representative to give a reliable picture of the situation.

A first conclusion that can be made is that the study of the mechanical properties of ceramics attracts a notable part (> 10%) of the attention of the scientific community. In particular the studies on new, flaw tolerant materials are of great interest. Some lack of interest is detected regarding a direct approach to typical engineering problems such as proof testing, SPT diagrams and reliability. Nevertheless if we want to substitute metals with ceramics for applications, the designer must appreciate the difference between ceramics and metals and the behaviour of ceramics in terms of engineering parameters and lifetime. To assess the mechanical properties and the structure reliability, procedures similar to those reported (2) in Figure 2 are generally followed.

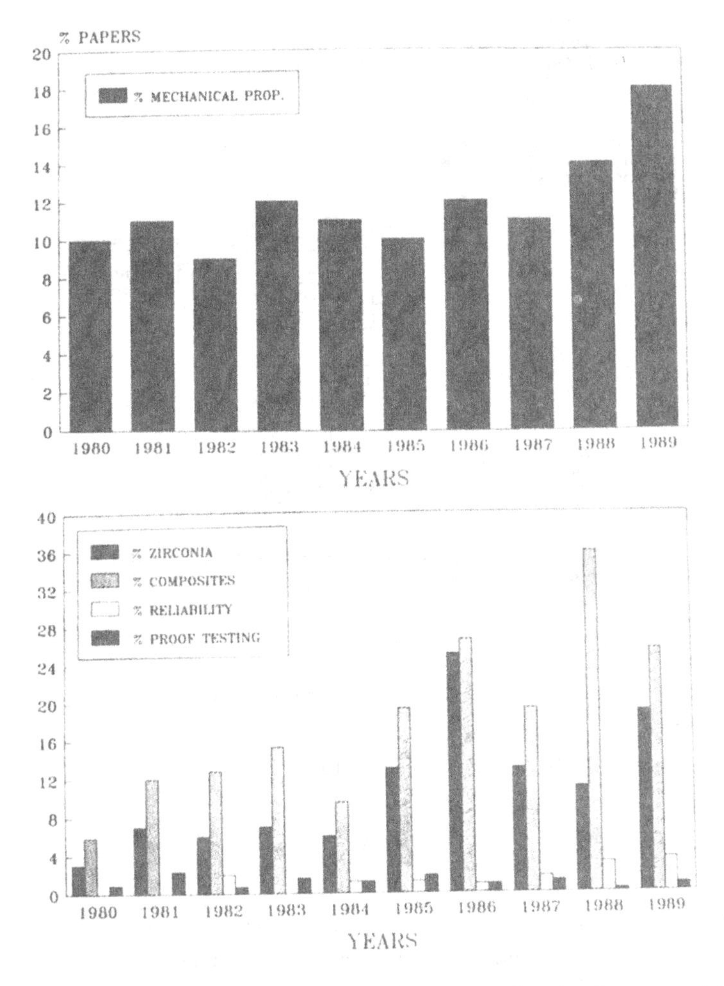

Figure 1. Above: Percentage of the papers on mechanical properties vs. the total number of the papers reviewed by the Am. Ceram. Soc. Abstracts. Below: Percentage of different topics vs. the total number of papers on mechanical properties.

FAILURE BEHAVIOUR

The main distinction between ceramics and metals is brittleness - the tendency to fail catastrophically by the growth of a single crack that originates from a very small defect. Generally these defects range from 1 µm or less to 100 µm and are significantly lower than the critical flaw size of several millimetres in metals. In ceramics, fracture occurs essentially by bond rupture at the tip of a sharp crack, whereas in metals some ductility accompanies fracture.

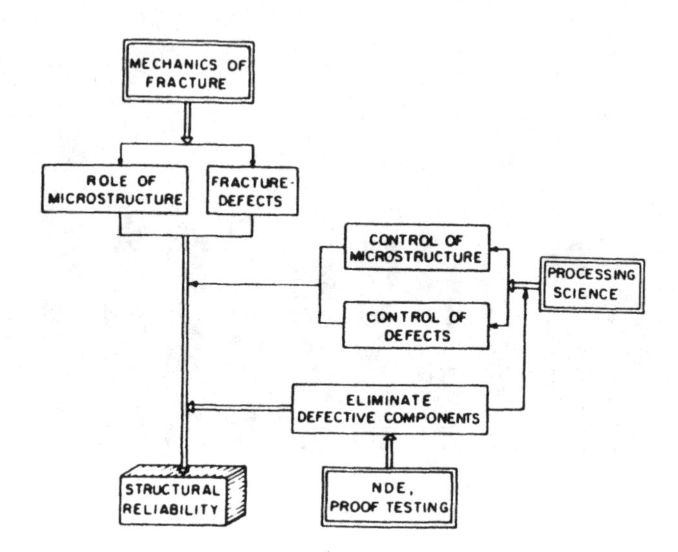

Figure 2. Schematic illustration of the relationship between mechanics of fracture, processing science, evaluation testing and structural reliability of ceramic components (after Evans (2)).

Fracture occurs in a brittle manner and originates from flaws present in the material. The tensile fracture strength of a ceramic can be understood in terms of Griffith (3) equation:

$$\sigma = YK_C/c^{\frac{1}{2}} \qquad (1)$$

where Y is a geometrical constant, c the crack length and K_C is the stress intensity factor or fracture toughness. As the fracture toughness of ceramics is relatively low it is clear that to retain a good strength the flaw size must be very small.

Figure 3 summarizes the relationship between critical defect size, fracture toughness and strength in some ceramic materials (4), while Figure 4 reports the comparison between fracture toughness of ceramics and other materials.

From equation (1), any increase of toughness (γ or K_C) should lead to an increase in strength. Nevertheless in some cases this is not true (e.g. some composites). In addition this equation assumes that the surface energy (γ) or fracture toughness (K_C) of the material is constant so that if the flaw size is known the fracture strength can be calculated. Unfortunately, it has been verified that these basic assumptions are not often valid and (γ) and K_C are not independent of crack size and time.

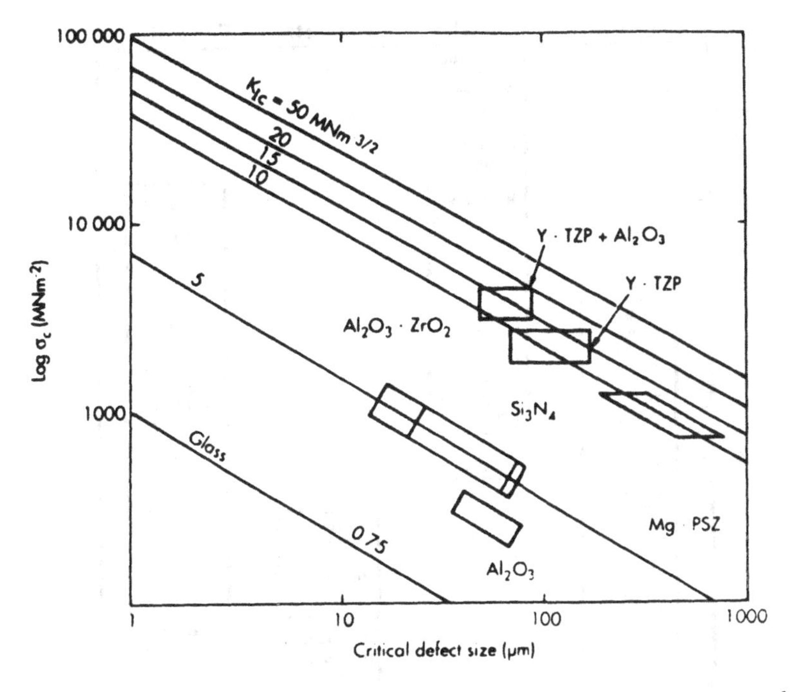

Figure 3. Relationship between critical defect size, fracture toughness and strength in ceramic materials (after Wilson et al. (4)).

The fracture of brittle solids can occur either by direct activation of non-interacting, pre-existent flaws or by generation of defects and coalescence processes as summarized (5) in Figure 5. The first mode generally pertains to fine grained ceramics at relatively low temperature (≤ 1100°C) and is characteristic of most of high strength engineering materials. In more coarse ceramics the failure arises from the generation of subcritical microcracks which form under stress with a contribution of very large localized residual stresses.

At high temperature new environmentally induced flaws are present. Generally the failure is due to the formation of cavitation which eventually grows into defects or grain boundary facets which can then link together by creep processes (6), Figure 6.

ORIGINS OF FLAWS

In ceramics, defects can be divided into three principal categories: surface cracks, fabrication flaws and environmentally induced flaws. However some complications arise from the non-uniqueness of the effects of the flaw size

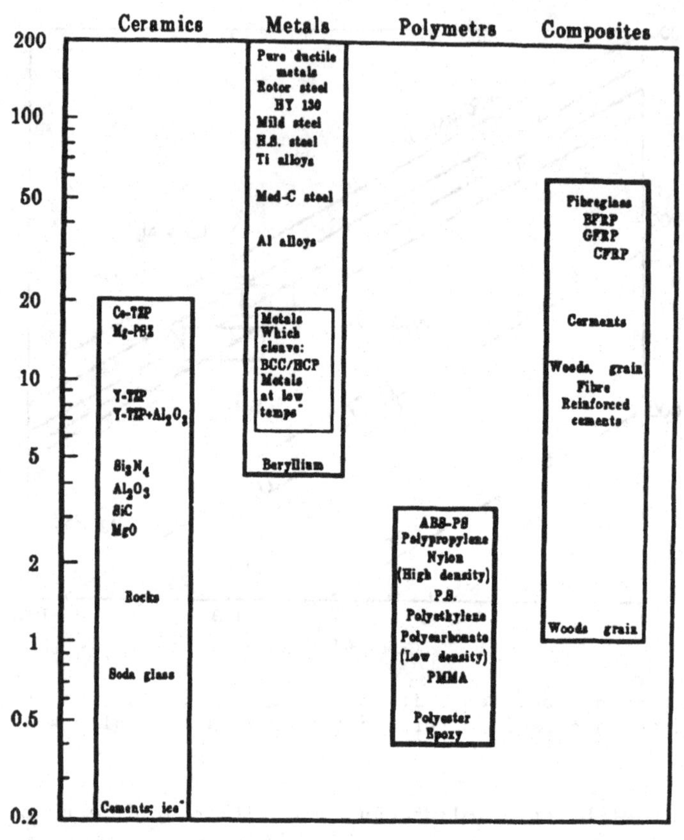

Figure 4. Toughness of some ceramics compared with other materials.

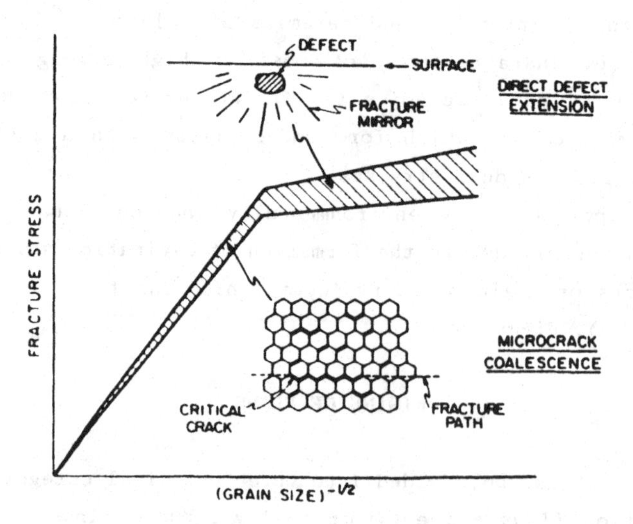

Figure 5. Schematic of failure processes in fine and coarse grained ceramics at relatively low temperature (< 1100°C), (after Evans et al. (5)).

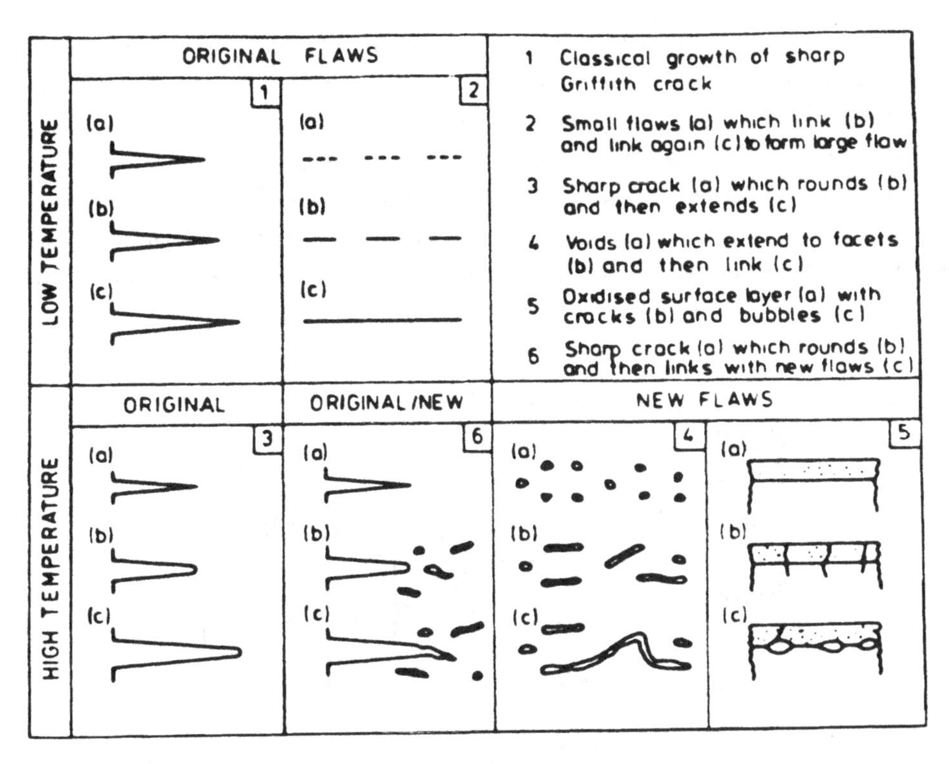

Figure 6. Schematic of crack nucleation and behaviour at low and high temperature (after Davidge (6)).

on fracture stress. This is expressed (7-9) by the effect of voids and inclusions in Si_3N_4 in Figure 7. It is clear that an inclusion of Si causes a more severe strength degradation that one of WC.

Surface Cracks

The most severe degradation is caused by surface cracks. Thus much effort has been concentrated on developing understanding of the behaviour of crack surfaces through the creation of models consisting of the generation of cracks formed on the surface of ceramics by loading a sharp, hard indenter (i.e. Vickers or Knoop pyramids). This sharp body plastically penetrates the brittle solid, leaving a residual contact impression inversely proportional to the hardness of the ceramic (10-13).

The localized plastic deformation generates residual stresses which lead to the formation of half-penny surface cracks and eventually subsurface lateral cracks that are approximately parallel to the surface (Figures 8 and

9). The crack normal to the surface (radial) determines the strength properties, while the crack parallel to the surface is responsible for material removal. Similar defects can occur when a hard particle impacts a surface (Figure 9) as the case of erosion (14), even if microstructural effects can modify the characteristics of the damage (15,16).

Experimental results confirmed the detailed fracture mechanics analysis (10,13) developed to describe this type of contact. Central forces

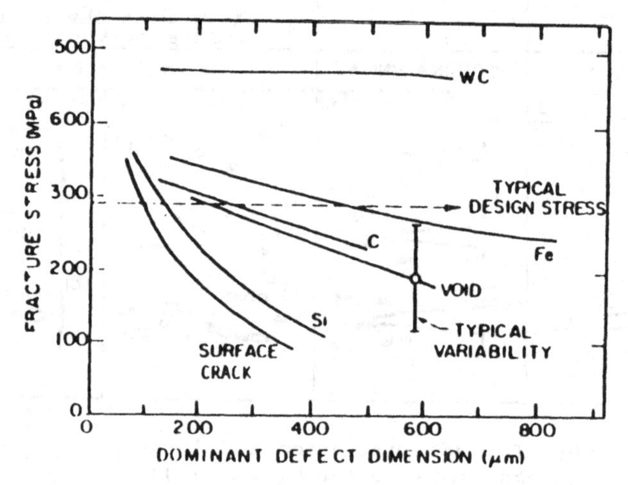

Figure 7. Relationship between strength and defect size for various types of inclusions in Si_3N_4 (after Evans (7,8); Wiederhorn et al. (9).

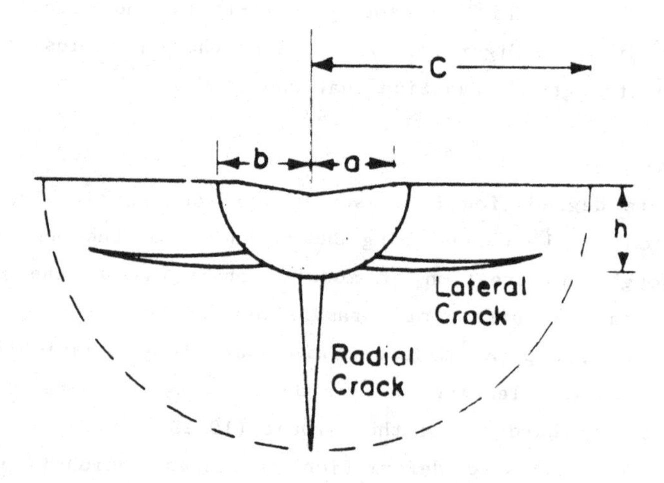

Figure 8. Schematic of cross-section of indentation flaw showing deformation zone and different crack types (after Marshall (10)).

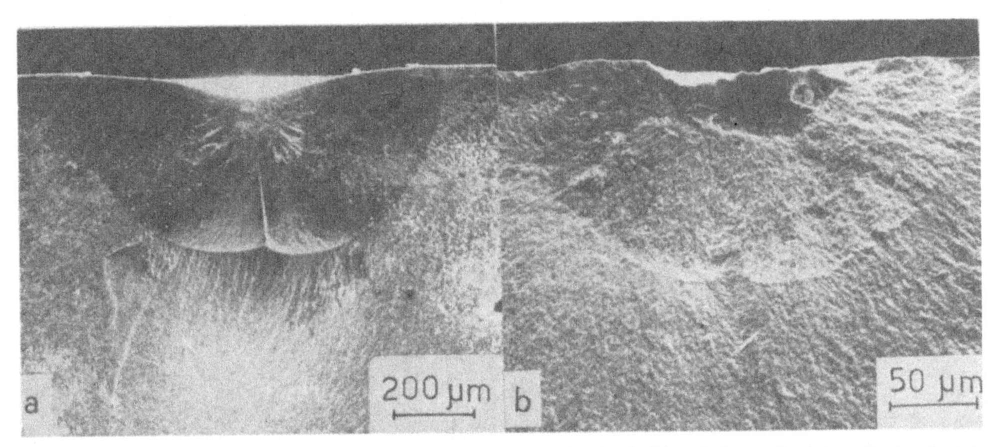

Figure 9. Cross-section of cracks in Y(3 Mol.%)-TZP. Left: Indentation crack at peak load P = 50 kg; Right: Impact crack at impact kinetic energy $U_k = 426*10^{-6}$ J.

imposed by the plastic zone generate a crack with a radius given by:

$$c = \zeta(E/H)^n P^{2/3} K_{IC}^{-2/3} \quad (2)$$

where ζ is a material-independent constant, n is an exponent ≈ ½, E is Young's modulus, H is the hardness, P is the external peak load exerted by the penetrating body and K_{IC} is the fracture toughness. This analysis defines the requirement on contact conditions and the material properties to minimize the damage. As E/H is generally a material independent parameter, fracture toughness appears to be the most significant material parameter.

A sharp contact exhibits a threshold load above which a well developed crack "pop-in" occurs with crack length given by equation (2). It has been verified (17) that the threshold load increases with the fracture toughness. So it is possible to inhibit the formation of cracks and, if the threshold is exceeded, to minimize the crack size by maximizing the fracture toughness. The nature of sub-threshold damage is relatively unknown and further investigations on this problem are needed.

Sliding/rolling contacts or particle impacts as well as machining of ceramics can induce surface cracks (Figure 10) and the analysis of contact stresses can be modelled by Vickers indentation. In the latter case, however, some discrepancies due to the aspect-ratio between length and depth of the crack should be taken into account (18).

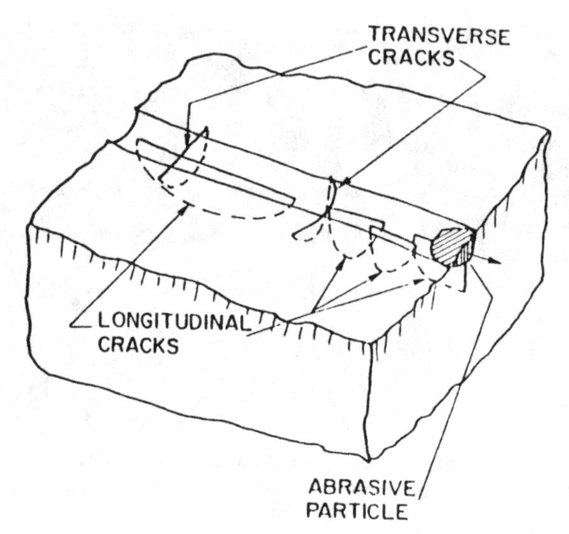

Figure 10. Possible crack formation during machining of ceramics (after Rice et al. (18)).

From equation (1) it is evident that strength-determining parameters are the toughness and crack length. The presence of residual stresses generates an additional stress intensity factor which decreases as the crack extends promoting a stable crack growth.

For indentation cracks the relation expressing the strength at failure in terms of fracture toughness and parameters characterizing the residual stress (19), is:

$$\sigma = 2(H/E)^{1/6} K_C^{4/3} P^{1/3} \tag{3}$$

where the symbols have the same meaning as in equation (2).

It should be noted that for a given initial crack size the residually stressed flaw is more severe than a stress-free flaw.

Fabrication Defects

Defects induced by fabrication processes that can act as failure origins include large voids, large grains, inclusions and shrinkage cracks. Some of the defects originate during the initial stages of fabrication; in particular powder purity, particle size distribution and morphology must be kept under control.

High contracting high modulus inclusions tend to detach from the matrix and produce a defect comparable, in principle, to a void. When an inclusion has a thermal contraction larger than the matrix, the inclusion experiences residual tension and cracking occurs at the boundary with the matrix or in the inclusion (20) (Figures 11a and 11b). When the thermal contraction of the inclusion is lower than that of the matrix, compression stresses are exerted on the inclusion and radial cracks form in the matrix (20), (Figure 11c). In this case the strength degradation is more severe and is comparable to that imposed by machining damage.

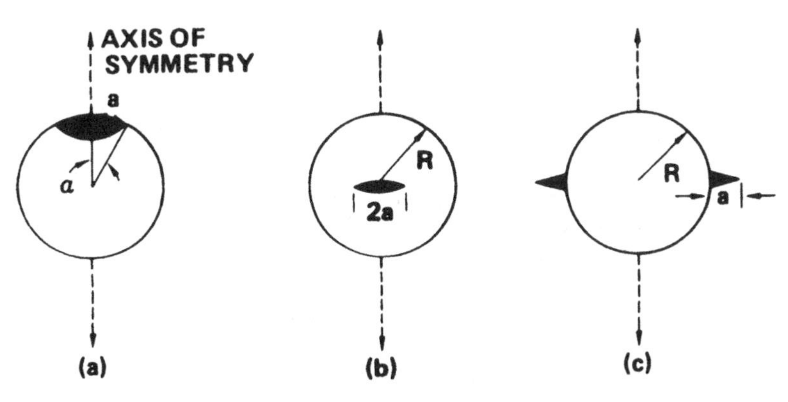

Figure 11. Effect of spherical inclusions in an infinite matrix. Microcracks are generated at the interface (a), at the centre of the inclusion (b), and radially from inclusion (after Green (20)).

There is similarity between this situation and the sharp indentation model, although the residual stresses are expressed in terms of thermal mismatch rather than elastic plastic deformation properties.

The fabrication defects such as pores or microcracks effect also the elastic constants (21-23), and an example of this dependence for Young's modulus has been observed in zirconia toughened alumina (24) (Figure 12).

STATISTICAL ANALYSIS

Strength values for a given material generally exhibit a large variation. In this situation it is important to identify a mathematical expression that allows a probablistic prediction of the performance.

The mathematical tool is the cumulative distribution function proposed by Weibull (25). This is based on the weakest-link-hypothesis which means that the strength is controlled by the most serious flaw in the specimen.

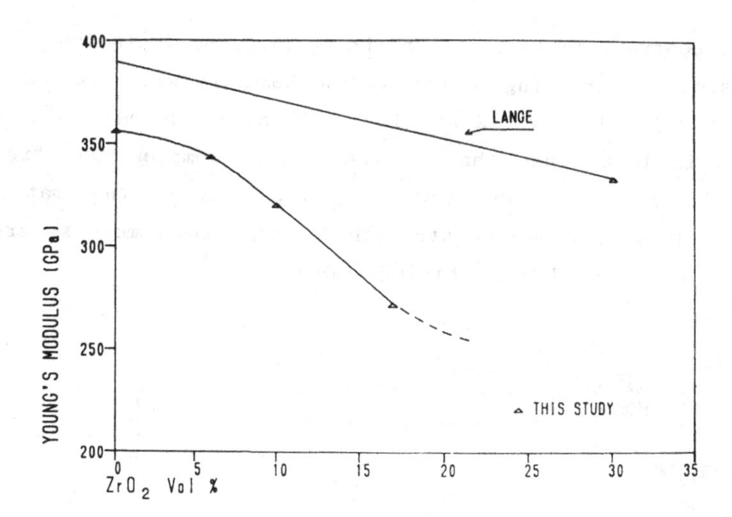

Figure 12. Effect of microcracking on Young's modulus in zirconia toughened alumina (ZTA) (after De Portu et al. (24)).

Based on this concept the probability of survival of a sample under a stress is:

$$P_s = \exp\left[-V((\sigma-\sigma_u)/\sigma_0)^m\right] \tag{4}$$

where V is the volume of the specimen, σ_u is a threshold stress below which the failure probability is zero, σ_0 is a scaling parameter and m is known as the Weibull modulus. Providing that $\sigma_u = 0$, as usually accepted, equation (4) can be written:

$$\ln \ln P_s^{-1} = \ln V + m\ln\sigma - m\ln\sigma_0 \tag{5}$$

Plotting equation (5) in terms of $\ln \ln P_s^{-1}$ and $\ln \sigma$ it is possible to determine the value of m - the slope of the straight line that fits the experimental points (Figure 13). This theory helps an engineer to estimate the failure probability of a component exposed to a fixed stress. However it should be taken into account that this approach is empirical and P_s is not independent of the volume of the specimen.

ENVIRONMENTAL AND TIME EFFECTS

In the previous paragraphs we considered crack formation and growth only

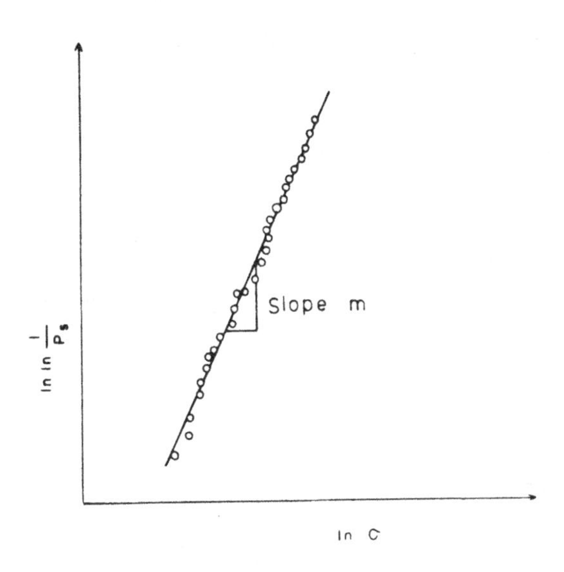

Figure 13. Example of a plot using Weibull statistical analysis.

under conditions where the crack is stationary if the stress intensity factor (K_I) is smaller than a critical value (K_{IC}), and rapidly propagating for $K_I \geq K_{IC}$.

If a ceramic fails at a certain stress in a given time it can fail at a lower stress after a longer time. This means that the ceramic undergoes a degradation with time, especially under aggressive atmospheres (26-27). This is particularly true for oxides in wet environments (e.g. some bioceramics), or at high temperature for all ceramics. Consequently we are not within the ideal conditions described by the Griffith equation.

As K_{IC} is a point on the crack velocity (v) - toughness (K_I) curve it is very important to analyse the curve itself because it is necessary for a prediction of the lifetime of a structure under stress. An idealized diagram (28) (Figure 14) shows three distinct regions for the crack growth velocity/stress intensity factor relationship. In region I the rate of the crack growth is reaction rate controlled and the relationship between the crack velocity (v) and the stress intensity factor can be expressed by the following equation (29-31)

$$v = AK_I^{\ n} \tag{6}$$

In region II the crack velocity does not depend on the stress intensity factor and is controlled by the diffusion of corrosive species to the crack

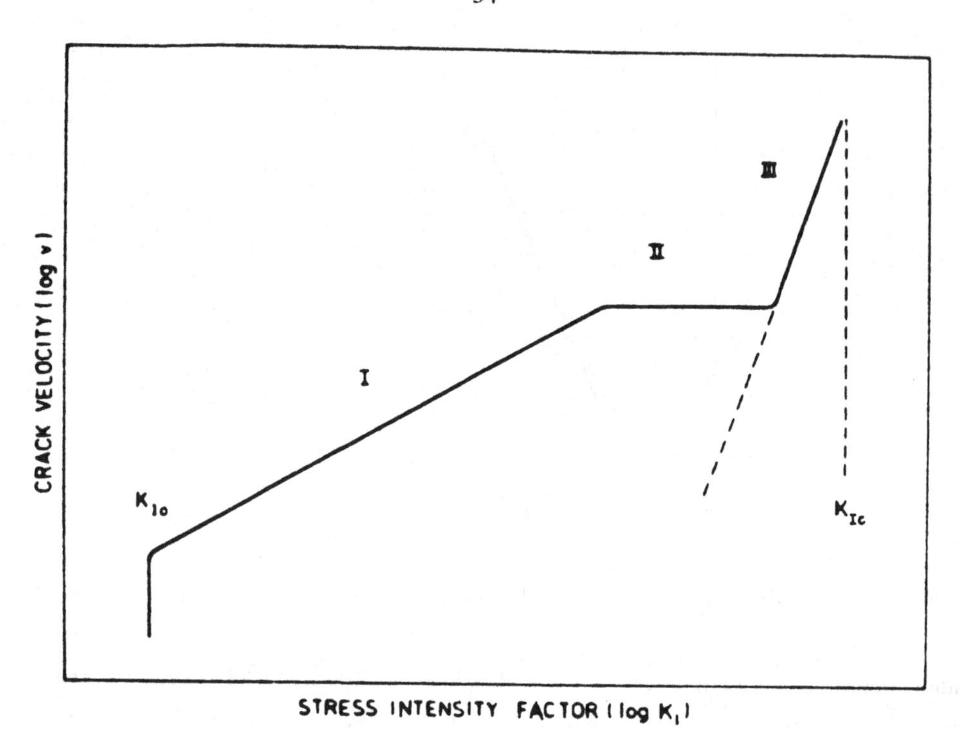

Figure 14. Idealized diagram for the crack velocity-stress intensity factor relationship (after Davidge (28)).

tip. In region III the crack propagation is environment independent and the behaviour is similar to region I but the slope is steeper. When the stress intensity factor reaches a critical value (K_{IC}) fracture occurs.

Generally the important effects concerning the time dependence of strength are related to region I. This means that the key parameter in predicting the time dependent behaviour of ceramics is the parameter n, as expressed in equation (6) which provides a measure of the sensitivity of the ceramics to slow crack growth. Several techniques using the direct observation of crack extension (for example the double cantilever beam or the double torsion test) have been used to obtain data for generating K/v curves.

Another possibility to evaluate n is through the strain-rate dependence of strength. In this case the ratio of the stresses σ at two different strain-rates $\dot{\varepsilon}$ is expressed by:

$$\left(\frac{\sigma_{\dot{\varepsilon}_1}}{\sigma_{\dot{\varepsilon}_2}}\right)^{n+1} = \frac{\dot{\varepsilon}_1}{\dot{\varepsilon}_2} \tag{7}$$

This relation is commonly obtained by simple short-term strength tests at different strain rates. However a problem connected with this technique is that the data do not represent crack growth at very low crack velocity (32) and here the crack growth parameter can be obtained by delayed failures tests (32).

In a delayed-fracture test the ratio between the failure time t_1, t_2 and the applied stresses σ_1, σ_2 is expressed by the following expression:

$$(\sigma_1/\sigma_2)^n = t_2/t_1 \tag{8}$$

For failure at the same stress level, the time to fracture ($t_{\dot{\varepsilon}}$) in a test at constant strain rate, is related to the time to fracture (t_σ) in a test at the maximum constant stress, by the following:

$$t_{\dot{\varepsilon}} = (n+1)\ t_\sigma \tag{9}$$

In this case, for each specimen fractured in a constant strain rate test, can be ascribed a failure time under constant stress conditions. These relationships permit the calculation of a strength-probability-time diagram.

Combining the results obtained from tests performed in strain rate or constant stress conditions it is possible to evaluate the parameter n. As the value of n obtained from these tests is related to the crack growth from inert flaws it includes both time and microstructural effects. Once the parameter n (typically > 10) and A have been measured for a material (equation (6)) it is possible to predict the lifetime for a given initial flaw size and loading conditions. However as equation (6) is empirical and the fundamental mechanisms involved have not been identified so far, extrapolation outside the data range should be avoided. In fact it has been verified (30) that several functional relations fit the available range of data but diverge outside this range.

Strength Probability Time (SPT) Diagrams

The SPT diagram (33) is used to summarize in a simple way the time dependent properties and statistical strength features. Once the parameters n and m are obtained the STP diagram can be produced. Through equation (8), after normalization of the failure stresses to a constant failure time (e.g. 1 sec), it is possible to construct failure lines for increasing time. From the diagram it is possible to determine the maximum stress allowed for a high survival probability for a fixed lifetime of a component.

Figure 15 shows a typical diagram obtained for biograde alumina (6). It is evident that a high reliability and long lifetime of bioalumina are obtained when it is considered that the maximum working stresses is limited to 70 MPa.

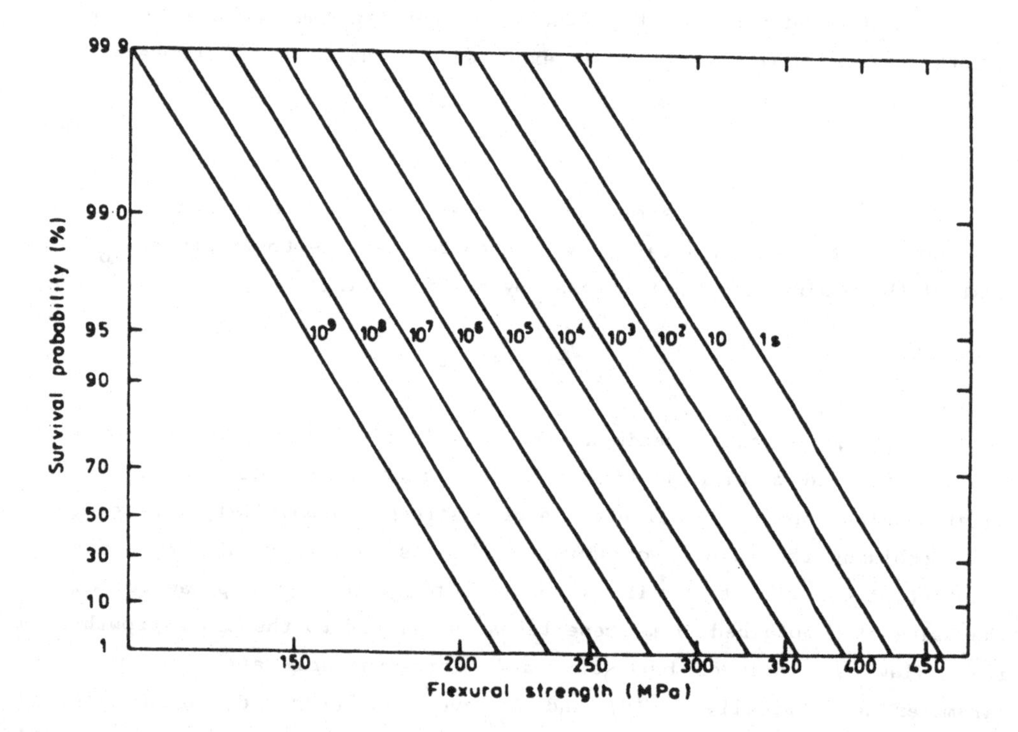

Figure 15. Strength-Probability-Time diagram for biograde alumina (after Davidge (6)).

PROCESS IMPROVEMENT

One way extensively explored to improve the mechanical properties of ceramics is the identification of better fabrication processes (34-35). Chemical routes to produce pure monodispersed fine particles have been defined (36). The role of the agglomerates has been well understood so that methods such as freeze-drying or direct consolidation from the colloidal state have been developed (34). New techniques for consolidating ceramics such as pressure sintering (37) or hot isostatic pressing (38-39) (HIP) are under investigation. Undoubted performance improvements have been obtained and this is a fruitful way to proceed. However these methods are, in some respects, complicated and expensive, and do not solve entirely the problem.

If we observe (40) (Figure 16) how the best flexural strength of alumina varied over a twenty year period from 1950 to 1970. The result of a simple extrapolation of the data suggest that in 1979 the strength should have been 3 GPa, which is approximately the ultimate theoretical value. If we consider that this value is at least three or four times higher than the value currently obtained it is clear that there is a natural limit above which a particular process cannot guarantee further spectacular improvement.

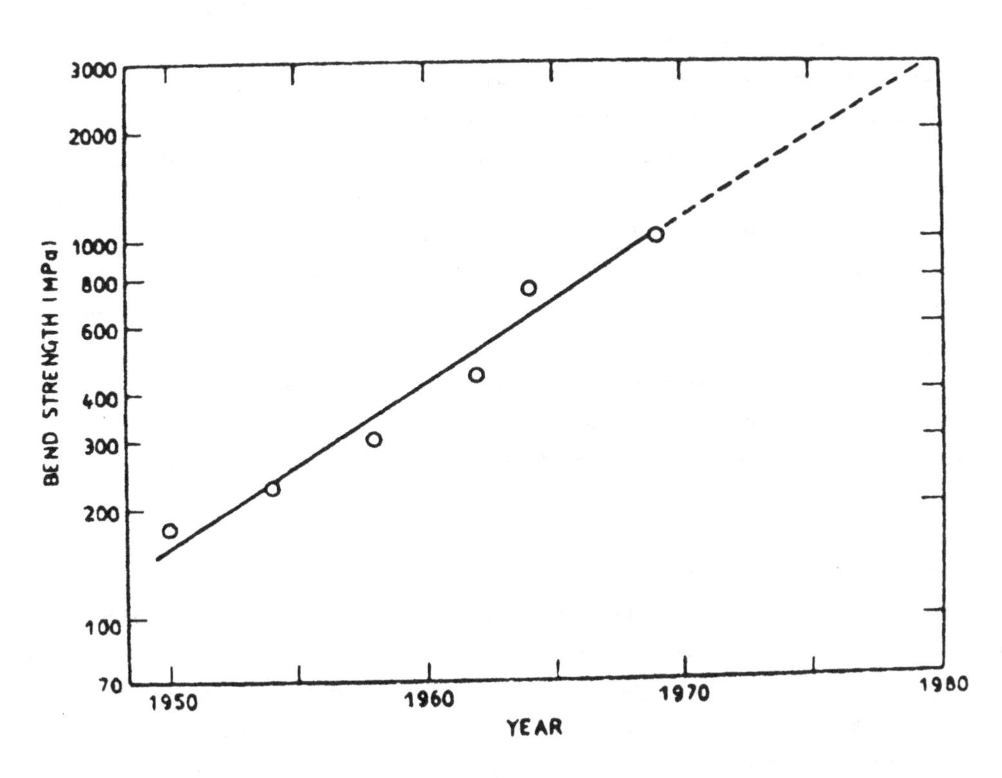

Figure 16. Improvement of bend strength of alumina in the period 1950-1970 (after Boy et al. (40)).

TOUGHENING OF MONOLITHIC CERAMICS

So far we have analyzed the fracture mechanics of ceramics in terms of basic understanding of the flaw size strength relationship, and design requirements for such materials. This knowledge can help us to live with their brittleness (low K_C), so that the reliability can be assessed by developing statistical analyses and/or non destructive methods. Another approach is to identify the source of defects and to reduce their size by developing new processing methods. In this case high strength ceramics with high Weibull modulus could be produced. Both of these approaches have been substantially developed in the past decade, however, they do not deal with the environmental or machining induced flaws.

The most satisfactory approach to the problem of reliability is the development of flaw-tolerant ceramics (monolithic or composite) having microstructures associated with a higher resistance to fracture. From this point of view the development of zirconia-based toughened ceramics can be considered a real breakthrough.

When 15 years ago Garvie et al. (41) proposed use of the phase transformation from the tetragonal to monoclinic symmetry to improve the fracture toughness of zirconia based materials, there was great excitement in the scientific community. Consequent improvements in toughness have been achieved by creating microstructures containing a second phase that can undergo transformation (42-50). Several mechanisms have been proposed (42-50) that can act independently or simultaneously to account for this phenomenon.

Although quite advanced fracture mechanics analyses have been developed for many of these mechanisms, the discussion on their effectiveness is still open. The main mechanisms identified are: stress induced transformation, microcracking, crack deflection, crack branching and crack bridging.

Stress Induced Transformation

Several investigations have been carried out on this subject (42-51). This kind of toughening mechanism involves the martensitic transformation from tetragonal to monoclinic structure. This transformation occurs on cooling, as the monoclinic phase is the stable structure at room temperature, and is accompanied by a volume increase of ≈ 4% and shear distortion of ≈ 7%. The transformation can be prevented by addition of oxides such as Y_2O_3, CeO_2, MgO, CaO or by a constraint of a surrounding stiff matrix. A grain size effect on the transformation and the existence of a critical size above which the spontaneous transformation occurs has been verified (46,50).

The stress induced transformation pertains to particles within a limited zone (process zone) near the tip of propagating crack. The toughening then derives from a crack shielding process due to the expansion of the transforming particles which causes a reduction in crack tip stresses. Two independent types of approach have been followed to describe this mechanism, a mechanical model (46) and a thermodynamic model (53-55). Fortunately both these analyses lead to the same conclusion (54,55), that the maximum toughness achievable has the form:

$$K_{IC} = K_{IC}^{m} + 0.22E\, e^{T}\, V_f\, h^{\frac{1}{2}}/(1-\nu) \tag{7}$$

where K_{IC}^{m} is the toughness of the transformed products in the process zone ahead of the crack, h is the transformation zone width, E is the Young's modulus, ν is the Poisson's ratio, V_f is the volume fraction of the transformed product and e^T is the transformation volume change (transformation strain). Figure 17 represents (53) the formation of the process zone and the related R-curve behaviour.

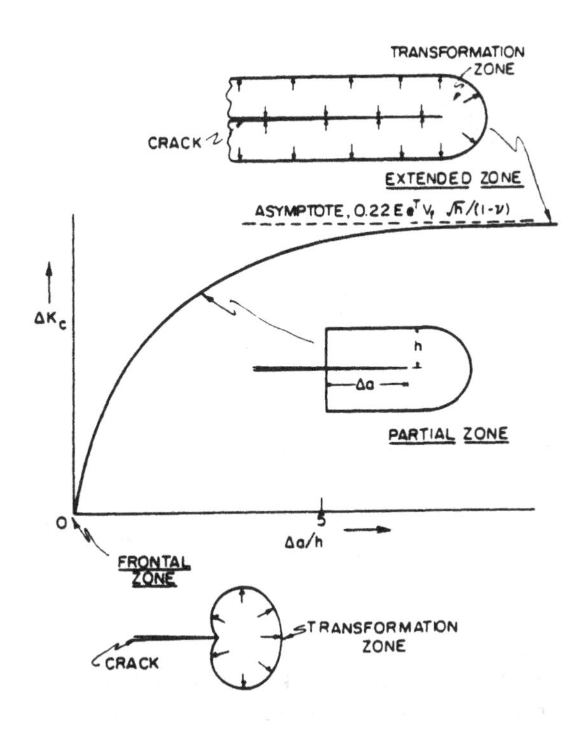

Figure 17. A schematic showing the transformation zone shape and the associated R-curve with the relative asymptote (after Evans (53)).

Independent measurements of the transformation zone indicate that most of the toughness is due to dilatation as described by equation (7) with a contribution from deflection effects. Study of these measurements reveals that the micromechanics model expressed in terms of the zone width parameter explains the basic character of transformation toughening.

The relation between the zone width and the criteria for transformation are not completely clear. It is open to discussion whether the transformation conditions are determined by thermodynamical processes or by nucleation. However the nucleation explanation argument seems to be preferred (56-58).

As mentioned above, the conditions for optimizing transformation toughening resides in the microstructure and thus in fabricating materials with varying particle size, shape and chemical composition to control the transformation zone width, etc. A possible way for optimizing the microstructure is represented (59) in Figure 18.

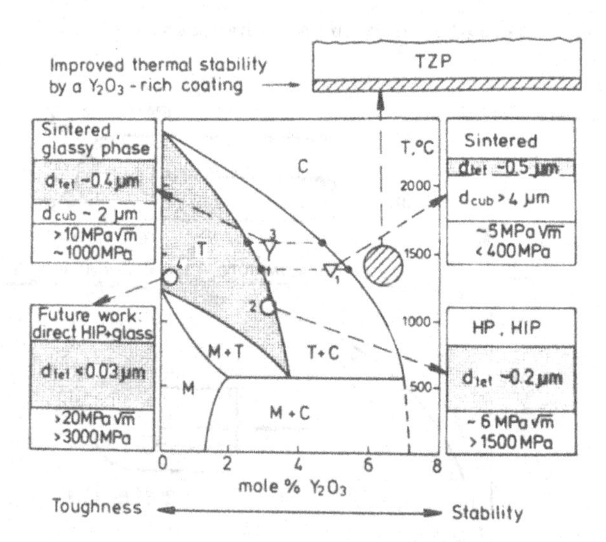

Figure 18. Possible compositions and procedures for improving mechanical properties of yttria-tetragonal zirconia polycrystals (Y-TZP) ceramics (after Claussen (59)).

Microcracking

Thermal expansion anisotropy, or the presence of a secondary phase with different thermal expansion or elastic properties, or inclusions that undergo phase transformation can induce localized stresses. The latter is the case for zirconia based materials in which the transformation from tetragonal to monoclinic symmetry on cooling induces shear stress and hence microcracking near the zirconia particles (20). If a critical grain size is exceeded by the metastable particle, spontaneous transformation occurs, inducing microcracking in the surrounding material. Below this critical size microcracking can be induced by applied tension (20) (stress induced microcracking). The microcracking mechanism stimulates toughening by the reduction of the modulus of the microcracked material within the process zone, and by dilatation induced by microcracks. The latter component provides the larger contribution to toughening (53-60). To be stimulated, the microcracking mechanism needs particular conditions. As only particles or grains within a narrow size range contribute to the toughening there are strict requirements on microstructure and uniformity for obtaining the largest toughening possible with this mechanism (47,60) (Figure 19). The microcrack toughening is less potent than transformation toughening, however it has been observed (42,61) that a combined effect of stress induced transformation and microcracking is beneficial.

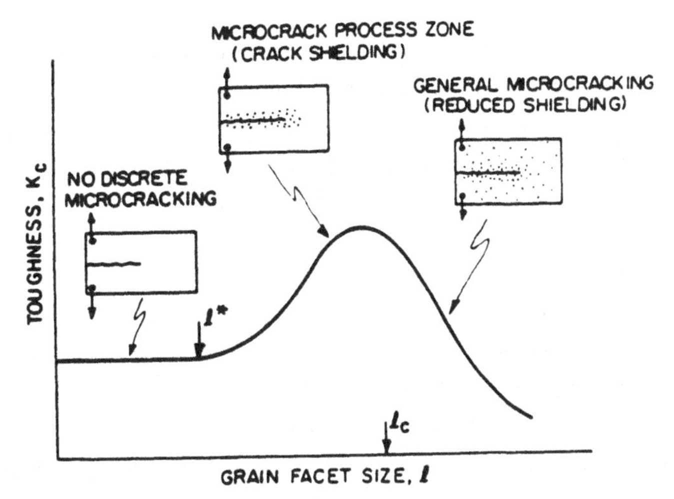

Figure 19. Effect of grain size on the development of microcracking mechanisms and relative trends in toughness (after Evans et al. (60)).

Crack Deflection

Deflection toughening pertains to phenomena occurring at the crack tip and is stimulated by the presence of a second phase. Analysis (62) suggests that the maximum toughening effect is related to the aspect ratio of the deflecting phase (Figure 20). This observation stimulated the development of whisker-containing composites. The predictions of the analysis have been verified experimentally (23,63,64) on ZrO_2-toughened Al_2O_3, Si_3N_4 and SiC.

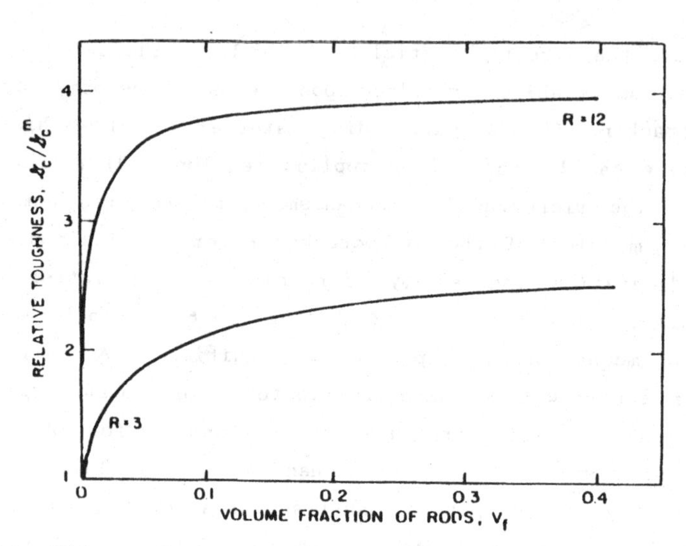

Figure 20. Effect of the particle aspect ratio on deflection toughening. R is the aspect ratio of the rod-shaped particles (after Evans et al. (63)).

Although this mechanism is sufficiently defined, further study of the second phase characteristics needed to induce deflection (i.e. residual strain, cleavage, resistant planes, etc.), are of great interest. In fact, since this toughening mechanism is still active at high temperature it can be utilized to optimize high temperature toughness.

Crack Bridging

All the mechanisms so far discussed occur in the vicinity of the crack tip.

Recently it has been verified that in frictionally bonded fibre composites (65) or large grained Al_2O_3 (66,67) another mechanism is present at large distances from the crack tip. The proposed mechanism for reducing the stress intensity factor, involves closure forces on crack surfaces dictated by mechanical interlocking of fibre ligaments or protruding grains on a rough fracture surface. Some problems in the analysis arise from the difficulties of determining the force/displacement relation for the crack bridging forces. As a wide range of ceramics such as whisker-containing composites or ceramics with elongated grains can potentially exhibit a similar mechanism, considerable effort to develop the analysis can be expected.

CRACK GROWTH RESISTANCE (R-CURVE)

As already mentioned (Figure 17) tough ceramics exhibit a crack growth resistance curve (53) (R-curve behaviour). Theoretical predictions have been confirmed by several experiments on zirconia based materials (68-70) (Figure 21) or other ceramics (71,72).

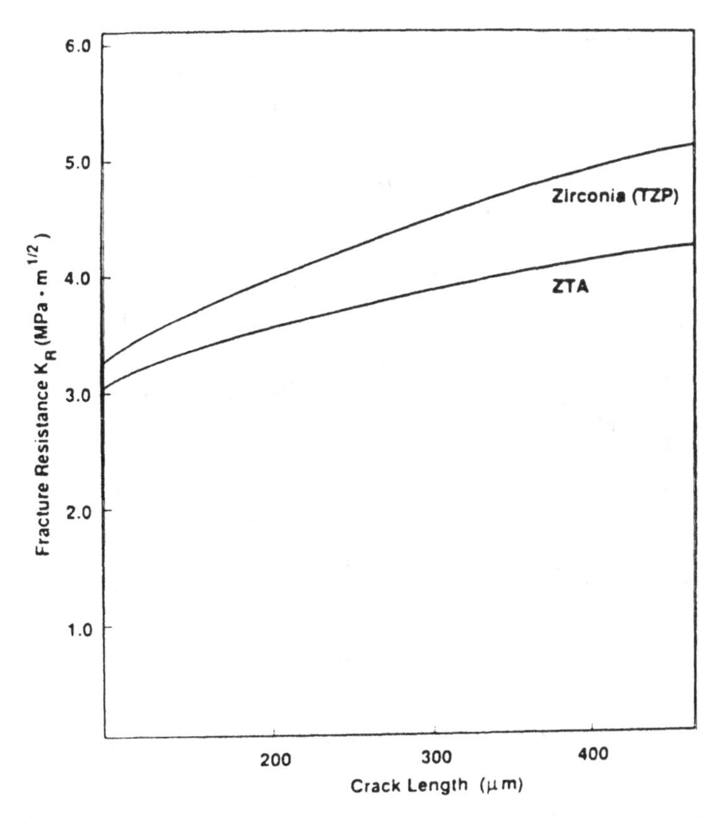

Figure 21. Predicted fracture resistance curves in zirconia based ceramics (after Ritter et al. (70)).

The resistance to crack growth in such materials increases as the crack length increases. The direct consequence of this behaviour is that the strength is no longer related uniquely to an initial flaw size. R-curve behaviour can explain the existence of a maximum in the strength-toughness relationship in different zirconia toughened ceramics (73) (Figure 22). Very high strengths (> 2 GPa) can be obtained in Y-TZP + Al_2O_3, but the toughness is ≈ 5-8 MPa $m^{\frac{1}{2}}$. On the contrary the tougher material (peak-aged

Mg-PSZ with K_{IC} approaching 15 MPa $m^{\frac{1}{2}}$) is not the strongest one, but this material is very damage-tolerant (strength is relatively crack-size insensitive). This evidence suggests that lower strength materials which exhibit strong R-curve behaviour will also exhibit a higher Weibull modulus, but a better knowledge of microstructural requirements for an appropriate R-curve must be achieved.

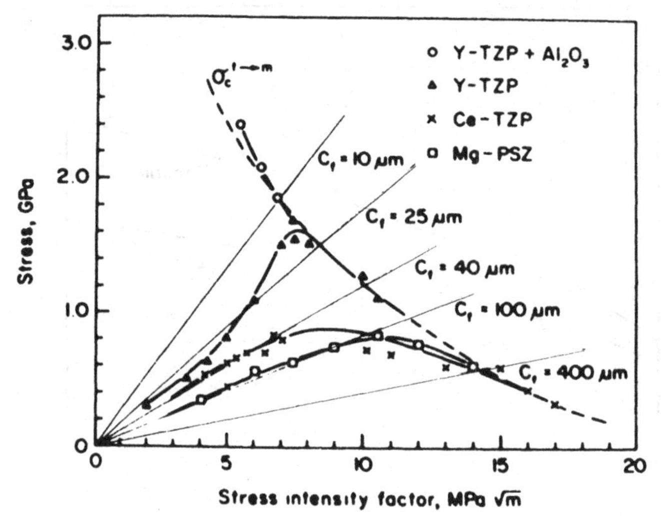

Figure 22. Plot of strength as a function of toughness for transformation-toughened zirconia based materials (after Swain (73)).

CONCLUDING REMARKS

The effort made by the scientific community to develop understanding of mechanical behaviour of ceramics has been demonstrated by a statistical analysis of the literature. It appears that some areas deserve more consideration and we can expect interesting advancement in these.

The mechanical properties of monolithic ceramics at room temperature are quite well defined. Failure occurs as a result of growth of pre-existing defects or environmentally produced flaws that grow from subcritical to critical size. The characteristics of various defects are relatively well understood and a fracture mechanics framework exists for structural design.

Several toughening mechanisms have been identified and a fracture mechanics analysis has been established to describe transformation toughening, microcracking and crack deflection in terms of microstructural parameters. The existence of R-curve behaviour exhibited by very tough ceramics is a very exciting discovery, which although needs better theoretical understanding, opens new prospects for utilization of structural ceramics.

ACKNOWLEDGEMENTS

The authors thank G. Farolfi for the statistical analysis of the literature and S. Guicciardi for the helpful discussion of the manuscript.

REFERENCES

1. Davidge, R.W., "Mechanical Properties of Ceramics - Perspectives Past and Future", pp. 439-54, in Proceedings of the Tenth International Conference "Science and Ceramics", Edited by H. Hausner, Deutsche Keramische Geselleschaft, 1980.

2. Evans, A.G., "Aspects of Reliability of Ceramics for Engine Applications", pp. 364-402, in Fracture in Ceramic Materials, Edited by A.G. Evans, Noyes Publications, 1984.

3. Griffith, A.A., "The Phenomena of Rupture and Flow in Solids", Philos. Trans. Roy. Soc. Lond., 1920, **A221**, 163.

4. Wilson, M.C. and Dover, R.J.,"Ceramic Matrix Composites", Metals and Materials, 1988, 752-56.

5. Evans, A.G. and Fu, Y., "The Mechanical Behaviour of Alumina: A Model Anisotropic Brittle Solid", pp. 56-98, in Fracture in Ceramic Materials, Edited by A.G. Evans, Noyes Publications, 1984.

6. Davidge, R.W., "Engineering Performance Prediction for Ceramics", in ASFM - Fracture Mechanics on Nonmetallic Materials, Ispra, October 14-18, 1985.

7. Evans, A.G., "Structural Reliability: A Processing-Dependent Phenomenon", J. Am. Ceram. Soc., 1982, **65**, 127-37.

8. Evans, A.G., "Engineering Property Requirements for High Performance Ceramics", Mater. Sci. Eng., 1985, **71**, 3-21.

9. Wiederhorn, S.M. and Fuller Jr, E.R., "Structural Reliability of Ceramic Materials", Mater. Sci. Eng., 1985, **71**, 169-86.

10. Lawn, B.R., Evans, A.G. and Marshall, D.B., "Elastic/Plastic Indentation Damage in Ceramics: The Median/Radial Crack System", J. Am. Ceram. Soc., 1980, **63**, 574-81.

11. Anstis, G.R. Chantikul, P., Lawn, B.R. and Marshall, D.B., "A Critical Evaluation of Indentation Techniques for Measuring Fracture Toughness: I, Direct Crack Measurements", J. Am. Ceram. Soc., 1981, **64**, 533-38.

12. Lawn, B.R., "The Indentation Crack as a Model Surface Flaw", pp. 1-25, in Fracture Mechanics of Ceramics, Vol. 5, Edited by R.C. Bradt, A.G. Evans, D.P.H. Hasselman and F.F. Lange, Plenum Press, NY, 1983.

13. Marshall, D.B., "Controlled Flaws in Ceramics: A Comparison of Knoop and Vickers Indentation", J. Am. Ceram. Soc., 1983, **66**, 127-31.

14. De Portu, G. and Ritter, J.E., Unpublished work.

15. Ritter, J.E., "Erosion Damage in Structural Ceramics", Mater. Sci. Eng., 1985, 71, 195-201.

16. Ritter, J.E., Jakus, K., Viens, M. and Breder, K., "Effect of Microstructure on Impact Damage of Polycrystalline Alumina", pp 55,1-55,6, in Proceedings of the Seventh International Conference on Erosion by Liquid and Solid Impact, Edited by J.E. Field and J.P. Dear, Cavendish Laboratory, Cambridge, UK, 1987.

17. Lawn, B.R. and Evans, A.G. "A Model for Crack Initiation in Elastic/ Plastic Indentation Fields", J. Mater. Sci., 1977, **12**, 2195-2199.

18. Rice, R.W. and Mecholsky Jr., J.J., "The Nature of Strength Controlling Machining Flaws in Ceramics", pp. 351-78, in Proceedings of the 2nd Symposium on the Science of Ceramic Machining and Surface Finishing, The National Bureau of Standards, Gaithersburg, Maryland, 1978.

19. Lawn, B.R., Marshall, D.B., Chantikul, P. and Anstis, G.R., "Indentation Fracture: Application in the Assessment of Strength of Ceramics", J. Aust. Ceram. Soc., 1980, **16**, 4-9.

20. Green, D.J., "Microcracking Mechanisms in Ceramics", pp. 457-78, in Fracture Mechanics of Ceramics, Vol. 5, Edited by R.C. Bradt, A.G. Evans, D.P.H. Hasselman and F.F. Lange, Plenum Press, NY, 1983.

21. De Portu, G. and Vincenzini, P., "Young's Modulus Porosity Relationship for Alumina Substrates", Ceramurgia International, Short Communications, 1979, **5**, 165-67.

22. De Portu, G. and Vincenzini, P., "Young's Modulus of Silicon Nitride Hot Pressed with Ceria Additions", Ceramurgia International, 1980, **6**, 129-32.

23. Budiansky, B. and O'Connell, J., "Elastic Moduli of Cracked Solid", Int. J. Solid Struct., 1976, **12**, 81.

24. De Portu, G., Fiori, C. and Sbairero, O., "Fabrication, Microstructure and Properties of ZrO_2-toughened Al_2O_3 Substrates", pp. 1063-73, in Advances in Ceramics, Vol. 24B, Edited by S. Somiya, N. Yamamoto and H. Yanagida, Am Ceram. Soc. Inc., Columbus OH, 1988.

25. Weibull, W., "A Statistical Distribution Function of Wide Applicability", J. Appl. Mech., 1951, **18**, 293-97.

26. Wiederhorn, S.M., "Subcritical Crack Growth in Ceramics", pp. 613-46, in Fracture Mechanics of Ceramics, Vol. 2, Edited by R.C. Bradt, D.P.H. Hasselman and F.F. Lange, Plenum Press, NY, 1974.

27. Krausz, A.S. and Krausz, K., "Time Dependent Failure of Ceramic Materials in Sustained and Fatigue Loading", pp. 333-40, in Fracture Mechanics of Ceramics, Vol. 8, Edited by R.C. Bradt, A.G. Evans, D.P.H. Hasselman and F.F. Lange, Plenum Press, NY, 1986.

28. Davidge, R.W., "Mechanical Behaviour of Ceramics", pp. 140, Cambridge University Press, Cambridge, 1979.

29. Ritter, J.E., "Engineering Design and Fatigue Failure of Brittle Materials", Fracture Mechanics of Ceramics, Vol. 4, 1978.

30. Wiederhorn, S.M. and Ritter, J.E., "Application of Fracture Mechanics Concepts to Structural Ceramics", pp. 202-14, in Fracture Mechanics Applied to Brittle Materials, ASTM STP -678, Edited by S.W. Freiman, ASTM, Philadelphia, PA, 1979.

31. Evans, A.G. and Wiederhorn, S.M., "Proof Testing of Ceramic Materials: An Analytical Basis for Failure Prediction", Int. J. Frac., 1974, **10**, 379.

32. Freiman, S.W., "A Critical Evaluation of Fracture Mechanics Techniques for Brittle Materials", pp. 27-45, in Fracture Mechanics of Ceramics, Vol. 6, Edited by R.C. Bradt, A.G. Evans, D.P.H. Hasselman and F.F. Lange, Plenum Press, New York, 1983.

33. Davidge, R.W., McLaren, J.R. and Tappin, G., "Strength Probability Time (SPT) Relationship in Ceramics", J. Mater. Sci., 1973, **8**, 1699.

34. "Forming of Ceramics", Edited by A. Mengels and G.L. Messing, Am. Ceram. Soc. Inc., Columbus, OH, 1984.

35. "Emergent Process Methods for High-Technology Ceramics", Edited by R.F. Davis, H. Palmour III and R.L. Porter, Plenum Press, NY, 1984.

36. "Science of Ceramic Chemical Processing", Edited by L.L. Hench and D.R. Urlich, John Wiley & Sons, NY, 1986.

37. Lange, D., Thaler, H. and Schwetz, K.A., "Gas-Pressure-Sintering of Silicon Nitride", pp. 1.329-1.335, in Euro-Ceramics, Vol. 1, Processing of Ceramics, Edited by G. de With, R.A. Terpstra and R. Metselaar, Elsevier Applied Science, London, 1989.

38. Larker, H.T., "Recent Advances in Hot Isostatic Pressing Processes for High Performance Ceramics", Mater. Sci. Eng., 1985, **71**, 329-32.

39. Butler, N.D., Hepworth, M.A., Iturriza, I. and Castro, F., "HIPing Glass-Encapsulated Silicon Nitride to Full Density", 1.304-1.308, in Euro-Ceramics, Vol. 1, Processing of Ceramics, Edited by G. de With, R.A. Terpstra and R. Metselaar, Elsevier Applied Science, London, 1989.

40. Boy, G. and Pincus, A.G., "Ceramic Age", 1970, **86**, 41.

41. Garvie, R.C., Hannink, R.H.J. and Pascoe, R.T., "Ceramic Steel?", Nature, 1975, **258**, 703-4.

42. Green, D.J., Hannink, R.H.J. and Swain, M.V., "Transformation Toughening of Ceramics", CCR Press Inc., Boca Raton, Florida, 1989.

43. "Advances in Ceramics Vol. 3, Science and Technology of Zirconia I", Edited by A.H. Heuer and L.W. Hobbs, Am. Ceram. Soc. Inc., Columbus, OH, 1981.

44. "Advances in Ceramics Vol. 12, Science and Technology of Zirconia II", Edited by N. Claussen, M. Ruhle and A.H. Heuer, Am. Ceram. Soc. Inc., Columbus, OH, 1984.

45. "Advances in Ceramics Vol. 24A-24B, Science and Technology of Zirconia III", Edited by S. Somyia, N. Yamamoto and H. Yanagida, Am. Ceram. Soc. Inc., Columbus, OH, 1988.

46. Evans, A.G. and Heuer, A.H., "Review-Transformation Toughening in Ceramics: Martensitic Transformation in Crack-Tip Stress Fields", J. Am. Ceram. Soc., 1980, **63**, 241-48.

47. Evans, A.G., Burlingame, N., Drory, M. and Kriven, W.M., "Martensitic Transformation in Zirconia - Particle Size Effects and Toughening", Acta Metall., 1981, **29**, 247-56.

48. McMeeking, R.M. and Evans, A.G., "Mechanics of Transformation - Toughening in Brittle Materials", J. Am. Ceram. Soc., 1982, **65**, 242-46.

49. Proceedings of the Second US - Australian - West German Transformation Toughening Workshop, J. Am. Ceram. Soc., 1986, **69**, 169-298.

50. Ibid, 1986, **69**, 511-584.

51. Gupta, T.K., "Role of Stress - Induced Transformation in Enhancing Strength and Toughness of Zirconia Ceramics", pp. 877-89, in Fracture Mechanics of Ceramics, Vol. 4, Edited by R.C. Bradt, D.P.H. Hasselman and F.F. Lange, Plenum Press, NY, 1978.

52. Lange, F.F., "Transformation Toughening, Part 1. Size Effect Associatd with the Thermodynamics of Constrained Transformation", J. Mat. Sci., 1982, **17**, 225-34.

53. Evans, A.G., "Toughening Mechanisms in Zirconia Alloys", pp. 193-192, as in Ref. 44.

54. Budiansky, B., Hutchinson, J. and Lanbroupolos, J., "Continuum Theory of Dilatant Transformation Toughening in Ceramics", Int. J. Solid Struct., 1983, **19**, 337-55.

55. Marshall, D.B., Evans, A.G. and Drory, M., "Transformation Toughening in Ceramics", pp. 289-307, in Fracture Mechanics of Ceramics, Vol. 6, Edited by R.C. Bradt, A.G. Evans, D.P.H. Hasselman and F.F. Lange, Plenum Press, NY, 1983.

56. Ruhle, M. and Heuer, A.H., "Phase Transformation in ZrO_2-containing Ceramics: II, The Martensitic Reaction in t-ZrO_2", pp. 14-32, as in Ref. 44.

57. Chen, I.W. and Chiao, Y.H., "Theory and Experiment of Martensitic Nucleation in ZrO_2-containing Ceramics and Ferrous Alloy", Acta Metall., 1985, **33**, 1827.

58. Heuer, A.H. and Ruhle, M., "On the Nucleation of the Martensitic Transformation in Zirconia (ZrO_2)", Acta Metall., 1985, **33**, 2101-12.

59. Claussen, N., "Microstructural Design of Zirconia-Toughened Ceramics (ZTC)", pp. 325-51, as in Ref. 44.

60. Evans, A.G. and Faber, K.T., "Crack Growth Resistance of Microcracking Brittle Materials", J. Am. Ceram. Soc., 1984, **67**, 255-60.

61. Ruhle, M., Claussen, N. and Heuer, A.H., "Transformation and Microcrack Toughening as Complementary Processes in ZrO_2-toughened Al_2O_3", J. Am. Ceram. Soc., 1986, **69**, 195-97.

62. Faber, K.T. and Evans, A.G., "Crack Deflection Processes - I Experiment", Acta Metall., 1983, **31**, 565-76.

63. Faber, K.T. and Evans, A.G., "Crack Deflection Processes - II Experiment", Acta Metall., 1983, **31**, 577-84.

64. Faber, K.T. and Evans, A.G., "Intergranular Crack Deflection Toughening in Silicon Carbide", J. Am. Ceram. Soc., 1983, **66**, C94-C96.

65. Marshall, D.B. and Evans, A.G., "Failure Mechanisms in Ceramic-Fibre/ Ceramic Matrix Composites", J. Am. Ceram. Soc., 1985, **68**, 225-31.

66. Steinbrech, R., Khehans, R. and Schaarwachter, W., "Increase of Crack Resistance during Slow Crack Growth in Al_2O_3 Bend Specimens", J. Mat. Sci., 1983, **18**, 265-70.

67. Swanson, P.L., Fairbanks, C.J., Lawn, B.R., Mai, Y.W. and Hockey, B.J., "Crack-Interface Grain Bridging as a Fracture Resistance Mechanism in Ceramics", J. Am. Ceram. Soc., 1987, **70**, 279-89.

68. Swain, M.V. and Hannink, R.H.J., "R-Curve Behaviour in Zirconia Ceramics", pp. 225-39, as in Ref. 42.

69. Breder, K., De Portu, G., Ritter, J.E. and Fabbriche, D.D., "Erosion Damage and Strength Degradation of Zirconia-Toughened Alumina", J. Am. Ceram. Soc., 1988, **71**, 770-75.

70. Ritter, J.E., De Portu, G., Breder, K. and Fabbriche, D.D., "Erosion Damage in Zirconia and Zirconia Toughened Alumina", pp. 171-76, in Materials Science Forum, **34-36**, Ceramic Developments: Past, Present and Future, Edited by C.C. Sorrell and B. Ben-Nissan, Trans. Tech. Publications Ltd., Switzerland, 1988.

71. Krause, R.F., "Rising Fracture Toughness for the Bending Strength of Indented Alumina Beams", J. Am. Ceram. Soc., 1988, **71**, 338-43.

72. Li, C.-W. and Yamanis, J., "Super-Tough Silicon Nitride with R-Curve Behaviour", Cer. Eng. and Sci. Proc. 1989, **10**, 532-45.

73. Swain, M.W., "Transformation Toughening: An Overview", J. Am. Ceram. Soc., 1986, **69**, iii.

FRACTURE MECHANICS OF CERAMICS

D. MUNZ
University and Nuclear Research Centre,
Postfach 3640, 7500 Karlsruhe 1,
Germany.

ABSTRACT

The failure of ceramic materials is caused by the extension of small flaws. Therefore, linear-elastic fracture mechanics can be applied to describe the failure behaviour. The main problem in the application of the simple fracture mechanics relation is the existence of a rising crack growth resistance curve, which is caused by crack bridging forces behind the advancing crack tip or by transformations in front of the crack tip. The increasing crack growth resistance leads to problems in the transformation of results from specimens with macrocracks to components with natural cracks.

INTRODUCTION

Failure of ceramic materials in most cases is caused by the extension of small flaws 10-500 µm in size. These flaws may be pores, inclusions, cracks or the surface roughness introduced during fabrication or grinding. At high temperatures failure may be caused by a creep process. Pores are created - often at grain boundaries - and can link to form microcracks and, finally, to a macrocrack.

The fracture process depends on the load history. Under a rapidly increasing load spontaneous unstable extension of the flaw may occur. In some materials a stable extension of the flaws under increasing load precedes unstable flaw extension. This effect is described by the term "R-curve behaviour".

Subcritical extension of the flaws may occur under constant or cyclic loading. Under constant loading at room temperature and moderately elevated temperatures this extension is termed environmentally enhanced crack extension. Breaking of atomic bonds is the elementary process responsible for this effect (1-7). At sufficiently high temperatures subcritical flaw extension is caused by creep phenomena (8-11).

Under cyclic loading the flaw extension is caused by a combination of environmental effects - as under constant loading - and special cyclic effects.

The description of the extension of flaws in ceramics is based on the principles of fracture mechanics. Accordingly, a flaw is assumed with a sharp tip and with a zero tip radius. Such flaws are called cracks. The application of a crack model to the description of the behaviour of real flaws may be doubtful. In the following sections this problem will be discussed in some detail.

UNSTABLE CRACK EXTENSION

Macrocrack Behaviour

A macrocrack is a crack artificially introduced into a specimen. Its depth and crack-front length are usually in the order of some millimetres. It is not easy to introduce such a crack. Different methods have been applied, some of them described in the next section.

In an ideally brittle material the stress intensity factor increases during loading until a critical value K_{Ic} is reached (Figure 1a). In a load controlled test unstable crack extension and sudden fracture occur. In a displacement-controlled test the crack extension can be stable under decreasing load (Figure 1c). The stability of the crack extension depends on the compliance of the loading system, i.e. the size of the specimen and the crack and the stiffness of the loading system. During crack extension $K_I = K_{Ic}$, independent of the amount of crack extension (Figure 1a).

In many materials a different behaviour is observed. Instead of the "flat" crack-growth resistance curve of Figure 1a an increase in K_I - now called K_{IR} - with increasing crack extension is observed, possibly leading to a plateau value (Figure 1b). Three main effects can be responsible for such a "rising" crack-growth resistance curve. Steinbrech et al. (12) have shown that for some materials, especially Al_2O_3 with large grain sizes, the crack faces behind the extending crack tip are not completely separated and that interaction forces are present between the crack surfaces. Thus, the actual stress intensity factor at the crack tip K_I is smaller than K_{Iappl} calculated from the applied stress by a value K_{Ibr}, which is caused by the bridging stresses:

$$K_I = K_{Iappl} - K_{Ibr} \tag{1}$$

with

$$K_{Iappl} = \sigma \sqrt{a}\, Y \tag{2}$$

The bridging stresses are dependent on the surface roughness. For a given roughness they are a function of the crack-opening displacement.

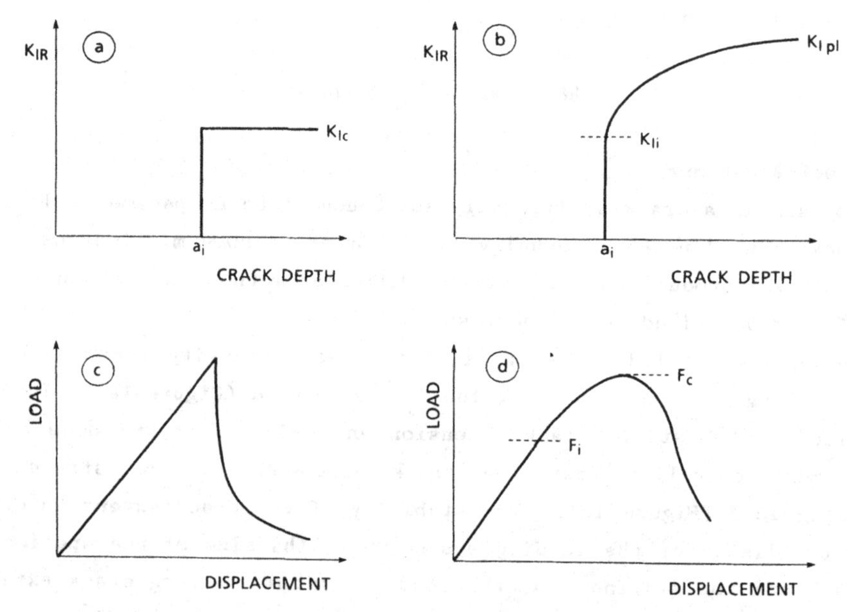

Figure 1. Flat (a) and rising (b) crack growth resistance curve with corresponding load displacement curves (c) and (d).

If a crack extends from a sharp notch, introduced by a saw, K_{Ibr} increases with the crack extension Δa. If the bridging stresses are only responsible for the rising R-curve effect, then during crack extension $K_I = K_{Io}$ and because of $K_{Iappl} = K_{IR}$

$$K_{IR} = K_{Io} + K_{Ibr} \tag{3}$$

ΔK_{br} has been calculated by several authors (13-15). Fett and Munz (15) have shown that the K_{IR}-Δa relation is not a unique material property, but is dependent on the geometry and size of the crack and of the component. This is in agreement with experimental results (12). Figure 2 shows the K_{IR}-Δa relation for different initial crack sizes in a bending specimen for a given relation between bridging stress and crack-opening displacement.

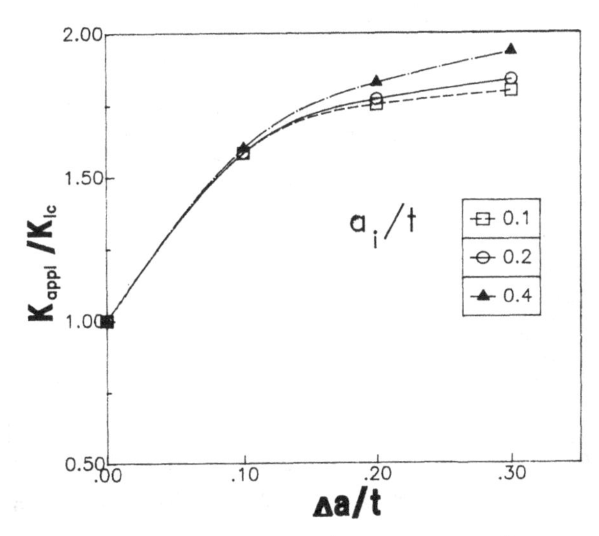

Figure 2. Influence of initial crack size a_i on R-curve calculated from bridging stresses (15).

Another reason for a rising R-curve can be an enlargement of the process zone ahead of the crack tip. This may be due to crack branching, especially in coarsely grained materials. A third effect occurs in metastable ceramics in which stress-induced transformations may occur. This effect is observed especially in ZrO_2 ceramics. In these materials transformation occurs ahead of the growing crack. The transformation-induced strains change the stress distribution and reduce the effective stress intensity factor. A steady state, and thus a plateau value of K_{IR}, is reached after some crack extension if the crack advances from a notch. A theoretical treatment of R-curves in metastable ZrO_2 ceramics can be found in references (16) and (17).

In non-precracked material cracks can develop locally within transformed areas (18). Then the fracture starts not necessarily from the natural flaws, but from the newly created cracks. For these materials the R-curves for the transformation induced cracks differ from the R-curves from notched specimens. Figure 3 shows an example. K_{IR} is plotted versus the crack length for the transformation-induced cracks because there is no initial flaw and versus the crack extension for the notched specimen. Transformation-induced residual stresses were discussed by the authors as one possible reason for the difference in behaviour of micro- and macro-cracks.

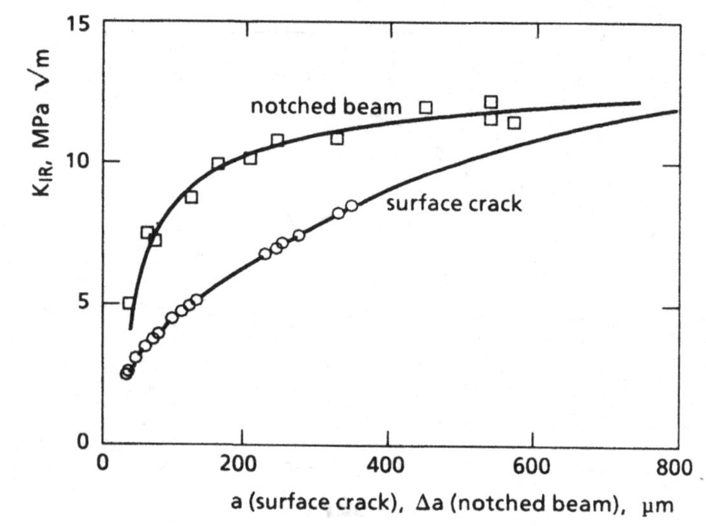

Figure 3. R-curves for magnesia-partially-stabilized zirconia (18).

Different R-curves for specimens with natural cracks and for notched specimens have also been observed by Steinbach and Schmenkel (19) for coarse grained Al_2O_3.

If all possible effects are combined, the stress intensity factor at the crack tip is given by

$$K_I = K_{Iappl} + K_{Ires} - K_{Ibr} - K_{It} , \quad (4)$$

where K_{Iappl} is the stress intensity factor calculated from the externally applied load, K_{Ires} from the residual stress field (can be positive or negative), K_{Ibr} from the bridging stresses; K_{It} is the reduction due to transformation. The critical stress intensity for crack extension is called K_{Io}. Then K_{IR} in terms of K_{Iappl} is given by

$$K_{IR} = K_{Io} - K_{Ires} + K_{Ibr} + K_{It} \quad (5)$$

The effect of crack branching leads to an increase in K_{Io} with crack extension. Unstable crack extension with load control is given by the relations

$$K_{Iappl} = K_{IR} \tag{6a}$$

$$\left(\frac{K_{Iappl}}{\partial a}\right)_{\sigma} = \frac{dK_{IR}}{da} \tag{6b}$$

Even if a K_{IR}-Δa relation independent of the initial crack size exists, the critical K_I at instability K_{Iinst} would depend on the crack size. It is important to realize that the material with the maximum plateau-value has not necessarily the largest K_{Iinst}. The critical value depends rather on the shape of the R-curve.

Determination of Fracture Toughness

General remarks: For the determination of fracture toughness different procedures and test specimens have been used. For a material with a flat crack-growth resistance curve all methods should lead to the same value, provided that the crack tip radius is small enough. For a material with a rising crack-growth resistance curve the measured value is dependent on the amount of crack extension at critical load.

The fundamental behaviour is shown in Figure 4. For simplicity, a crack-growth resistance curve independent of the initial crack length is assumed. Figure 4a shows the R-curve starting from a notch depth a_N. During pre-cracking the crack is extended to depth a_i, which is the starting crack depth for the fracture toughness test. The stress intensity factor obtained is K_{Ip}. Figure 4b shows the remaining part of the R-curve (curve B) and K_{Iappl} at the critical load. Instability occurs at $K_{Iappl} = K_{Ip}$, which then is the value obtained for K_{Ic}. Curve A in Figure 4b shows the situation in the absence of precracking. Then, during the fracture toughness test, the crack extends from $K_{Iappl} = K_{Ic}$ until instability at the tangent point. Usually, K_{Ic} is calculated from the load at instability and the initial crack length a_i instead of the real crack length $a_i + \Delta a_c$. It can be seen that precracking may lead to a K_{Ic} value near the plateau value of the R-curve. The consequence for the different method of K_{Ic} determination will be shown below.

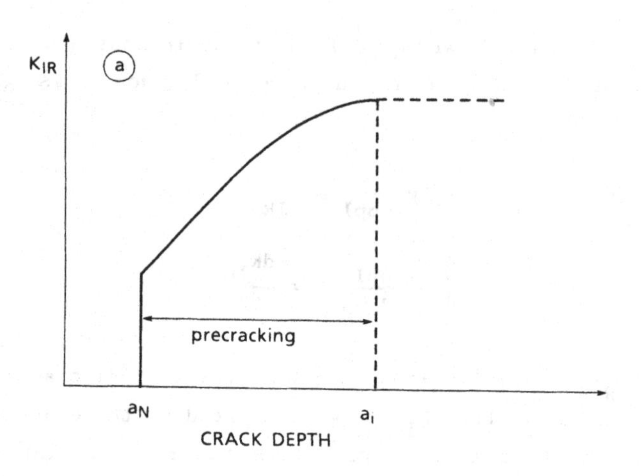

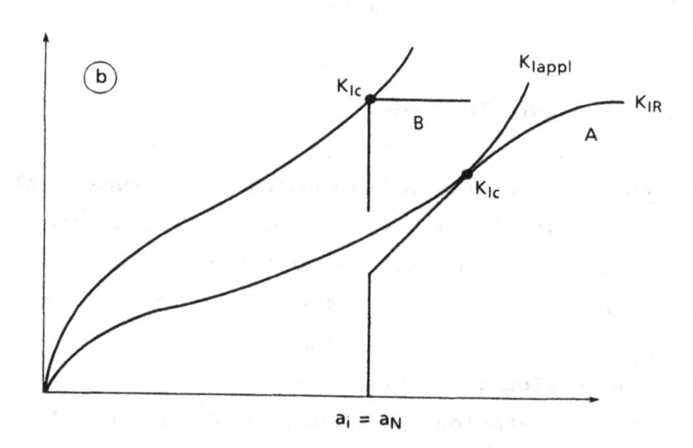

Figure 4. Effect of precracking on K_{Ic}. a: R-curve, b: K_{Ic} for precracked (B) and not precracked (A) specimens.

Bend specimen with through-the-thickness cracks: Usually, four-point bend tests are carried out. The "crack" is often a narrow slit of about 100 µm width. K_{Ic} is independent of the width s below a critical value s_c. This value is dependent on the material. Therefore, it has to be ensured that the applied s is below s_c. Curve A in Figure 4b is relevant to notched specimens and, therefore, K_{Ic} is a value at the beginning of the R-curve.

Sharp cracks can be introduced by different methods. Warren and Johannesson (20) applied a technique, according to which cracks are initiated from a Vickers or Knoop indentation by loading with a 'bridge'.

Suresh et al. (21) introduced a sharp crack from a notch by cyclic loading in compression. So far no information has been available on R-curve behaviour in fatigue crack growth.

Specimens with triangular (Chevron) notches: Three different types have been used so far (Figure 5): short bar, short rod and four-point bend specimens. A crack is initiated at the tip of the triangular notch during the loading of the specimen. During a crack extension the load passes through a maximum.

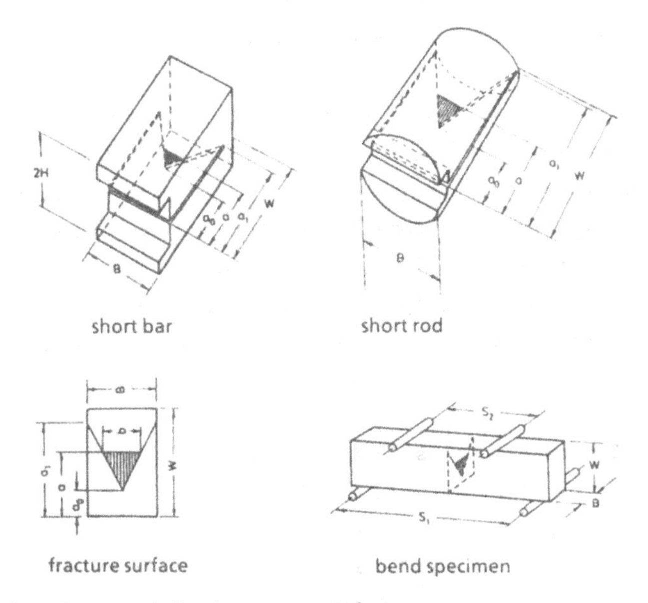

Figure 5. Specimens with chevron notches.

The relation between K_{IR} and the load is (22):

$$K_{IR} = \frac{F}{B\sqrt{W}} \left[\frac{1}{2} \frac{dC^*}{da} \cdot \frac{a_1 - a_o}{a - a_o} \right]^{\frac{1}{2}} = \frac{F}{B\sqrt{W}} \cdot Y^* \qquad (7)$$

with $a = a/W$, $a_1 = a_1/W$, $a_o = a_o/W$.

The geometric parameters B, W, a_o, a_1 are given in Figure 5. C* is the dimensionless compliance given by C* = EBC, where C is the compliance of the specimen with a trapezoidal crack front (configuration as the crack proceeds through the triangular ligament of the chevron notch).

Y* first decreases with increasing crack extension and then increases again. K_{Ic} is calculated from the maximum load and the minimum of Y*:

$$K_{Ic} = \frac{F_m}{B \sqrt{W}} Y^*_m \qquad (8)$$

For a flat R-curve Y_m corresponds to the maximum load F_m and eq. (8) is correct. For a rising R-curve the maximum load does not occur exactly at the crack length where $Y^* = Y_m$. Nevertheless, eq. (8) leads to a value of K_{Ic}, which is on the K_{IR}-Δa curve. Because of the relatively large crack extension up to maximum load a relatively high value of K_{Ic} near the plateau value is obtained. Also an effect of specimen size on K_{Ic} is inherent with this method for a material with a rising R-curve.

Specimen with Knoop cracks: With a Knoop-indenter semi-elliptical surface flaws can be introduced in a bend specimen. The depth of the cracks is < 1 mm and therefore approaches the size of real flaws. The main problem with this method is the residual tensile stresses at the crack tip which are introduced during precracking. These stresses can be released by removing a small surface layer through grinding or annealing. As in other precracked specimens high K_{Ic} values and an increase in K_{Ic} with increasing crack size can be expected for materials with an R-curve behaviour.

Conclusion: The main conclusion is that a unique method does not exist for K_{Ic} determination and that for materials with a rising crack growth resistance curve the measured K_{Ic} depends on the amount of crack extension at maximum load. For specimens with a chevron notch crack extension before occurrence of maximum load is unavoidable. The same is true for the Knoop cracks. For four point bend specimens with through the wall cracks, the precrack length depends on the precracking procedure.

There is a need for standardization of the K_{Ic} evaluation method. It is necessary to obtain a value of K_{Ic}, which is determined at a specified amount of crack extension, preferably at the beginning of crack extension.

Microcrack Behaviour

The strength of a ceramic material containing a crack of depth a can be calculated from the fracture toughness K_{Ic} applying the basic relation

$$\sigma_c = \frac{K_{Ic}}{\sqrt{a}\, Y} \qquad (9)$$

In a $\log \sigma_c$ - log a plot a straight line with a slope of -0.5 is predicted by eq. (9). For small cracks deviations from this straight line have been observed (23-25), as shown schematically in Figure 6.

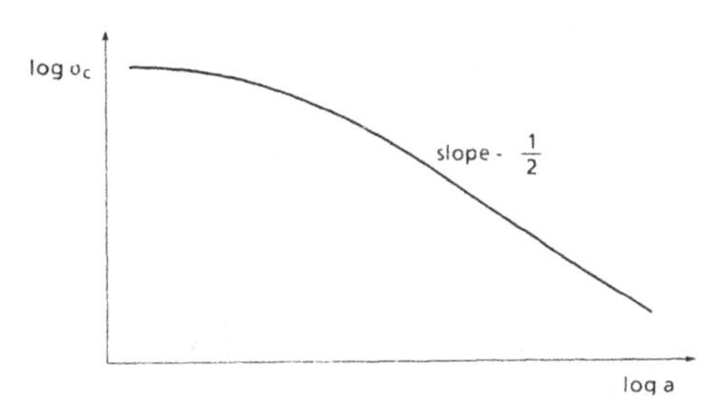

Figure 6. Fracture stress versus flaw size according to results of Usami et al. (25).

There are several possible reasons for such a deviation and, generally, several problems opposing the application of eq. (9) to small natural flaws in ceramics:

- flaw geometry
- microstructure
- breakdown of linear elastic fracture mechanism
- R-curve behaviour
- residual stress.

Flaw geometry: The application of eq. (9) requires that the flaw be described as a crack with a sharp tip. Many flaws in ceramic materials are three-dimensional such as pores or inclusions. Nevertheless, these flaws very often are described as crack-like defects. This can be assumed as a conservative assessment of the flaws because a sharp crack is more dangerous than a pore. However, starting from a pore or an inclusion a sharp crack can be created which leads to special configurations.

The most conservative procedure is to describe the flaws as internal elliptical cracks or semi-elliptical surface cracks with the crack plane orientated perpendicularly to the stress axis. The stress intensity factor varies along the crack front. Relations are given by Newman and Raju (26) for surface cracks and by Fett and Mattheck (27) for internal cracks.

If cracks are initiated at pores a pore-crack configuration can be described as sphere or hemisphere with a circumferential crack (28). The stress intensity factor - or Y in eq. (9) - depends on the ratio of crack length a to pore radius R (Figure 7). Relations are given by Baratta (29). The crack length should be of the order of the grain size (28). However, it has been found that the experimental results could in some cases be described only by application of a much larger crack length (30,31).

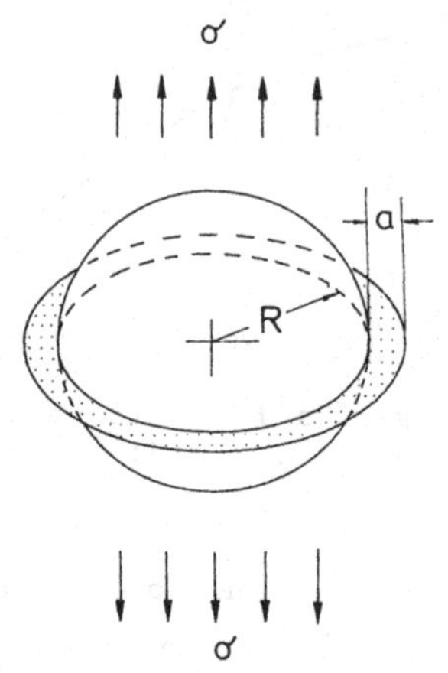

Figure 7. Pore with circumferential crack.

Inclusions with a perfect bonding to the matrix lead to stress concentrations due to differing elastic constants. In addition, thermal stresses may develop during cooling after processing of the material due to different thermal expansions. Different cracks can develop, dependent on these physical properties: the size of the inclusions and the strength of the inclusions, the matrix and the interface; cracks in the inclusion; circumferential cracks in the matrix; and interfacial cracks.

<u>Microstructure and application of linear-elastic fracture mechanics</u>: Linear-elastic fracture is a continuum mechanics theory neglecting the microstructure of the material. Fracture mechanics (linear-elastic or elasto-plastic) cannot be applied if the crack is in the order of the microstructure (grain size) of the material. The fracture stress then is dependent on the orientation of the crack with respect to the grain configuration.

A second requirement for application of linear-elastic fracture mechanics is that the size of the process zone ahead of the crack tip is small compared to the crack size. The process zone is the region ahead of the crack tip where damage, e.g. microcracking, occurs before the onset of unstable fracture. Usually, the size of this zone is not known. As a first approximation eq. (9) can be modified for small cracks, if the crack length a is replaced by an effective crack length.

$$a_{eff} = a + \rho/2 \tag{10}$$

where ρ is the size of the process zone:

$$\sigma_c = \frac{K_{Ic}}{(a + \rho/2)^{\frac{1}{2}}\, Y} \tag{11}$$

Applying eq. (11) to results of Cook et al. (23) leads to values of ρ between 20 μm and 300 μm to ratios of ρ to grain size between 2 and 25, provided that all other effects are neglected. The results of Usami et al. (25) can be described with eq. (11) and ρ = 36 μm.

R-curve behaviour: As already mentioned, eq. (9) cannot be applied to materials showing a rising R-curve. Fracture stresses have to be calculated from eq. (6). This requires the knowledge of the K_{IR}-Δa curve, which may be dependent on the initial crack size.

The effect of the R-curve on the relation between fracture stress and flaw size is shown in Figure 8 for two R-curves with the same initial value (K_{Io} = 5 MPa √m and plateau value (K_{Ip} = 10 MPa √m) but different slopes. It can be seen that the decrease in strength with increasing flaw size is less than predicted for a flat crack growth resistance curve with K_{IR} = const. If the prediction is made with the plateau value, which may be obtained from a fracture toughness test with specimens containing macro-cracks (K_{Ic} = 10 MPa √m), the fracture stress is over-estimated. It can also be seen from Figure 8b that in a log σ_c - log a plot deviations from the straight line at large cracks may be caused by the R-curve effect.

Residual stresses: Ceramic materials may have residual stresses resulting from the fabrication process. The stresses may have steep gradients at the surface. Residual stresses and the stresses caused by external loading can be superimposed linearly. Therefore, compressive residual stresses increase the tensile strength and tensile residual stresses decrease the strength. For steep gradients large flaws may be only partially influenced by the residual stresses.

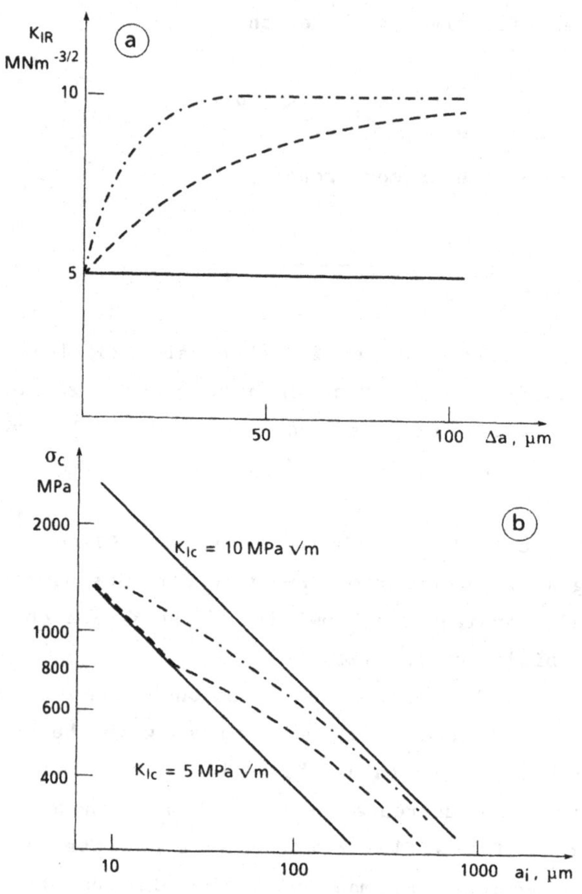

Figure 8. Two R-curves (a) and corresponding relation between fracture stress and flaw size (b).

SUBCRITICAL CRACK EXTENSION UNDER CONSTANT LOAD

General Relation

Subcritical crack extension for a material in a given environment is described by a relation between the crack growth rate da/dt and the stress intensity factor K_I:

$$v = da/dt = v\ (K_{Ic}) \qquad (12)$$

If Figure 9 the general behaviour is shown schematically in a log-log plot. A lower bound K_{Ic} may exist below which no crack extension occurs.

Such a threshold was detected especially for glass (32,33). For ceramics, the existence of a threshold has not been proved so far. In some publications a threshold is indicated (25,34); however, the data base seems to be insufficient. Within several decades a linear relation has been found for several ceramics and, therefore, a power law relation between v and K_I exists:

$$v = AK_I^n \tag{13}$$

At large crack growth rates a plateau region may exist before unstable fracture occurs at $K_I = K_{Ic}$.

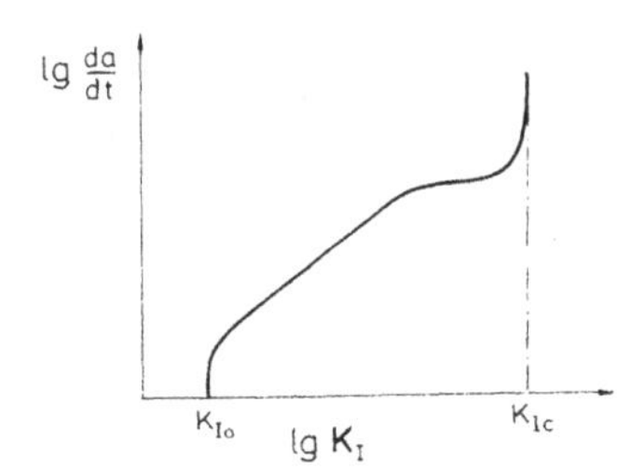

Figure 9. Typical v-K curve.

For lifetime evaluation eq. (13) can be used because the transition in the plateau region occurs at very high crack growth rates.

From eq. (13) and the general relation

$$K_I = \sigma \sqrt{a}\, Y \tag{14}$$

the general lifetime relation for time dependent stresses can be written:

$$\int_0^{t_f} \left[\sigma(t)\right]^n dt = a \frac{2}{AY^n(n-2)} \left[a_i^{\frac{n-2}{2}} - a_c^{\frac{n-2}{2}}\right] \tag{15}$$

In this relation it is assumed that Y is not dependent on the crack length. Because of the small amount of crack extension this assumption may be valid.

The initial crack size a_i and the final crack size a_c can be replaced by

$$a_i = \left[\frac{K_{Ic}}{\sigma_c \; Y} \right] \qquad (16)$$

$$a_c = \left[\frac{K_{Ic}}{\sigma_f \; Y} \right] \qquad (17)$$

σ_c is the inert fracture strength, which can be measured in a test with high loading rate where any subcritical crack growth is avoided. Equation (16), however, can be applied only if the strength and the delayed failure are caused by the same flaw population. σ_f in eq. (17) is the actual stress at failure.

Equation (15) then can be rewritten as

$$\int_0^{t_f} \left[\sigma(t) \right]^n dt = B \left[1 - \frac{\sigma_f}{\sigma_c}^{n-2} \right]. \qquad (18)$$

with

$$B = \frac{2}{AY^n (n-2) K_{Ic}^{n-2}} \qquad (19)$$

Because of the high values of n found, very often $(\sigma_f/\sigma_c)^{n-2} << 1$ and then

$$\int_0^{t_f} \left[\sigma(t) \right]^n dt = B\sigma_c^{n-2} \qquad (20)$$

Determination of the v-K_I Relation for Specimens with Natural Flaws

The v-K_I relation, especially the parameters A and n in eq. (13), can be determined directly, from specimens with macrocracks (see below), or indirectly from specimens with natural flaws.

Dynamic bending method: The most currently used method so far has been the so-called dynamic bending test. Specimens are loaded in four-point bending at different loading rates. For a constant stress rate $\dot{\sigma}$ eq. (20) leads to the fracture strength σ_f:

$$\sigma_f^{n+1} = B\sigma_c^{n-2}\,\dot{\sigma}(n+1)\left[1 - (\sigma_f/\sigma_c)^{n-2}\right] \tag{21}$$

This relation is shown in Figure 10 in a log σ_f - log $\dot{\sigma}$ plot. For high loading rates σ_f approaches the inert fracture strength. For low loading rates, there is

$$\sigma_f^{n+1} = B\sigma_c^{n-2}\,\sigma(n+1) \tag{22}$$

The parameter n can be obtained from the slope of the log σ_f - log $\dot{\sigma}$ plot and the quantity $B\sigma_c^{n-2}$ and thus A from relation (19) is obtained from the location of the straight line.

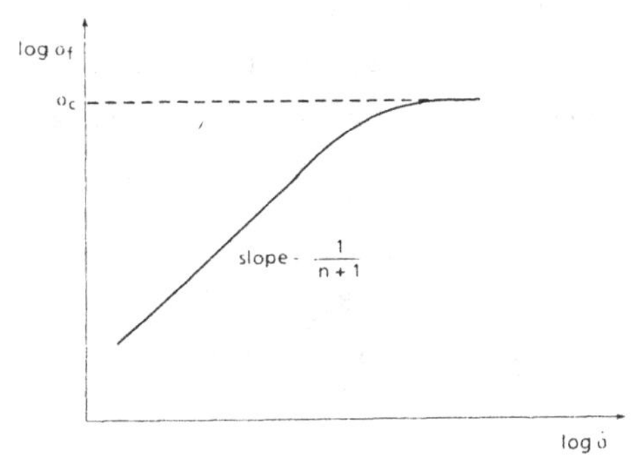

Figure 10. Fracture stress versus loading rate according to eq. (21).

To apply eq. (22) it has to be ensured that all data points occur within that section of the log σ_f - log $\dot{\sigma}$ curve which can be described by a straight line. Figure 11 shows results for hot-pressed silicon nitride at elevated temperatures; the transition in the plateau region can be seen.

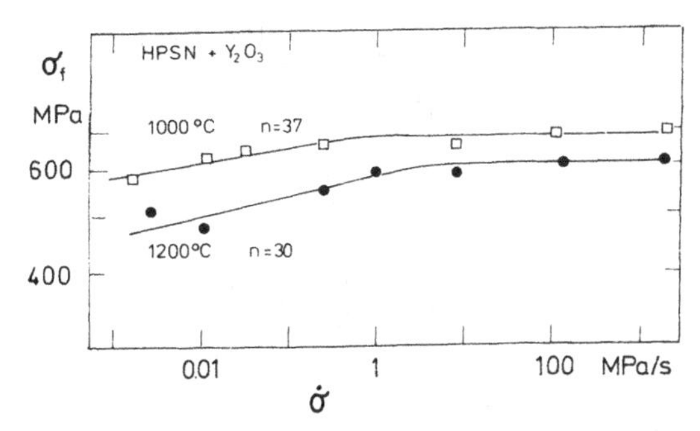

Figure 11. Bending strength versus loading rate for hot-pressed silicon nitride (35).

Lifetime method: From eq. (20) the lifetime for constant stress is given by

$$t_f = \frac{B\sigma_c^{n-2}}{\sigma^n} \tag{23}$$

From a log σ - log t_f plot the parameter n can be obtained from the slope and the quantity $B\sigma_c^{n-2}$ from the position of a straight line connecting the data points. Application of this method calls for many specimens because of the large scatter in lifetime, and it is therefore very time-consuming.

Modified lifetime method: A modified lifetime procedure is described in (36). It relies on the scatter of the lifetime and does not require that a power law relation between crack growth rate and stress intensity factor is assumed. From eqs. (12) and (13) the lifetime for constant stress is given by

$$t_f = \frac{2}{\sigma^2 Y^2} \int_{K_{Ii}}^{K_{Ic}} \frac{K_I}{v(K_I)} dK_I \tag{24}$$

K_{Ii} is the initial stress intensity factor given by

$$K_{Ii} = \sigma \sqrt{a_i}\, Y \tag{25}$$

Differentiation with respect to K_{Ii} leads to the crack growth rate at the beginning of the test

$$v(K_{Ii}) = - \frac{2K_{Ii}^2}{t_f \sigma^2 Y^2} \frac{d[\log K_{Ii}]}{d[\log (t_f \sigma^2 Y^2)]} \tag{26}$$

Introducing

$$K_{Ii} = \frac{\sigma}{\sigma_c} K_{Ic} \tag{27}$$

leads to

$$v(K_{Ii}) = -\frac{2K_{Ic}^2}{t_f\sigma_c^2Y^2}\frac{d[\log \sigma/\sigma_c]}{d[\log (t_f\sigma^2)]} \tag{28}$$

For an evaluation of eq. (28) the relation between σ/σ_c and $t_f\sigma^2$ has to be obtained. One possibility would be to introduce cracks (e.g. Knoop cracks) of the same size into specimens and to measure the lifetime t_f for different stresses σ, while the strength is measured in a separate test. Another method - which is recommended here - is to utilize the natural scatter of the lifetime for tests with constant stress and the scatter in strength. The basic assumption for this procedure is that the inert fracture strength and the lifetime are determined by the same flaw population. A number of n specimens are tested under constant stress and the lifetimes obtained range from t_{f1} to t_{fn}. The same number of specimens are used for the measurement of fracture strength and the results also range of σ_{c1} to σ_{cn}. Then log σ/σ_{ci} is plotted versus log t_{fi} σ^2, assuming that both specimens with σ_{ci} and t_{fi} have the same initial flaw size. For a given combination t_{fi}, σ_{ci} the crack growth rate is calculated from

$$(K_{Ii}) = -\frac{2K_{Ii}^2}{t_{fi}\sigma_{ci}^2Y^2}\, m \tag{29}$$

where m is the slope of the log σ/σ_{ci} - log t_{fi} σ^2 plot. This value is related to the stress intensity factor

$$K_{Ii} = \frac{\sigma}{\sigma_{ci}} K_{Ic} \tag{30}$$

The advantage of this method is that no assumption has to be made on the form of the v-K_I relation and that very low crack-growth rates can be obtained. An example of the evaluation procedure is shown in Figure 12.

Specimens with Macrocracks

Several specimens with through the thickness cracks have been used to measure crack growth rates. The double-torsion specimen was widely used in the past. A detailed description of the method was given by Fuller (38). Direct measurements of the crack extension can be made or the crack length can be obtained from the compliance of the specimen. The double cantilever beam specimen is another specimen which was used (39).

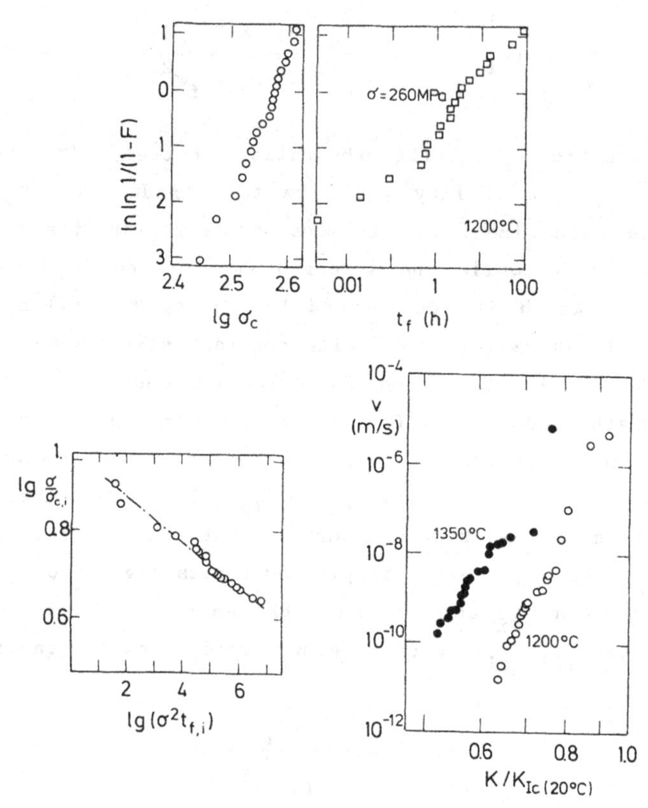

Figure 12. Evaluation procedure for v-K curve for hot-pressed silicon carbide (for 1350°C only v-K curve is shown) (37).

The measurements with these specimens are restricted to crack growth rates down to about 10^{-8} m/sec, whereas with the modified lifetime method crack growth rates of 10^{-12} m/sec can be obtained.

However, the main problems in these tests are R-curve effects. This is shown in Figure 13. For Al_2O_3 the crack growth rate was measured in 3-point bending using the compliance method. For each specimen the stress intensity factor increased with increasing crack length. The crack growth rate, however, first decreased and then increased again. Figure 13 also shows the $v-K_I$ curve obtained indirectly with specimens having natural flaws. The initial crack growth rates of the specimens with macrocracks was the same as for the specimens with microcracks. The decrease of the crack growth rates can be related to the R-curve behaviour. In principle, the same relation as for stable crack extension under increasing load - eq. (4) - can be applied.

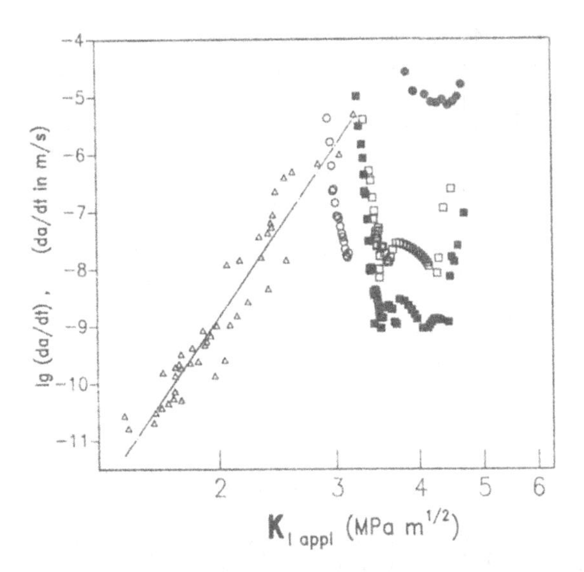

Figure 13. v-K_I curves of Al_2O_3 for macrocracks and for natural cracks (Δ) (40).

The crack growth rate is a function of the stress intensity factor at the crack tip: for a power law relation and considering only the bridging stresses,

$$v = v(K_I) = AK_I^n = A[K_{Iappl} - K_{Ibr}]^n = A^* \left[\frac{K_{Iappl} - K_{Ibr}}{K_{Ic}} \right]^n \qquad (31)$$

In Figure 14 calculated v-K_{Iappl} curves are shown for an assumed relation between K_{Ibr} and crack extension of

$$K_{Ibr} = K_I^* \left[1 - \exp \frac{a - a_i}{a^*} \right] \qquad (32)$$

Comparing Figure 13 and Figure 14 it can be seen that the macrocrack behaviour can be explained by the R-curve effect. A detailed evaluation will be given elsewhere.

Figure 13 shows clearly that no unique relation exists between crack growth rate and stress intensity factor - at least not for the macrocracks. Therefore, the macrocrack method cannot be recommended for the determination of the v-K_I relation for lifetime prediction.

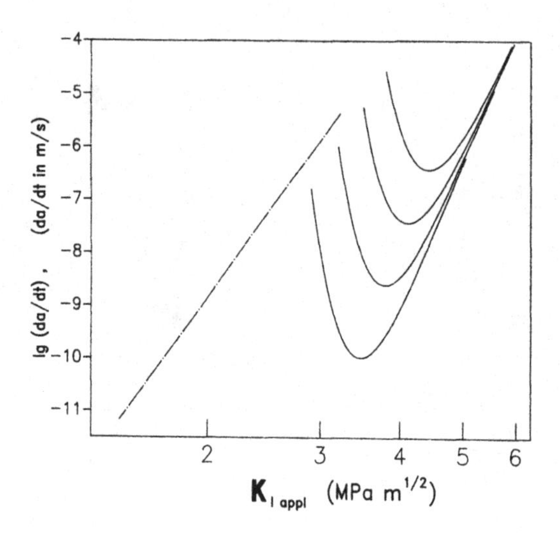

Figure 14. Calculated v-K_I curves from eqa. (31) and (32) according to results of Figure 13 ($n = 19$, $A^* = 7.02 \cdot 10^{-5}$ m/s, $K_{Ic} = 4$ MPa $\sqrt{m}$, $K_I^* = 1.945$ MPa $\sqrt{m}$, $a^* = 0.17$ mm).

SUBCRITICAL CRACK EXTENSION UNDER CYCLING LOADING

The lifetime under cyclic loading can be predicted from the results of static loading, applying eq. (15), under the assumption that no additional cyclic fatigue effect occurs. For a periodic stress cycle with the period T

$$\sigma = \sigma_m + \sigma_a . f(t), \qquad f(t) = f(t+T) \tag{33}$$

the lifetime is obtained from eq. (15):

$$t_{fz} = \frac{B\sigma_c^{n-2}}{\sigma_m^n} \cdot \frac{1}{g(n, \sigma_a/\sigma_m)} , \tag{34}$$

where the function g is given by

$$g = \frac{1}{T} \int_o^T \left[1 + \frac{\sigma_a}{\sigma_m} f(t) \right]^n dt \tag{35}$$

If the lifetime is plotted versus the amplitude σ_a or the maximum stress $\sigma_{max} = \sigma_m + \sigma_a$ with constant $R = \sigma_{min}/\sigma_{max}$, then eq. (34) has to be replaced by

$$t_{fz} = \frac{B\sigma_c^{n-2}}{\sigma_a^n} \cdot \frac{1}{g} \left[\frac{1 - R}{1 + R}\right]^n \quad (36)$$

$$t_{fz} = \frac{B\sigma_c^{n-2}}{\sigma_{max}^n} \cdot \frac{1}{g} \frac{2^n}{(1 + R)^n} \quad (37)$$

If the crack growth rate is measured during cyclic loading, the results can be represented as the crack-growth rate per cycle da/dN as a function of the range of the stress intensity factors ΔK or, alternatively, as the average crack-growth rate during one cycle $\bar{v} = \Delta a/T$ as a function of the mean value K_{Im} of the cycle. If no additional cyclic effect occurs, then the relations

$$\bar{v} = A\, K_{Im}^n \cdot g \quad (38)$$

and

$$\frac{da}{dN} = \left[\frac{1 + R}{2(1 - R)}\right]^n g\, T\, A\, (\Delta K)^n \quad (39)$$

should be valid.

In Figure 15 results of Evans and Fuller (41) are shown for porcelain, for which the predicted and measured crack growth rates are in good agreement.

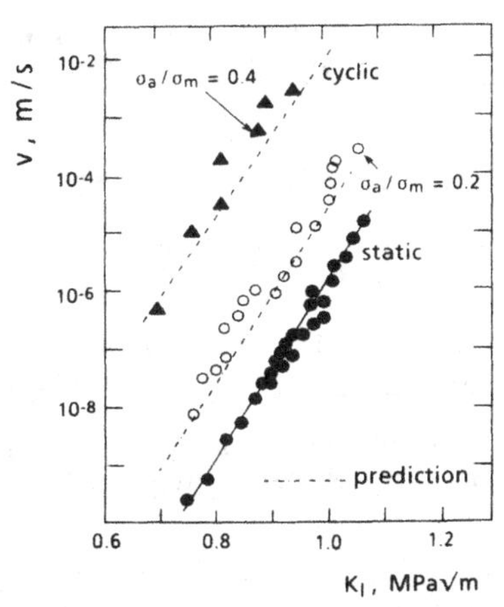

Figure 15. Crack growth rate for static and cyclic loading for porcelain; dashed lines are predictions from static loading (41).

This result, however, is not typical of ceramics. In many investigations it was shown that the lifetime under cyclic loading is much shorter than predicted from static loading (34,37,42,43,44). Figure 16 shows some results.

As for subcritical crack growth under static loading, the question about the effect of R-curves and the possible discrepancy between macrocrack and microcrack behaviour arises.

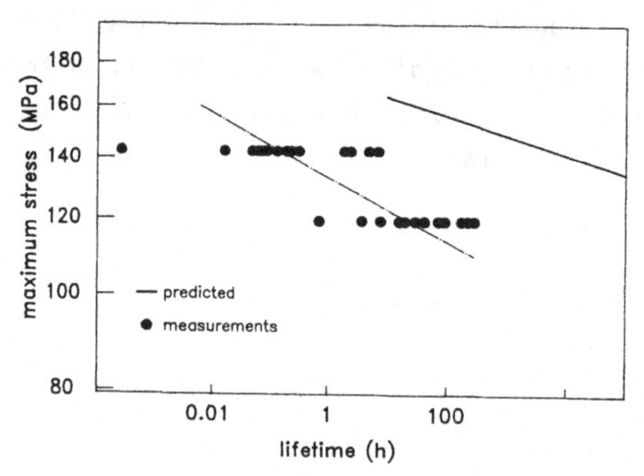

Figure 16. Lifetime for cyclic loading and prediction from static loading for alumina (44).

REFERENCES

1. Wiederhorn, S.M., Johnson, H., Diness, A.M. and Heuer, A.H., Fracture of glass in vacuum, J. Amer. Ceram. Soc., 1974, **57**, 336-341.

2. Wiederhorn, S.M. and Bolz, L.H., Stress corrosion and static fatigue of glass, J. Amer. Ceram. Soc., 1970, **53**, 543-548.

3. Fuller, E.R. and Thomson, R.M., Lattice theories of fracture, in: Fracture Mechanics of Ceramics IV, Plenum Press, 1978, 507-548.

4. Lawn, B.R., An atomistic model of kinetic crack growth in brittle solids, J. Mater. Sci., 1975, **10**, 469-480.

5. Krausz, A.S. and Mshana, J., The steady state fracture kinetics of crack front spreading, Int. J. Fract., 1982, **19**, 277-293.

6. Fett, T. and Munz, D., Zur Deutung des unterkritischen Risswachstums in keramischen Werkstoffen, DFVLR-Forschungsbericht, FB 8307, Cologne, 1983.

7. Fett, T., A fracture mechanical theory of subcritical crack growth in ceramics, Accepted for publication in Int. J. Fract.

8. Evans, A.G. and Rana, A., High temperature failure mechanisms in ceramics, Acta Met., 1980, **28**, 129-141.

9. Porter, J.R., Blumenthal, W. and Evans, A.G., Creep fracture in ceramic polycrystals, Acta Met., 1981, **29**, 1899-1906.

10. Dalgleish, B.J., Johnson, S.M. and Evans, A.G., High temperature failure of polycrystalline alumina, J. Amer. Ceram. Soc., 1984, **67**, 741-750.

11. Page, R.A., Lankford, J., Chan, K.S., Hardmann-Rhyne, K. and Spooner, S., Creep cavitation in liquid phase sintered alumina, J. Amer. Ceram. Soc., 1987, **70**, 137-145.

12. Steinbrech, R., Knehans, R. and Schaarwächter, W., Increas of crack resistance during slow crack growth in Al_2O_3 bend specimens, J. Mat. Sci., 1983, **18**, 265-270.

13. Evans, A.G. and Faber, K.T., Crack growth resistance of microcracking brittle materials, J. Amer. Ceram. Soc., 1984, **67**, 255-260.

14. Mai, Y. and Lawn, B.R., Crack-interface grain bridging as a fracture resistance mechanism in ceramics: II, Theoretical fracture mechanics model, J. Amer. Ceram. Soc., 1987, **70**, 289-294.

15. Fett, T. and Munz, D., Influence of crack surface interactions on stress intensity factor in ceramics, J. Mat. Sci. Lett., 1990, **9**, 1403-1406.

16. Rose, L.F., Kinematical model of stres-induced transformation around cracks, J. Amer. Ceram. Soc., 1986, **69**, 208-212.

17. McMeeking, R.M. and Evans, A.G., Mechanics of transformation-toughening in brittle materials, J. Amer. Ceram. Soc., 1984, **65**, 242-246.

18. Marshall, D.B. and Swain, M.V., Crack resistance curves in magnesia-partially-stabilzied zirconia, J. Amer. Ceram. Soc., 1988, **71**, 399-407.

19. Steinbrech, R. and Schmenkel, O., Crack-resistance curves of surface cracks in alumina, Comm. Amer. Ceram. Soc., 1988, **71**, C271-C273.

20. Warren, R. and Johannsson, B., Creation of stable cracks in hard metals using 'bridge' indentation, Powder Met., 1984, **27**, 25-29.

21. Suresh, S., Ewart, L., Maden, M., Slaughter, W.S. and Nguyen, M., Fracture toughness measurements in ceramics: pre-cracking in cyclic compression, J. Mat. Sci., 1987, **22**, 1271-1276.

22. Barker, L.M., A simplified method for measuring plane strain fracture toughness, Eng. Fract. Mech., 1977, **9**, 361-366.

23. Cook, R.F., Lawn, B.R. and Fairbanks, C.J., Microstructure-strength properties in ceramics: I, Effect of crack size on toughness, J. Amer. Ceram. Soc., 1985, **68**, 604-615.

24. Hoshide, T., Furuya, H., Nagase, Y. and Yamada, T., Fracture mechanics approach to evaluation of strength in sintered silicon nitride, Int. J. Fract., 1984, **26**, 229-239.

25. Usami, S., Takahashi, I. and Machida, T., Static fatigue limit of ceramic materials containing small flaws, Eng. Fract. Mech., 1986, **25**, 483-495.

26. Newman, J.C and Raju, I.S., An empirical stress intensity factor equation for the surface crack, Eng. Fract. Mech., 1981, **15**, 185-192.

27. Fett, T. and Mattheck, C., Stress intensity factors of embedded elliptical cracks for weight function application, Int. J. Fract., 1989, **40**, R3-R11.

28. Evans, A.G. and Tappin, G., Proc. Br. Ceram. Soc., 1972, **20**, 275-297.

29. Baratta, F.I., Mode I stress intensity factors for various configurations involving single and multiple cracked spherical voids, Fracture Mechanics of Ceramics, Vol. 5, 1983, 543-567.

30. Munz, D., Rosenfelder, O., Goebbels, K. and Reiter, H., Assessment of flaws in ceramic materials on the basis of non-destructive evaluation, Fracture Mechanics of Ceramics, Vol. 7, 1986, 265-283.

31. Heinrich, J. and Munz, D., Strengths of reaction-bonded silicon nitride with artificial pores, Amer. Ceram. Soc. Bull., 1980, **59**, 1221-1222.

32. Wiederhorn, S.M. and Bolz, L.H., Stress corrosion and static fatigue of glass, J. Amer. Ceram. Soc., 1970, **53**, 543-548.

33. Fett, T., Germerdonk, K., Grossmüller, A., Keller, K. and Munz, D., Subcritical crack growth and threshold in borosilicate glass, J. Mat. Sci., 1991, **26**, 253-257.

34. Kawakubo, T. and Komeya, K., Static and cyclic fatigue behaviour of a sintered silicon nitride at room temperature, J. Amer. Ceram. Soc., 1987, **70**, 400-405.

35. Keller, K., Ph.D. Thesis, University of Karlsruhe, FRG, 1989.

36. Fett, T. and Munz, D., Determination of V-K_I curves by a modified evaluation of lifetime measurements in static bending tests, Comm. Amer. Ceram. Soc., 1985, **68**, C213-C215.

37. Fett, T. and Munz, D., Subcritical crack extension in ceramics, MRS International Meeting on Adv. Mats., Vol. 5, 505-523, Materials Research Society, 1989.

38. Fuller, E.R., An evaluation of double-torsion testing, ASTM STP 678, 1979, 3-18.

39. Freiman, S.W., Murville, D.R. and Mast, P.W., Crack propagation studies in brittle materials, J. Mater. Sci., 1973, **8**, 1527-1533.

40. Fett, T. and Munz, D., Subcritical crack growth of macro-cracks in alumina with R-curve behaviour, submitted to J. Amer. Ceram. Soc.

41. Evans, A.G. and Fuller, E.R., Crack propagation in ceramic materials under cyclic loading conditions, Met. Trans., 1974, **5**, 27-33.

42. Grathwohl, G., Ermüdung von Keramik unter Schwingbeanspruchung, Mat.-wiss. u. Werkstofftechn., 1988, **19**, 113-124.

43. Dauskardt, R.H., Yu, W. and Ritchie, R.O., Fatigue crack propagation in transformation-toughened zirconia ceramics, J. Amer. Ceram. Soc., 1987, **70**, C248-257.

44. Fett, T., Martin, G., Munz, D. and Thun, G., Determination of da/dn-ΔK curves for small cracks in alumina in alternating bending tests, to be published in J. Mat. Sci.

WEAKEST-LINK FAILURE PREDICTION FOR CERAMICS

L. DORTMANS and G. de WITH
Centre for Technical Ceramics,
PB 595, 5600 AN, Eindhoven,
The Netherlands.

ABSTRACT

Detailed problems associated with measurement and interpretation of the bend strength of ceramics are presented. The relationship between 3- and 4-point bend and biaxial test results is discussed in the light of current theoretical understanding. There remain a number of unresolved aspects and a new, 'damage mechanics' approach is mooted for further consideration.

INTRODUCTION

Weakest-link failure prediction for components is usually based on data obtained from 3- or 4-point bend tests but several pitfalls are encountered. First, the bend tests themselves have problems, not only due to the limited volume tested but also to the test procedure and the quality of the specimen surface. Once these data are reliably generated, the next question is the transformation of these uni-axial data to multi-axial data, but no universally applicable relation is yet available. In the next section we discuss the problems associated with the bend test and the transformation of uni-axial to multi-axial data. Once this procedure has been carried out, further problems emerge; it is concluded that alternative approaches may be worthy of consideration. Here we propose a damage approach based on micro-structural insights and anisotropic deformation behaviour.

BEND TESTS

The problems associated with bend tests can be divided into three categories. Firstly, the problems directly related to the actual execution of the test:

- the application of simple beam theory
- friction effects at the supports
- local stresses at the supports
- unequal applied moments
- twisting and wedging of the specimen.

The problems associated with the performance of the test have been dealt with adequately (1-2). The general conclusion is that it is possible to obtain data reliable to about 1% if and only if all the above mentioned effects are taken into account properly.

Secondly, the problems related to the surface condition of the specimens. Here we refer to:

- roughness of the specimen surface
- damage introduced during grinding
- residual stresses in the specimen surface
- significance of the specimen surface for the actual component.

The situation here is less clear. Often for a component a certain (low) surface roughness is demanded. This is usually obtained by grinding and, if necessary, subsequent polishing. The surface damage introduced during grinding is not removed during polishing. Experiments with polished specimens show that by polishing the strength is increased somewhat but also that the Weibull modulus decreases. Moreover due to grinding, compressive residual stresses can occur which may disappear either partially or wholly during polishing.

Possibilities exist to improve the strength of specimens (and components) e.g. by ion implantation or by ductile or 'damage free' grinding. The former method (6) tries to remedy the damage done during grinding by introducing compressive stresses into the surface and, possibly, rounding off of the defect tips and healing of the defects. This has resulted in a higher strength (about 15%) and Weibull modulus (about 100%) for materials implanted with a dose of about 10^{17} cm^{-2} of noble gas ions. The latter method (4) is in principle much better suited since it minimizes the damage introduced during normal grinding. This can be realized by using an extremely stiff machine with 0.1 μm positioning capability and highly developed grinding wheel technology. Moreover, very low feed rates and cutting depths are used. Using this technique (4) for a particular type of sialon, an increase in strength from 480 for normally ground material to

875 MPa for ductile ground material was observed. Also the roughness decreased from 0.35 μm to 8 nm.

Residual stresses are important to consider. Unfortunately only for a few materials a more or less complete analysis of the residual stresses is performed, but these few studies indicate that a substantial influence is possible. For certain materials which show a phase transformation, e.g. ZrO_2 or $BaTiO_3$, the influence can be even larger due to the difference in thermal expansion coefficient and specific volume for the two phases involved.

Probably superfluously, it should be remarked that all predictions for components can only be done reliably if the surface of the component has received the same treatment as the specimens tested.

Thirdly, all aspects related to the interpretation of the test data. Here we can refer to:

- estimation of the failure probability
- extraction of the model parameters by different methods like least-squares, maximum-likelihood or method of moments.

A concise survey of these aspects has recently been given (7).

MULTI-AXIAL STRESSES

After the completion of the data collecting procedure and the identification of the type (surface or volume) and number of defects involved, a procedure is required to transform the uni-axial data into multi-axial data. For this transformation stress volume/surface integrals from the bend test have to be known. It has been shown (3) that the information from bend tests can be generated by using the analytical formulae for the stress integrals, within an error of ∿ 1%. Thus no finite element calculations are necessary, at least if the bend test specimen satisfies certain size requirements.

Unfortunately, once these integrals are given, no general prescription exists how to apply them to a multi-axial stress state. Even for a transformation to a bi-axial situation the situation is not clear (5). Well known are the models of Weibull (normal stress criterion), Stanley (independent stress criterion) and Lamon (maximum strain energy release rate criterion). On the assumption that 4-point bend data are available, these models do predict identical results for 3-point bend test. However, for the equi-biaxial stress state, e.g. as used in a ball-on-ring test, the situation is completely different. The estimated strength for the ball-on-ring test from 4-point data can differ as much as 10% for the three models, dependent on the value of the Weibull modulus, Figure 1. This may not seem large but it must be recognized that these differences lead to far larger differences in the failure probabilities, e.g. a factor 10. Experimental work to establish whether a certain model is applicable and under what conditions is necessary. Extension to multi-axial stress states is the next aspect to consider.

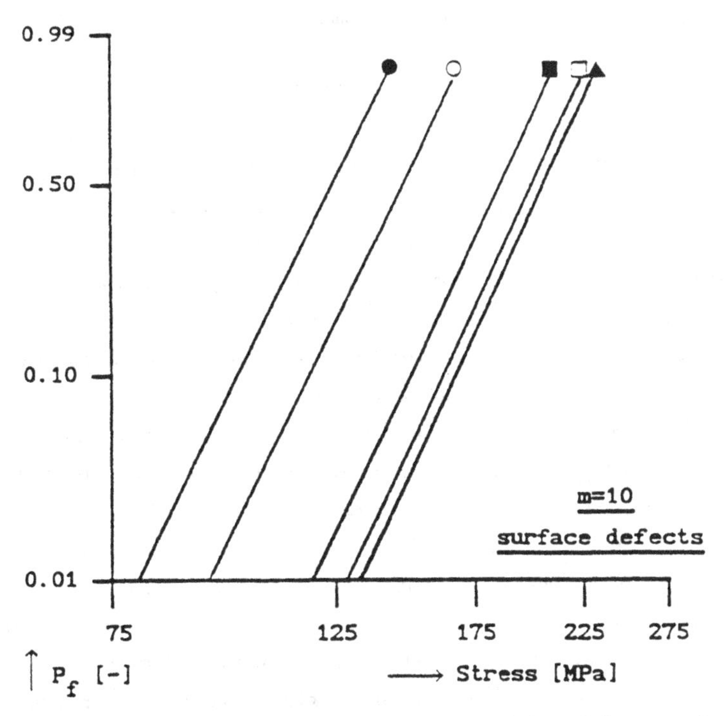

Figure 1. Weibull plot with predictions for models of Lamon, Stanley and Weibull for uni-axial (3- and 4-point) and bi-axial (ball-on-ring) bend tests: ● = 4-point bend test; O = 3-point bend test; □ = ball-on-ring test model of Lamon; Δ = ball-on-ring test model of Stanley; █ = ball-on-ring model of Weibull.

FURTHER PROBLEMS

Even if the problems with the statistical approach could be solved, some remaining items remain unaddressed. To mention just three:

- slow crack growth
- corrosion
- creep.

The first two phenomena change the nature of the defects continuously during loading in ways that are not completely defined. Introduction of these aspects into a statistical approach seems cumbersome. While slow crack growth and corrosion can operate at room temperature, creep operates only a high temperature. Again incorporation into the statistical approach seems rather difficult. Separate approaches are therefore necessary for mechanically loaded components when the above mentioned phenomena play an important role. From the problems encountered it seems fair to conclude that alternative solutions would be welcome.

A POSSIBLE ALTERNATIVE

The statistical approach is based on fracture mechanics applied to microscopic defects. It neglects to a large extent the information that is available from the microstructure of the tested specimens. A completely opposite approach is continuum damage mechanics (8). In this formalism, in its simplest form, a damage parameter is introduced describing the amount of isotropically distributed damage in the material. An associated damage evolution law is postulated. When a certain critical value for the damage parameter is reached, the material fails. Originally the theory was completely phenomenological; no clear statements were made about the nature of the defects or the basis of the damage evolution law; the information from microstructures was neglected. Recently, however, some micro-mechanical models have been put forward which can form a basis for the constitutive behaviour of brittle materials (9-10). A combination of the constitutive behaviour of these models with continuum damage mechanics could result in a 'damage mechanics' approach, taking into account microstructural data but nevertheless firmly rooted in an anisotropic continuum deformation theory. It is clear though, that the success of such an approach is critically dependent on the possibility of extracting from microstructural analysis the relevant parameters in a concise form.

REFERENCES

1. Newnham, R.C., Strength tests for brittle materials. Proc. Brit. Ceram. Soc., 1975, **25**, 281-93.

2. Hoagland, R.G., Marshall, C.W. and Duckworth, W.H. Reduction of errors in ceramic bend tests. J. Am. Ceram. Soc., 1976, **59**, 189-92.

3. Chau, F.S. and Stanley, P., An assessment of systematic errors in beam tests on brittle materials. J. Mater. Sci., 1985, **20**, 1782-86.

4. Shore, P., State of the art in 'damage free' grinding of advanced engineering ceramics. Brit. Ceram. Proc., 1990, **46**, 189-200.

5. Dortmans, L. and de With, G., Weakest-link failure prediction for ceramics using finite element post-processing. J. Eur. Ceram. Soc., Submitted.

6. McHargue, C.J., The mechanical properties of ion implanted ceramics - A review. Defect and Diffusion Forum, 1988, **57-58**, 359-80.

7. Bergman, B., How to estimate Weibull parameters. Brit. Ceram. Proc., 1987, **39**, 175-85.

8. Chaboche, J.L., Continuous damage mechanics - A tool to describe phenomena before crack initiation. Nucl. Engng. Des., 1981, **64**, 233-47.

9. Kachanov, M. and Laures, J.-P., Three-dimensional problems of strongly interacting arbitrarily located penny-shaped cracks. Int. J. Fract., 1989, **41**, 289-313.

10. Fabrikant, V.I., Close interaction of coplanar circular cracks in an elastic medium. Acta. Mech., 1987, **67**, 39-59.

CREEP OF NON-OXIDE CERAMICS CONSIDERING OXIDATIVE INFLUENCES

F. THÜMMLER
Institut für Werkstoffkunde II
University of Karlsruhe,
D7500 Germany.

ABSTRACT

Creep of silicon nitrides and silicon carbides is considered, including the effects of oxidation. Basic behaviour is defined, the relevant background equations presented, and the parallel microstructural mechanisms discussed. For porous silicon nitride the pore size rather than amount is critical in determining creep under oxidation. In dense sintered silicon nitride the presence or absence of viscous glassy phases is the main factor controlling behaviour. Silicon carbide ceramics show relatively low creep rates and are not so sensitive to oxidation effects.

GENERAL

Non-oxide ceramics are amongst the most important materials for prospective applications in engineering, especially at high temperatures. Si_3N_4- and SiC-based materials are the most common, each representing a group with a variety of compositions and properties. Their covalent bond provides the potential of excellent high temperature mechanical properties. However, due to their poor inherent sinterability they are densified, with a few exceptions, by using sintering additives, leading to multiphase microstructures. They are thermodynamically unstable in an oxidative environment, but can form protective SiO_2-rich surface layers, which are indispensablè for good oxidation behaviour.

For high temperature and long term applications, creep as well as oxidation and subcritical crack growth (fatigue) are the most important properties controlling damage and lifetime. They may be influenced strongly by oxidation processes. In this paper, the creep behaviour together with the influence of oxidation is outlined.

Several creep equations have been developed for the accumulated creep strain ε and creep rate $\dot{\varepsilon}$, characterizing transient as well as steady state creep, as shown in Table 1. For transient creep both the time exponent and stress exponent n are important, while for steady state creep the n-values essentially defines the creep mechanisms.

TABLE 1
Transient and steady state creep equations

Transient Creep	$\varepsilon_t = b.t^{1-C}$	$(b = \frac{A}{1-C})$; C = 0.5-0.8,
	$\dot{\varepsilon}_t = A.t^{-C}$	
	$\dot{\varepsilon}_t \sim e^{B\sigma}$	
	$\dot{\varepsilon}_t = A \cdot \frac{\sigma^n}{a^m} \cdot \exp \frac{-Q}{RT} \cdot t^{-C}$	
Steady State Creep	$\dot{\varepsilon}_s = \frac{A.D.G.b}{k.T} (\frac{b}{a})^p \cdot (\frac{\sigma}{G})^n$	D = effective Diff. Coeff.
	$\dot{\varepsilon}_s \sim \frac{D_\nu \cdot \sigma}{a^2}$	(Nabarro-Herring) D_v = Vol. Diff. Coeff.
	$\dot{\varepsilon}_s \sim \frac{D \cdot \sigma^n}{a^3}$	(Coble, Gifkins-Snowdon) n = 1: full accommodation without failure generation. n = 1-2: accommodation with failure generation. n = ≥ 4: dislocation motion or excessive failure generation.
Cumulative Creep Strain	$\varepsilon(t) = \varepsilon_0 + \beta (\frac{t}{t_0})^m + \dot{\varepsilon}_s \cdot t$	

For many years, creep of brittle materials such as ceramics was tested generally in the 3pt- or (preferably) 4pt-bending test, mostly in several countries with a standardized specimen, 3.5 x 4.5 x 79 mm^3. This test is easy to conduct due to the simple specimen geometry. But the complicated stress distribution across the bending bar during creep and the non-steady state character of stress distribution (1) does permit definition of a constant nominal stress during the experiments. Furthermore, the compression part of the sample above the neutral axis restricts the specimen from giving full creep response under stress on the tension side. Therefore, tensile creep testing became more important during recent years (2), in spite of the difficulties in reliable operation of testing machines including strain measurement at high temperatures, and the higher costs of the test specimen.

CREEP CURVES, MECHANISMS AND DAMAGE PROCESSES

The "classical" creep curve, ie the plot of creep stain vs time at constant stress and temperature, consists of three parts, namely the primary (transient), secondary (steady state) and the tertiary creep, the latter leading to creep fracture. Depending on the material, microstructure, stress and temperature, the creep curve of non-oxide ceramics often deviates from this form. The transient range can be very extended, followed directly by the tertiary creep and fracture, sometimes without any steady state regime, as shown in Figure 1. An equilibrium between "strengthening" and "weakening" factors may not exist or is reached only apparently during a short period. This seems to be characteristic for a microstructure having "strong", non-deformable grains and a "weak" intergranular, often amorphous phase. This grain boundary phase is decisive for the creep properties; deformation of the grains occurs only to a very small extent if at all. Below a certain stress and temperature, only transient creep occurs, reaching very low creep rates after longer times without fracture. This has been observed for several Si_3N_4 materials.

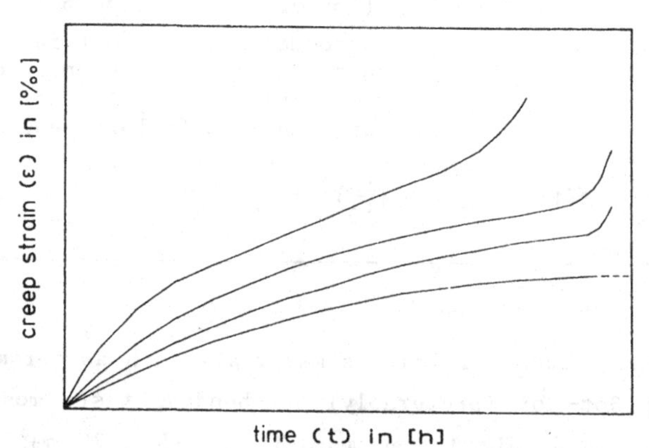

Figure 1. Different types of creep curve (ε-t).

Stress exponents n and activation energies Q are normally calculated in steady state regions, but can also be obtained, at constant time or after constant strain during transient creep. A linear dependence $\varepsilon \sim \sigma^n$ with n = 1 is characteristic of diffusional controlled creep <u>within</u> the grains, according to Nabarro-Herring, ie when a grain boundary phase is not present or does not control the creep process. This is the case in some RBSN and SSiC materials (Figure 2), as pointed out by Cannon and Langdon (1,2).

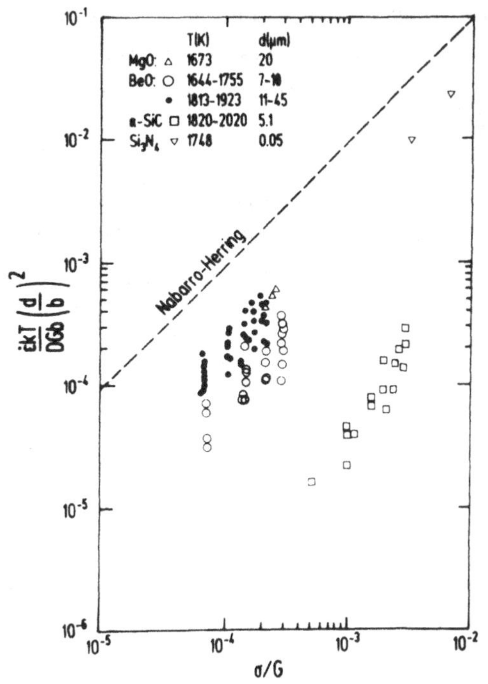

Figure 2. Normalized creep rate plotted against normalized stress for Nabarro-Herring creep (1,2).

When grain boundary processes are rate controlling, an exponent $n = 1$ can be obtained when no microstructural damage (pore or crack formation) occurs. This includes full geometrical accommodation of the microstructure by grain deformation, grain movement in the viscous phase or easy flow of the viscous phase. Newly developed pores and cracks, often occurring during creep, are measured as an additional strain in the material, leading to stress exponents $n > 1$. In practice, n-values between 1.5 and 2.5 are reported in many creep experiments (Table 2) with Si_3N_4 (3), obviously indicating the dominance of grain boundary processes combined with microstructural damage. Pore formation (cavitation), growth and coalescence, crack initiation at triple points and crack growth as a consequence of grain boundary sliding have been observed in many cases in oxide and non-oxide ceramics and is characteristic for temperatures between 0.4 and 0.7 T_m. n-values higher than 3, may indicate extensive microstructural damage as well as plastic behaviour with gross dislocation movement. The latter has been reported only rarely and is expected particularly at high temperatures and high stresses. Figure 3 shows cavitation with residual glass fibrils after tensile creep testing of SSN (Y_2O_3, MgO), while Figure 4 demonstrates creep porosity in the tension side of a bent specimen of SiSiC.

TABLE 2
n_p-values indicating the stress effect on the primary creep rate at constant times for SSN (Y_2O_3, MgO)

Temp. [°C]	n_p(t = const.)		
	t = 5 h	10 h	50 h
1200	3.5	3.6	4.0
1250	1.7	1.8	1.1
1300	2.0	2.3	-
1350	2.0	-	-

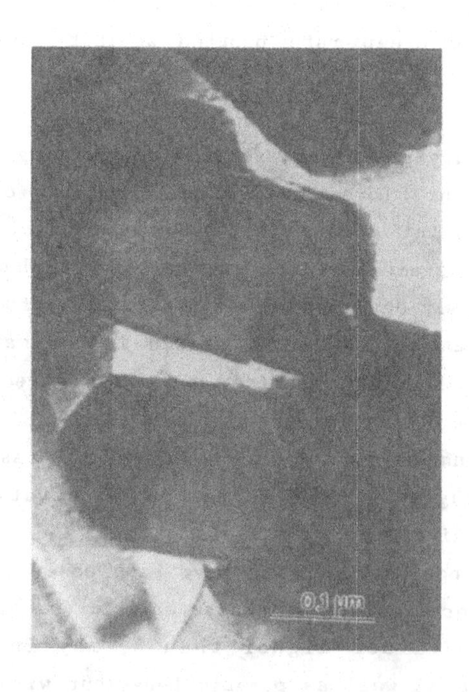

Figure 3. SSN (Y_2O_3, MgO), TEM micrograph (bright field) showing cavitation with residual glass fibrils after creep deformation (8).

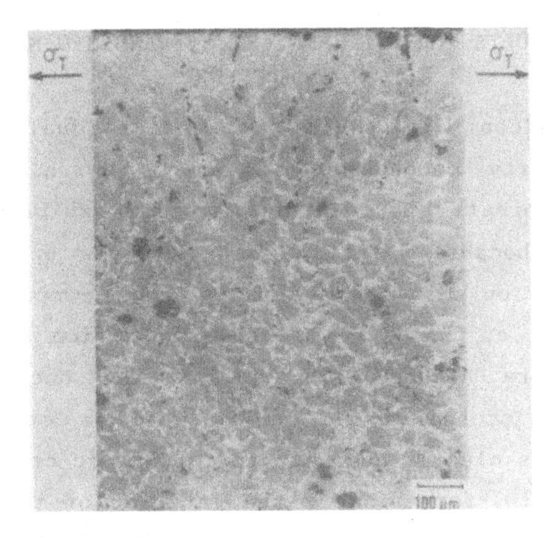

Figure 4. Development of creep porosity in the tensile region of a SiSiC bend specimen (10).

The growth of newly developed flaws under creep conditions competes with the slow crack growth (SCG) of pre-existing cracks. Under high stress, low temperature conditions, the material normally fails by SCG, according to the K_I-concept, while at lower stresses and high temperatures creep fracture normally occurs. Behaviour is even more complicated when surface failures (pits) are developed by oxidation processes, resulting from surface regions with low oxidation resistance, induced by impurities or inhomogeneities (Figure 5) (4,5). Stress concentrations may occur at these points, leading to an additional failure source. General views of creep deformation have been collected for several ceramics in deformation maps (6), where the different creep mechanisms are delineated in the stress temperature field.

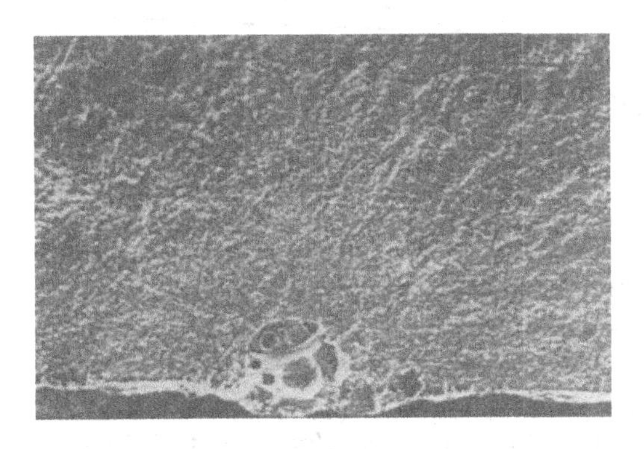

Figure 5 Oxidation pits as failure sources (4).

SPECIFIC RESULTS FOR DIFFERENT Si_3N_4- AND SiC-MATERIALS (7)

Reaction bonded silicon nitride (RBSN) has a residual porosity of 15-20%, and thus the density dependence of creep, correlated with oxidation processes is of interest. Low density material exhibits considerable high creep deformation, because internal oxidation via open porosity provides in situ SiO_2-formation in the interior. The competitive formation of sealing SiO_2 layers at the surface and penetration of the cross-section means that the pore channel diameter is the decisive parameter for resistance against internal oxidation as well as creep resistance. In Figures 6 and 7 it is clearly demonstrated, that the creep behaviour of RBSN is related to the SiO_2-formation in the interior of the specimen, measured by X-ray analysis as cristobalite (or by Rutherford backscattering of α particles as SiO_2-phases). A model of internal as well as surface oxidation of RBSN is shown in Figure 8. For this oxidation (and consequently creep) behaviour, the pore channel size, measured by Hg-intrusion porosimetry, is critical rather than the total amount of porosity. This is demonstrated by measuring the oxidation rate of some RBSN grades with a similar density (2.6 g/cm³) but different pore channel size distribution (Figure 9). High density (> 2.7 g/cm³) guarantees more or less an early sealing of surfaces in isothermal experiments, but most RBSN grades show degradation effects under severe cyclic conditions in ambient air.

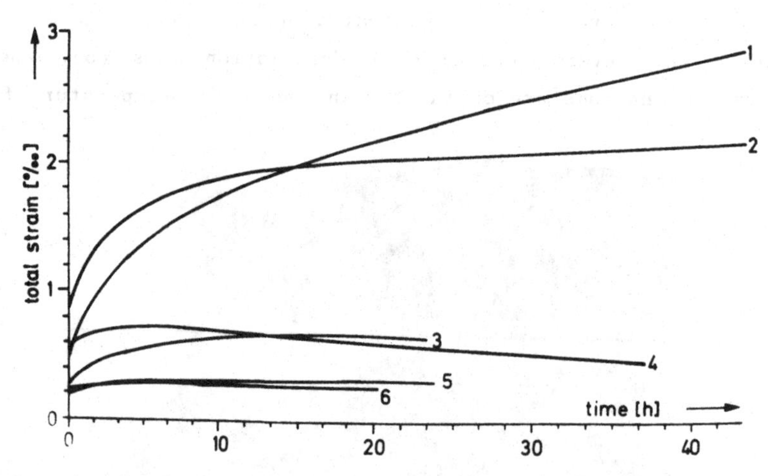

Figure 6. 4pt-bending creep curves (1300°C, 40 MN/m²) for different RBSN materials (7) (bulk densities are in g/cm³; 1 : 2.18; 2 : 2.14; 5 : 2.56; 6 : 2.45; 4 : 2.12 preoxidized; 5 : 2.13 SiC-coated).

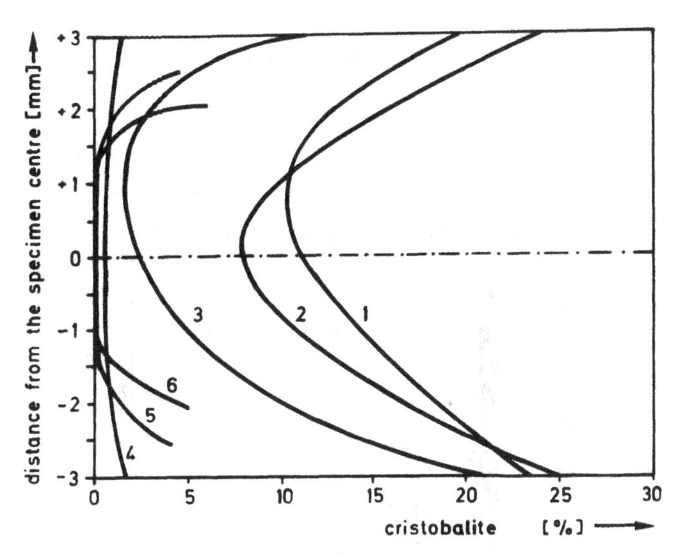

Figure 7. Profiles of cristobalite over the cross-section of RBSN samples after creep experiments (7).

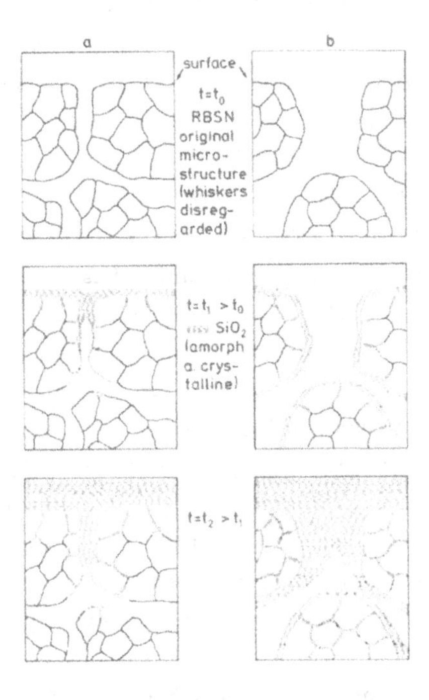

Figure 8. Model of RBSN oxidation (a) small, (b) large pore-channel sizes (7).

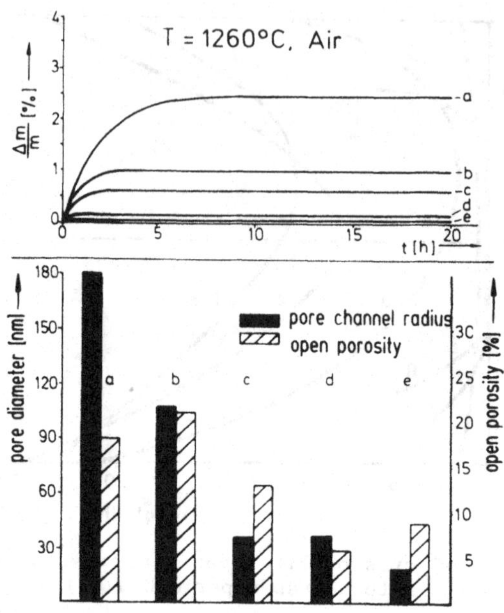

Figure 9. Oxidation of RBSN materials (ρ = 2.6 g/cm³) with different porosity characteristics (12).

Dense Si_3N_4 materials (HPSN, SSN, SRBSN) show correlations between grain boundary phase composition on one hand and creep as well as oxidation resistance on the other. There are "critical temperatures", above which both creep and oxidation are strongly enhanced. We have seen this in different Y_2O_3-, Y_2O_3-Al_2O_3-TiO_2- and Y_2O_3-Al_2O_3-MgO-containing Si_3N_4 materials, having several grain boundary phases and, generally, a glass phase. Good examples are SSN- and SRBSN-products, as shown in Figures 10 and 11. Their creep rate vs time plots change from pure transient to steady state or tertiary creep with subsequent fracture above a certain temperature (> 1225 and > 1300°C, respectively). At a very similar temperature, their oxidation rate constant increases discontinuously as demonstrated in Figure 12. Also the time exponent C of the transient creep law reaches its lowest value (0.45 to 0.55) at this critical temperature (Figure 13).

These phenomena are not fully understood, but obviously both processes are strongly dependent on viscosity of the grain boundary phases, the creep response by increased viscous flow and the oxidation rate by increased oxygen diffusion. This correlation has been observed for Y_2O_3- and Al_2O_3-, but not for MgO-containing Si_3N_4.

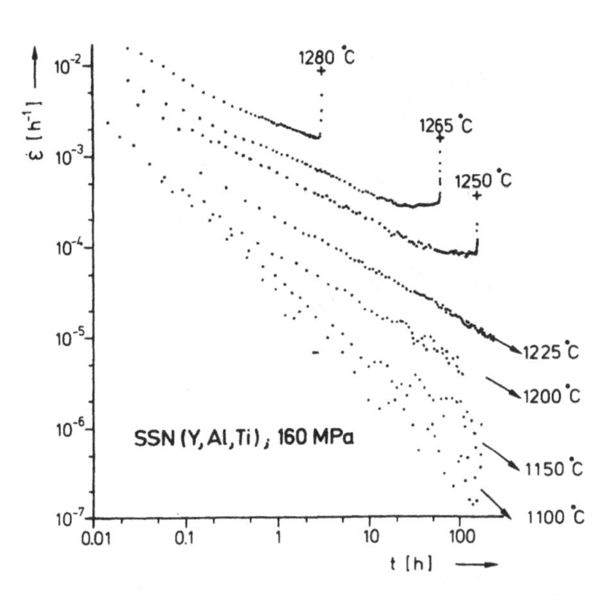

Figure 10. 4pt-bend creep curves of SSN in ambient air (11).

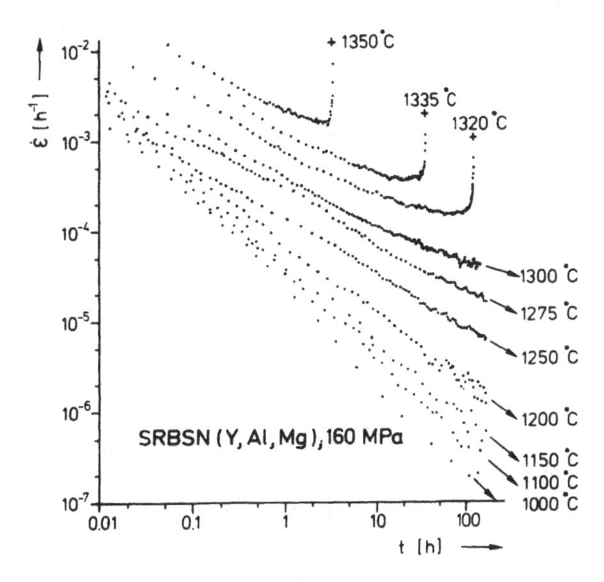

Figure 11. 4pt-bend creep curves of SRBSN in ambient air (11).

The creep properties of sintered Si_3N_4 can be changed considerably by pre-annealing in ambient air, as shown in Y_2O_3-MgO containing SSN (2,4). MgO is a constituent providing a low viscosity grain boundary phase and consequently a high creep strain under the experimental conditions according

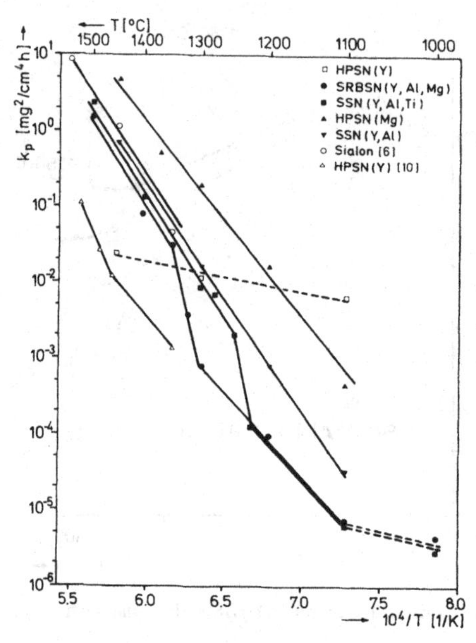

Figure 12. Arrhenius diagram of the parabolic oxidation rate constant of several Si_3N_4 materials (11).

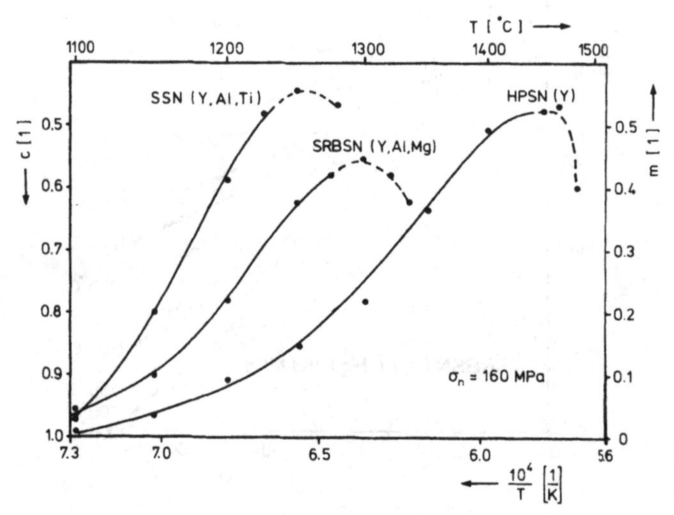

Figure 13. Temperature influence on the time exponent C of the transient creep equation (11).

to Figure 14. Long term pre-annealing provides a reduction of tensile creep strain of nearly one order of magnitude after 150 to 200 h, leading also to increased lifetime. The reason is the time-dependent increase of viscosity of the grain boundary phase by diffusion of species like Mg^{++} and Ca^{++} out to the surface. In the case of Si_3N_4 with Yb_2O_3 additive the grain boundary phase is fully crystallized and no easily diffusing cations could be found in the bulk. Microstructure and composition are more stable under annealing conditions and no influence of long term annealing could be observed (8). This material seems to have an excellent potential for high temperature applications, may be up to 1400°C. This demonstrates the well known fact that the adjustment of sintering additives is of major importance.

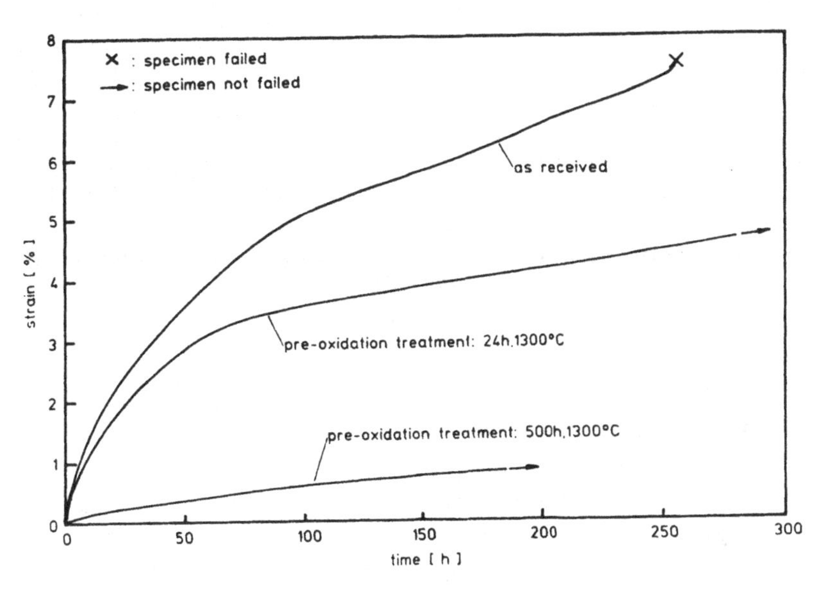

Figure 14. Tensile creep curves of SSN (Y_2O_3, MgO) at T = 1300°C and σ = 40 MPa after various pre-oxidation treatments (8).

SiC materials are generally very creep resistant, when the content of doping elements and impurities is not too high (7). No distinct influence of oxidation or pre-oxidation treatment on creep is found, besides a moderate reduction of the creep fracture strain. A characteristic for most sintered SiC(B,C) grades is a certain content of free carbon in the microstructure, being a residue of the carbon additive as sintering and de-oxidation aid (9) (Figure 15). Simetimes they contain some boron or even boron carbide as separate inclusions. Probably they have no distinct influence on creep, unless they form linear arrays, being potential nuclei for crack formation. SiC with Al-additive may have a somewhat lower creep resistance than SiC(B,C), which may not be critical for many actual and perspective applications. The influence of pre-oxidation treatment on creep (Figure 16) is not dramatic for the times investigated, but some embrittlement (reduction in lifetime) occurs. This is probably more serious in porous SiC products after long time annealing under load, as a consequence of oxidation induced subcritical crack growth.

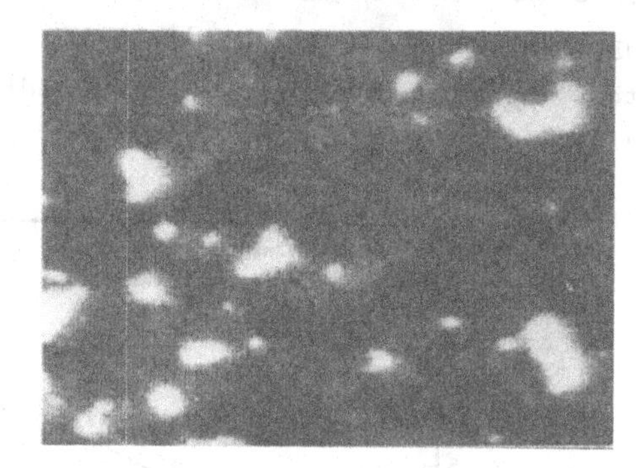

Figure 15. HRAES-distribution map of free carbon (on a fracture surface of (B_4C) doped SSiC (9).

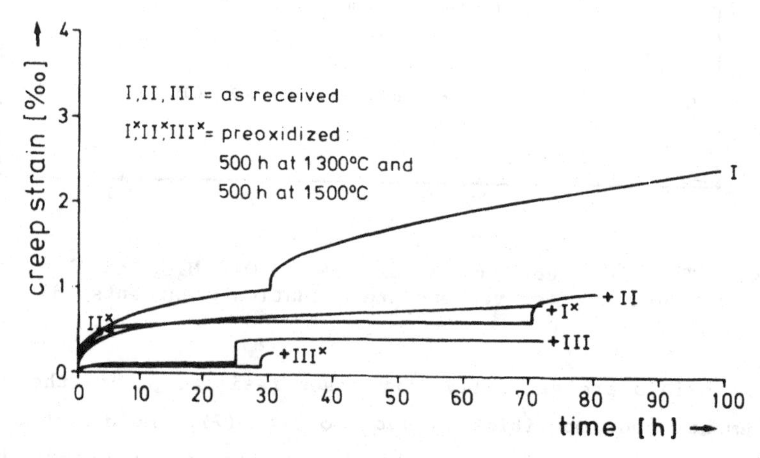

Figure 16. Creep of different SSiC materials in air. (T = 1400°C; σ = 130/160 MN/m^2).

Siliconized Silicon Carbide (SiSiC) does not show the extreme creep and oxidation resistance as high quality SSiC. Its microstructure contains many phase boundaries Si/SiC, with the free silicon as a more plastic "binding" phase at application temperatures. The Si/SiC boundaries act as diffusion paths for impurities (Ca, Al, Fe) influencing the composition of the surface scale. Nevertheless, SiSiC with adequate purity is a highly creep- and oxidation-resistant material, not strongly influenced by external or internal oxidation (10).

CONSEQUENCES AND CONCLUSIONS

The creep rate and creep strain of non-oxide ceramics is mainly controlled by the viscous behaviour of (oxide) grain boundary phases and the formation of microdefects. The deformation of the main ceramic phase is of little or no importance in most cases. Very high creep resistance can be achieved, when the grain boundary phase is minimal or when this phase is highly refractory. Many, but not all of the non-xoide ceramics considered show extended transient creep ranges, ie continuously decreasing creep rate in bending as well as in tensile experiments (7,8). A crystalline, nearly undeformable phase with a viscous phase at grain boundaries is the characteristic microstructure for this behaviour. The following features may explain the extended non-steady state creep:

- The grains are forming an increased number of solid contacts with increasing time, providing higher stiffness and inhibiting further deformation. This can be expected during bend creep and is accompanied by continuous stress redistribution during creep.
- Time dependent increase of viscosity of the grain boundary phase by outwards diffusion of impurity or additive species, like Mg, Ca, et al.
- Time dependent increase of viscosity of the grain boundary phase by continuous crystallisation.
- Formation of a time dependent stable network of microcracks, leading to apparently higher creep rates at shorter times. This may occur in microstructures without a glassy phase, as observed in Yb_2O_3-Si_3N_4 (8).

Extended primary creep occurs not only under bending but also under tension, so that the first point above cannot be dominating. The same is true for the third point, because the crystallisation of the SiO_2 phase takes place mainly in the initial creep period in the performed experiments. Thus, points two and four seem to be important, depending on the character of the grain boundary phases present.

Thus, microstructural instabilities during creep seem to be responsible for this type of creep curve. Si_3N_4 and SiC materials, which are microstructurally stable under creep conditions, exhibit a more "classical" creep behaviour with a clear steady state range.

It seems to be very important in practice to reach creep rates at a reference temperature (of say 1300°C) in the range of 10^{-5}/h or less. Only a small cumulated strain of a few tenths of a percent can be tolerated. When a small cumulated strain and a low minimum creep rate can be achieved, then the danger of creep fracture is not high even after long exposure times. Thus, a low creep rate from the onset is the best approach for practical applications. If this cannot be achieved, the occurrence of an extended transition range into the low creep rate mentioned above seems to be acceptable.

REFERENCES

1. Cohrt, H., Grathwohl, G. and Thümmler, F., Non-stationary stress distribution in a ceramic bending beam during constant creep load, Res. Mechanica., 1984, **10**, 55-71.

2. Gürtler, M. and Grathwohl, G., Tensile creep testing of sintered silicon nitride, in: "Creep and Fracture of Engineering Materials and Structures", Swansea, 1st-6th April 1990, (In print).

3. Cannon, R.W. and Langdon, T.G., Review: Creep of Ceramics, Pt. 1, Mechanical characteristics, J. Mat. Sci., 1983, **18**, 1-50; Pt. 2, An examination of flow mechanisms, J. Mat. Sci., 1988, **23**, 1-20.

4. Van der Biest, O., Weber, C. and Garguet, L.A., in: "Ceramic Materials and Components for Engines", (Ed. by V.J. Tennery), The American Ceramic Society, 1988, 729.

5. Grathwohl, G., Mikrostrukturelle Versagensphänomene beim Kriechen und bei der Ermüdung keramischer Werkstoffe; Fortbildungsseminar, "Mechanisches Verhalten keramisher Werkstoffe - Kennwerte, Werkstoffauswahl, Dimensionierung", Karlsruhe, 6th-9th March 1990.

6. Sargent, P.S. and Ashby, M.F., Deformation mechanism maps for silicon carbide, Scripta Metall., 1983, **17**, 951-57.

7. Thümmler, F. and Grathwohl, G., High temperature oxidation and creep of Si_3N_4- and SiC-based ceramics and their mutual interaction, MRS Internat. Meeting on Advanced Materials, Tokyo, May 1988.

8. Gürtler, M., Dissertation, Universität Karlsruhe, 1991.

9. Hamminger, R., Grathwohl, G. and Thümmler, F., Microanalytical investigations of sintered SiC, Pt. 1, J. Mat. Sci., 1983, **18**, 353-364; Pt. 2, 3154-3160.

10. Cohrt, H., Dissertation, Universität Karlsruhe, 1984.

11. Ernstberger, U., Grathwohl, G. and Thümmler, F., High temperature durability and limits of sintered and hot pressed silicon nitride materials, Int. J. High Technology Ceramics, 1987, **3**, 43-61.

12. Porz, F. and Thümmler, F., Oxidation mechanism of porous silicon nitride, J. Mat. Sci., 1984, **19**, 1283-95.

PRACTICAL DESIGNING ASPECTS OF ENGINEERING CERAMICS

J.T. van KONIJNENBURG, C.A.M. SISKENS and S. SINNEMA
Hoogovens Industrial Ceramics,
PB 10000, 1970CA Ijmuiden,
The Netherlands.

ABSTRACT

The application of engineering ceramics to practical engineering applications is discussed. The more important properties of ceramics are summarised. Calculation methods for failure criteria are presented and the working method described here shows clearly that model calculations can be of great help in designing with advanced ceramics.

The application of finite element analysis and software has permitted redesign of the wear-sensitive wall-ironing ring as a viable ceramic component for the deep drawing of metal cans. The balancing of forces has led to a design free of tensile stresses.

INTRODUCTION

Over the last decade engineering ceramics has become one of the key issues in materials research throughout the industrialized world. After the successful introduction of advanced ceramics in the field of electronics, research concentrated on applications of ceramic materials for structural components and a number of process industry installations.

The energy crises in the seventies triggered research in the field of energy saving. In the US and Japan extensive research programmes were initiated to develop engines for public transportation with improved fuel efficiency. This could be achieved by increasing the operating temperature of diesel engines or by using gas turbines working at high temperatures. For this goal materials had to be developed which could operate at temperatures < 1400°C. Only ceramics would qualify as candidate materials. In Japan a programme sponsored by MITI was commenced to develop a diesel engine made from ceramics. In the early eighties the engine was produced which triggered world-wide interest in the outstanding possibilities of advanced ceramics. Nearly a year later two gas turbines of around one hundred horse power were presented which were also made of ceramic. The excitement in the press was intense with many journalists predicting a new stone-age by the end of the twentieth century.

Now, almost a decade later, we know that application possibilities of advanced ceramics are still in many cases only a promise. The engines described never reached industrial production and most applications of advanced structural ceramics are still in their infancy. However, advanced ceramics have great possibilities in industry when used appropriately. It is necessary to design the material for a specific purpose, which will be different for each application. In this presentation some guidelines will be given for successful development of ceramic applications and it will be shown that successful design is possible.

SPECIFIC PROPERTIES OF ADVANCED CERAMICS

Engineering ceramics are used for structural applications according to their specific mechanical and chemical behaviour and Table 1 summarises the key properties.

TABLE 1
Specific properties of structural ceramics

- Refractoriness
- Brittleness
- Abrasion resistance
- High compressive strength
- High Young's modulus
- Relatively low expansion coefficient
- Broad range of thermal conductivity values
- Chemical stability

The values of the above properties vary significantly between the commonly used engineering ceramics, for example bend strength and

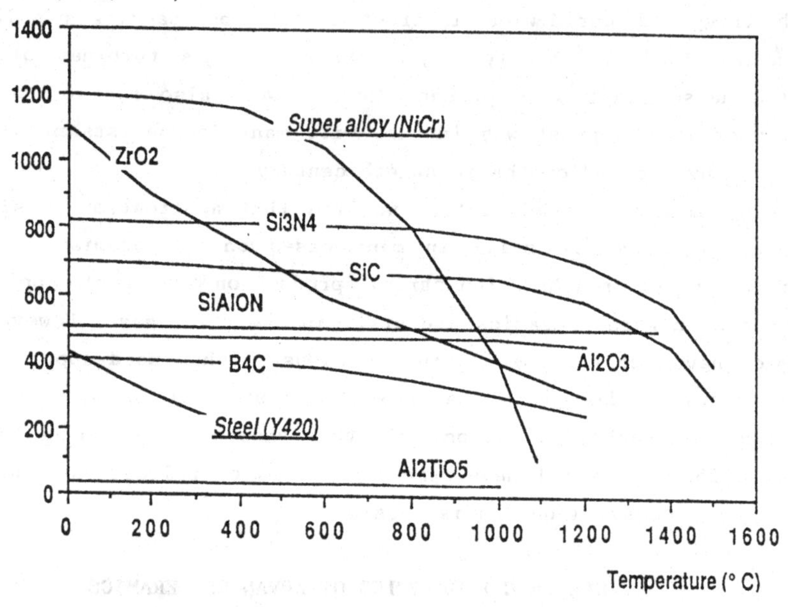

Figure 1. Strength and refractoriness of advanced ceramics.

refractoriness, Figure 1. Table 2 gives properties for a number of advanced ceramic materials.

TABLE 2
Important properties of various advanced ceramics (indicative values)

	Steel	Al_2O_3	ZrO_2	SiC	Si_3N_4	Al_2TiO_5
Density (kg/m^3)	7870	3800	6000	3100	3250	3200
Young's Modulus (GPa)	215	380	200	390	280	13
Modulus of Rupture (MPa)	600	300	900	350	600	30
K_{1C} (MPa$\sqrt{m}$)	-	3-4	10	4	5-8	-
Weibull Modulus	>35	4-8	>10	5-12	10-20	<5

In practice, as is well known, the properties of ceramics vary between different manufacturers and production routes due to variations in the powders and the complex production routes. It is necessary, therefore, to design ceramic components with a high safety factor. The variation in properties is expressed generally by the Weibull statistical method, in terms of failure probability, as indicated in Figure 2. Table 2 indicates the moduli for different ceramic materials which vary between 4 and 20. In comparison metals show Weibull moduli > 35.

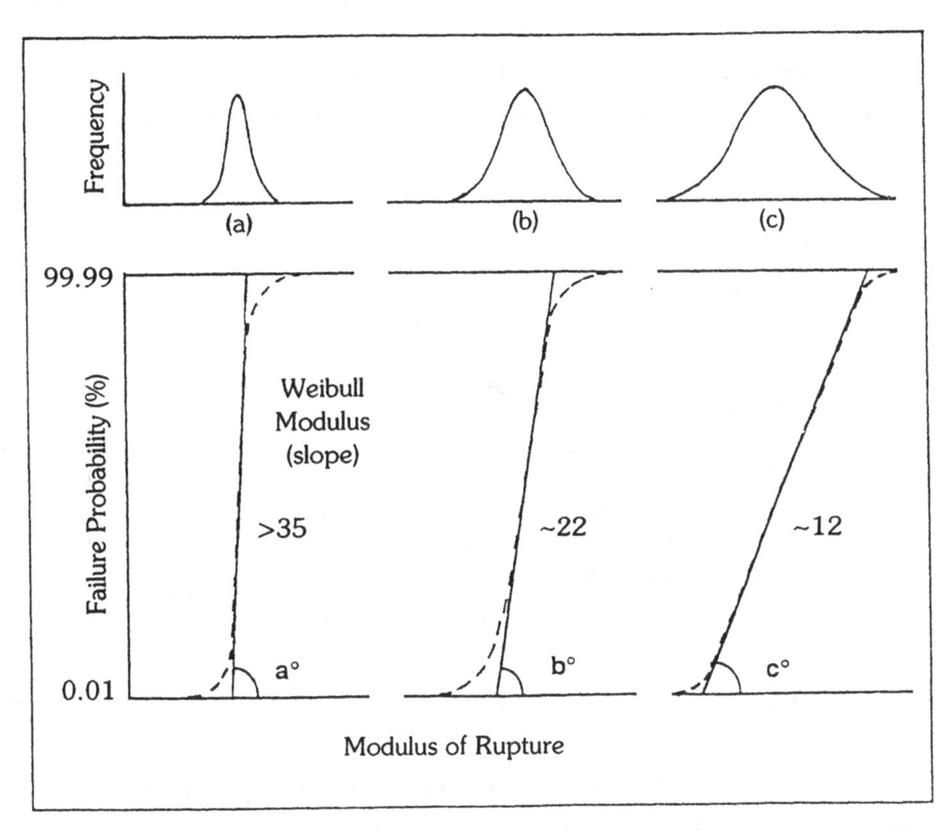

Figure 2. Failure probability expressed with Weibull statistics. Upper sketches indicate histograms of strength. Lower sketches show that when the data is plotted according to the Weibull function, the slope of the lines gives an estimate of the Weibull modulus.

In practice, maximum allowable tensile stresses have to be decreased significantly due to the finishing of the component surface by means of grinding and polishing. These procedures introduce surface cracks resulting in much lower strengths associated with the component surface compared with the bulk of the component.

When a successful design of a component is feasible, it is necessary to join it to the working installation or the engine. Usually temperature differences occur across the joint, and the installation in which the component is used will heat up and cool down periodically. This introduces the problem of the differential expansion of the ceramic-metal joint.

Figure 3 shows the thermal expansion curves of different ceramics in comparison with steel. The expansion of steel is much higher than the expansion of nearly all ceramics, therefore special care has to be taken in joining ceramics to metals.

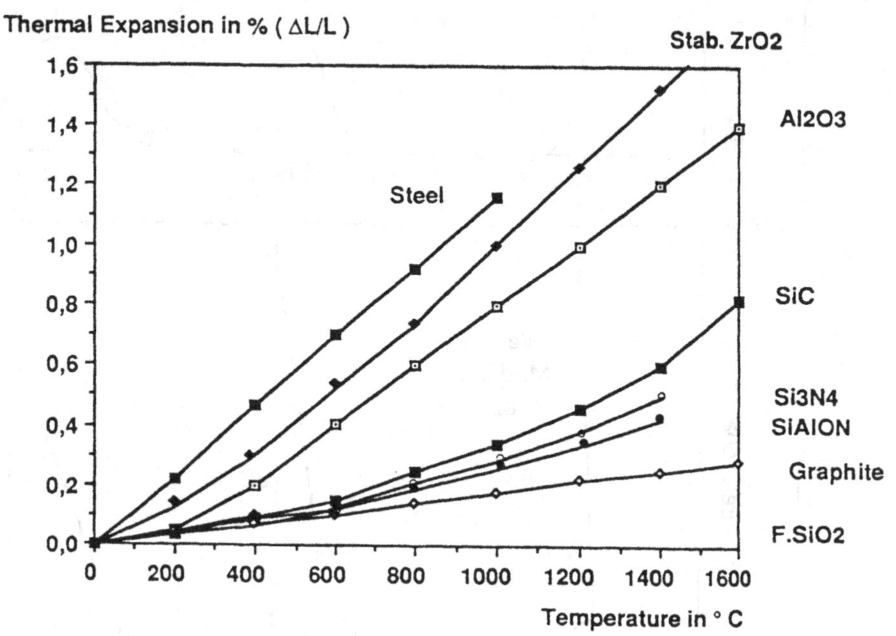

Figure 3. Thermal expansion of advanced ceramics in comparison with steel.

The brittle behaviour of ceramics makes it desirable to design a component such that the component is exposed to compressive forces, with tensile stresses kept to a minimum. Designing with ceramics is only possible when the ceramic properties are well known to the engineer and when the engineer is conversant with design in brittle materials, as were the engineers of a century ago when the brittle cast iron was one of the most important materials of construction.

The general remarks made here show clearly that it is necessary to carry out model calculations as part of the design of a component in ceramics.

CALCULATION METHODS

The finite-element method is a powerful tool for stress calculations in ceramic components. To predict stress distributions in complex geometries is essential during the design of ceramic components. However, care has to be taken during the mathematical analysis and the results are very sensitive to the choice of the elements, the fineness of the grid and the experimental methods by which the essential parameters have been determined. Furthermore, it has to be taken into account that the physical properties of ceramics may vary from component to component due to uncertainties in the manufacturing techniques, as indicated above. Some outlines of the methods used by Hoogovens Industrial Ceramics are described below.

Software: The model is based on standard finite-element software (NASTRAN), adapted to the specific properties of ceramics. In the past calculations have been made at Hoogovens for ceramic refractory materials used this method, and the model has been optimised with the help of simulated laboratory measurements. The technique was shown to be suitable for failure prediction on ceramic refractories. The same procedure is now followed to adapt the model for advanced ceramics.

Stress modelling: Since the stress-strain behaviour for ceramic materials is approximately linearly elastic, calculation of the stresses is more simple than for metals. Figure 4 shows the simple stress-strain behaviour for ceramics compared with ductile or tough materials.

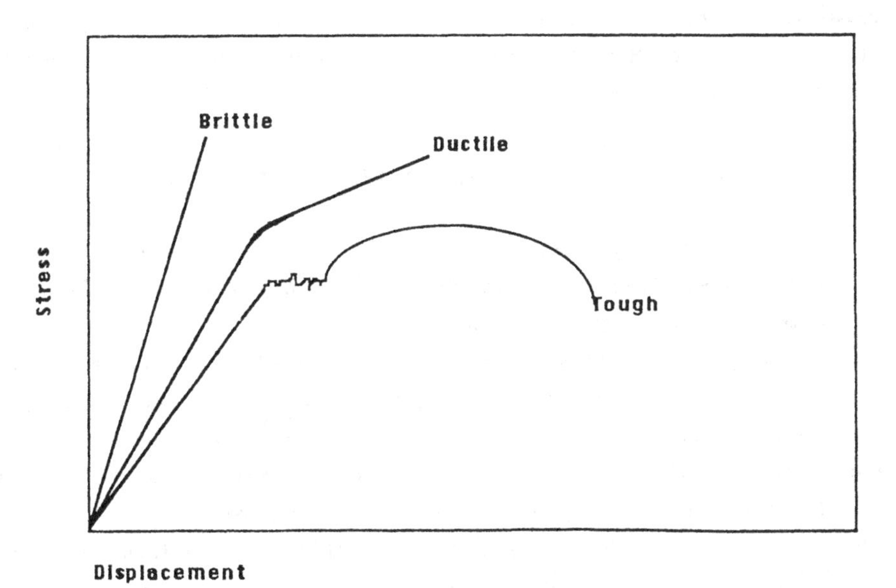

Figure 4. Stress-strain curves for various types of material.

Stresses are calculated for every element in a grid following the displacements of every node of the grid. In each node the stress situation can be described in cartesian co-ordinates with 6 stresses: σ_x, σ_y, σ_z, τ_{xy}, τ_{yz}, τ_{xz}. The first three stresses are tensile or compressive stresses with vector directions along the axes of the co-ordinate system. The other stresses are shear stresses working in the planes of the co-ordinate system. It is possible, however, by adding and subtracting of vectors, to simplify the description in such a way, that the stress situation can be described by three stresses only, to eliminate the shear stresses. By this manipulation the co-ordinate axes are rotated for each node. The overall xyz co-ordinate system changes to a local ABC co-ordinate system, which the principal stresses are σ_A, σ_B, σ_C.

Yield criterion: An additional problem exists of translating actual properties which are mainly measured uni-axially (e.g. bending and tensile test), to the multi-axial stresses as required by the model. In other words, the stress situation expressed in principal stresses has to be compared to a given equivalent yield stress. Techniques to calculate equivalent yield stresses for multi-axial states of stress for metals are the well-known methods of von Mises (maximum distortion energy theory) and Tresca (maximum shear theory). Values derived from these methods can then be compared directly to experimentally determined uni-axial strength data.

Consider the von Mises theorem as an example. The so called von Mises stress is calculated from:

$$\sigma_{\text{von Mises}} = [1/2\{(\sigma_A-\sigma_B)^2 + (\sigma_B-\sigma_C)^2 + (\sigma_C-\sigma_A)^2\}]^{\frac{1}{2}} \qquad (1)$$

The von Mises stress is dependent only on the difference between the principal stresses. Failure occurs when the von Mises stress is larger than the equivalent uni-axial yield strength.

Graphically, the yield surface can be presented by a cylinder, the axis of which is formed by the line where $\sigma_A = \sigma_B = \sigma_C$. Any contribution of the three principal stresses positioned outside the cylinder leads to failure. In both von Mises' and Tresca's theorems the failure stress value is numerically the same for compressive and tensile stresses.

However, for ceramics a significant difference exists between the tensile and the compressive failure stress. Thus both criteria are inappropriate for analysing stresses in ceramics. The difference between tensile and compressive stresses is taken into account in the criterion formulated by Drücker and Prager which was developed for solving problems in soil-mechanics. This criterion is thus more appropriate for ceramics than for instance the von Mises criterion.

The yield function for the criterion of Drücker and Prager is given by the following equation:

$$f = 1/3.\alpha\ (\sigma_A+\sigma_B+\sigma_C) + [1/2\{(\sigma_A-\sigma_B)^2 + (\sigma_B-\sigma_C)^2 + (\sigma_C-\sigma_A)^2\}]^{\frac{1}{2}} \quad (2)$$

For $\alpha = 0$ the criterion reduces to the von Mises criterion Equation (1). In ABC-space the yield function is represented by a cone. Again any contribution of the three principal stresses positioned outside the cone leads to failure. The yield cone is schematically given in Figure 5.

The above shows that for designing in ceramics compressive and tensile yield strengths have to be known. Since theoretical relations between tensile and compressive yield strengths are lacking, both parameters have to be determined experimentally. Proper experiments are hard to perform, very time-consuming and expensive.

From the engineers point of view it would be valuable if research could be carried out to find relationships (if they exist) between the compressive and the tensile strengths to determine a value for α, which could be related to bending strength measurements.

DESIGNING WITH ADVANCED CERAMICS: A CASE STUDY

Methods used for designing with advanced ceramics can be illustrated with the help of the experiments and calculations which are carried out for ceramic rings used for the production of cans for beverage packaging.

The production of a two-piece can is carried out in several steps. The first step is to form a tin plate or aluminium plate of 0.3 mm thickness into a cup. The wall of such a cup is elongated in three steps by means of three wall-ironing rings. This wall-ironing process is schematically illustrated in Figure 6. The ironing itself is carried out by a hard metal insert in the ring. A punch pushes the cup through these three rings and the wall thickness is reduced to 0.1 mm. Since a large number of cans have to be produced (240 cans/minute), interruptions to the process should be avoided as much as possible. To achieve this, a highly wear resistant hardmetal is used for the insert. This material however has certain

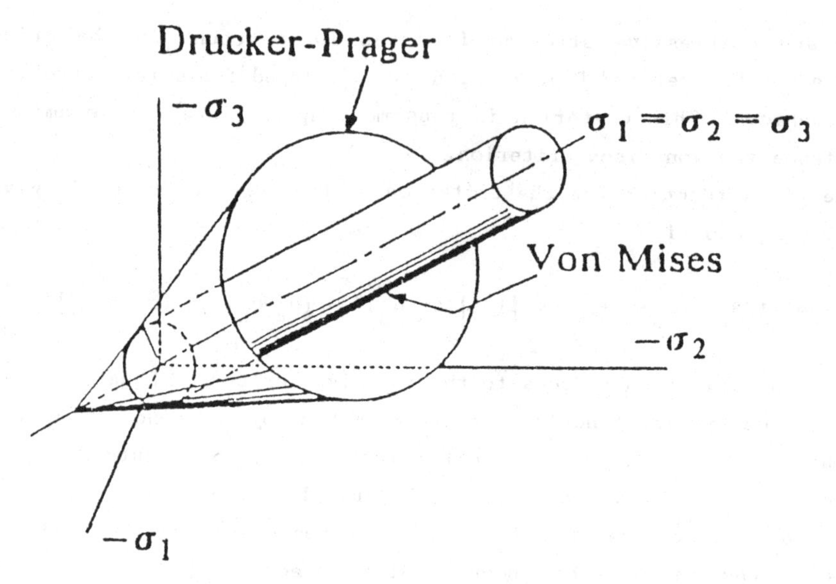

Figure 5. Schematical presentation of the difference between the von Mises criterion and the Drücker-Prager criterion.

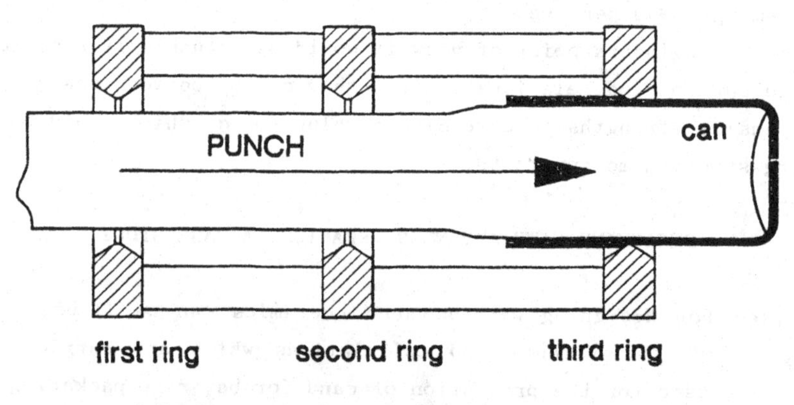

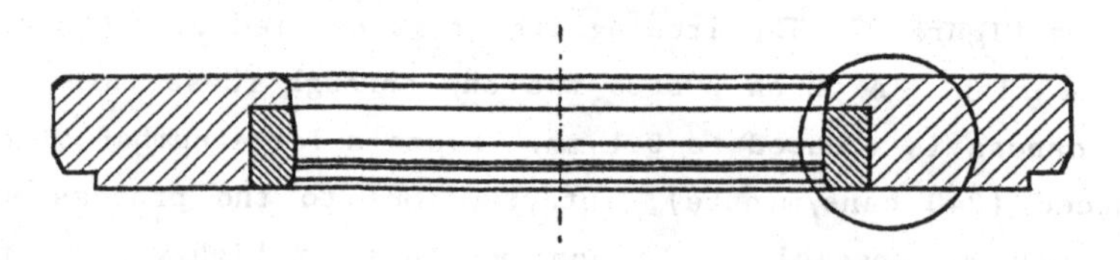

Figure 6. The wall ironing process and the design of the ring with the insert.

disadvantages and the possibility of using highly abrasion-resistant ceramics was thus studied.

This study was carried out in four phases. The first phase included stress calculations using the finite-element method as described earlier. These calculations facilitated a first materials choice and were used to redesign (if necessary) the inserted ring in such a way to eliminate tensile stresses in the ceramic insert during the wall-ironing process.

The grid used for the finite-element calculations of the ring and insert is shown in Figure 7. To minimise tensile stresses the insert is shrunk into the steel housing. The forces between the can and the hardmetal insert were determined in earlier calculations and experiments, but it must be kept in mind, that the results are valid only under the tribological conditions of the hardmetal.

However, in the second phase these values were used for an initial, approximate calculation for the ceramic insert, and stresses within the entire ring were calculated. First of all, the stresses were calculated after application of the wall-ironing load without the effect of shrinking, and the results are shown in Figure 8. After that, the effect of shrinking the insert in the metal holder was calculated in the unloaded situation. Finally, both conditions were combined, and the result is shown in Figure 9. In order to get an impression of the stresses in relation to the maximum strength of the ceramic the Drücker-Prager criterion was used (Figures 8 and 9) and it was thus possible to minimise the combined stresses and to find optimal shrinkage conditions.

The calculations show, that for optimal shrinkage conditions the stresses at the most crucial area (where the can wall touches the ring) have been significantly reduced. Furthermore our calculations show clearly that tensile stresses are eliminated, thus achieving the goal defined earlier. This gave confidence that it would be possible to machine the surface of the ring without the danger of cracking during the wall-ironing process, even though machining may cause small cracks and reduce locally the tensile strength below the value found during testing of the bulk material.

Based on these calculations the third phase of the study was the choice of the most suitable ceramic material(s) and the production of the actual inserts. Two ceramic materials were chosen and produced according to the desired design. The rings were shrunk into metal holders according to the predicted shrinking conditions. The assemblies were successfully tested under pilot can production conditions.

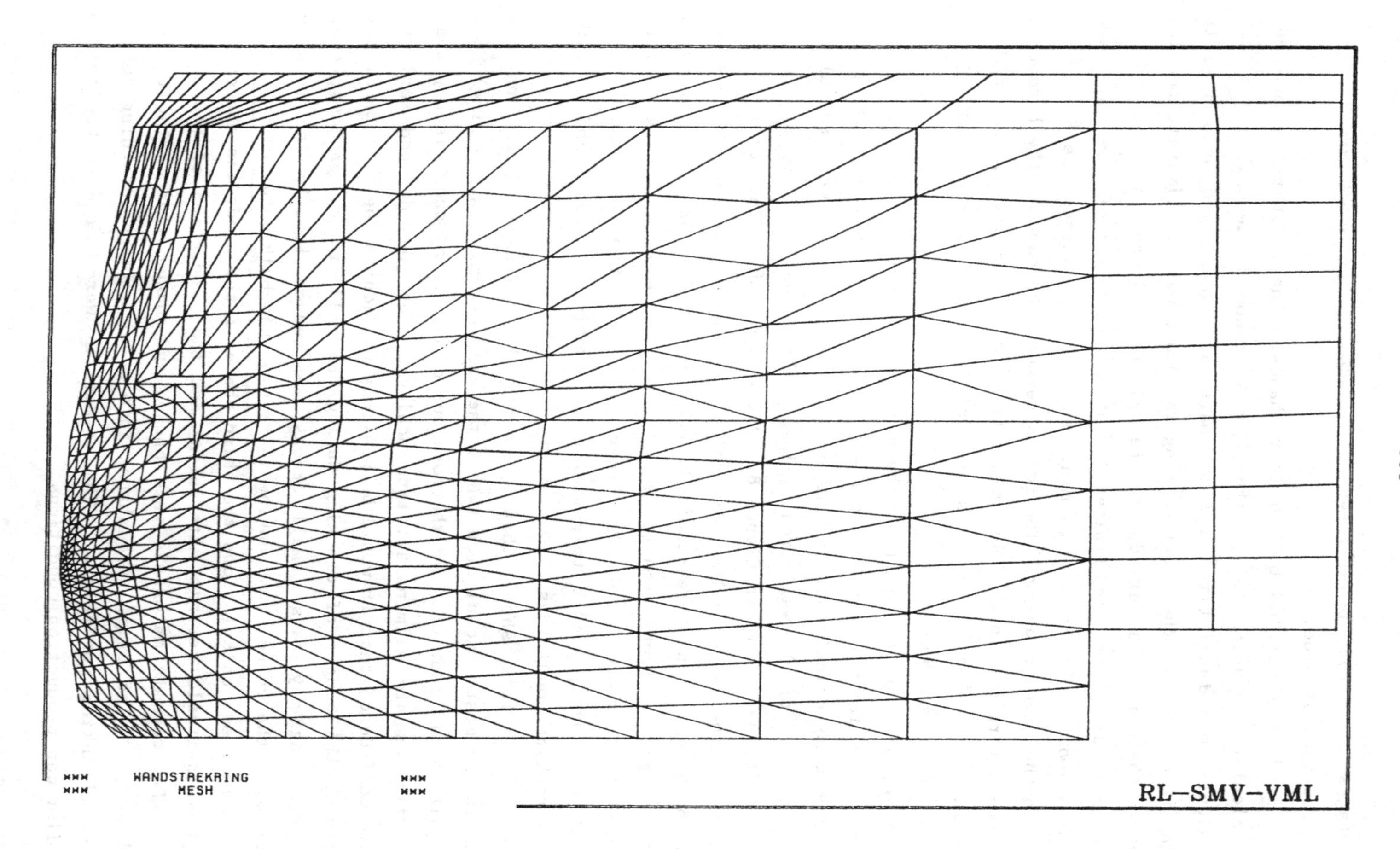

Figure 7. The grid used for the FE calculations of the ring and the insert.

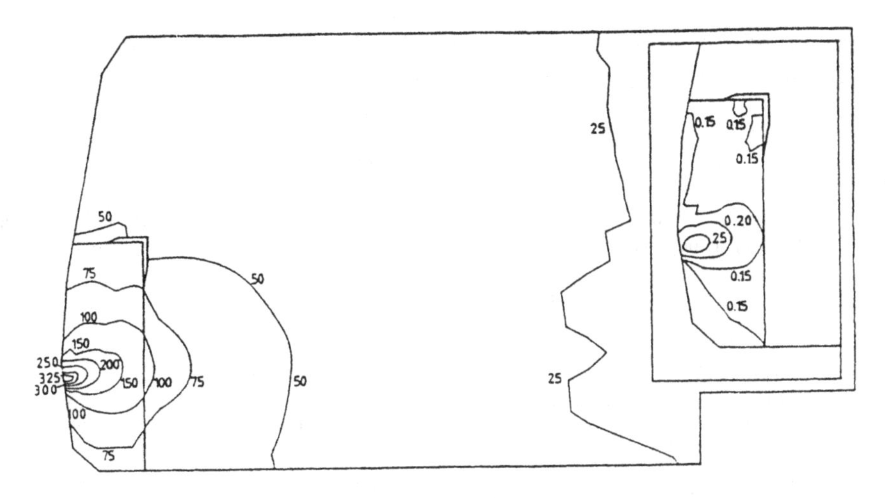

Figure 8. Huber-Hencky stresses (MPa) after load (without the effect of shrinking). The values of the stress/strength ratio according to the Drücker-Prager criterion are given in the insert.

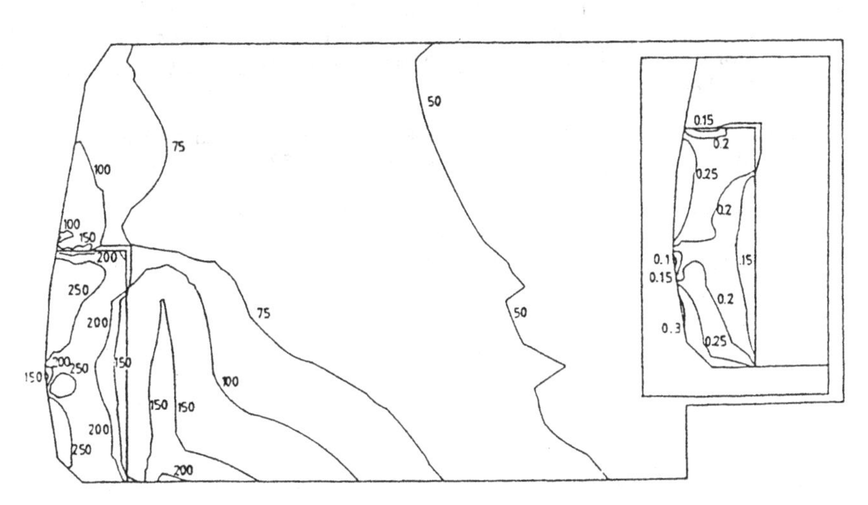

Figure 9. Huber-Hencky stresses (MPa) after load and shrinkage. The values of the stress/strength ratio according to the Drücker-Prager criterion are given in the insert.

The fourth and final step was to produce several hundred cans using the ceramic insert. The cans were produced without damaging the insert and showed excellent surface conditions. None of the ceramic inserts tested failed. The experimental data for the process-generated forces are currently under investigation. With these data the stress model will be refined and the design will be optimised if necessary. After making these adjustments, experiments will be carried out under normal production conditions in order to determine the lifetime using ceramic materials as inserts for the wall-ironing process.

REFERENCES

1. Davidge, R.W., Mechanical behaviour of ceramics. Cambridge University Press, Cambridge 1979.

2. Rieger, W., Konstruieren mit Keramik ein Problem? Chimia, 1989, **43**.

3. Grathwohl, G., Current testing methods - A critical assessment in mechanical testing of engineering ceramics at high temperatures, 1988, p. 31. Dyson, B.F., Lohr, R.D. and Morrell, R., editors, Elsevier, London.

4. Butter, J.A.M. and Rengers, J., Temperature distribution and thermal stresses in refractory material. Inst. Metals, Sutton Coldfield, October 1987.

5. Ras, H.B, van den Hoeven, J.A. and Jansen, E., A numerical calculation of stress distribution during wall-ironing. Proceedings of the 2nd Int. Conf. of Numerical Methods in Ind. Forming Processes, 1986, p. 299-304.

6. Drücker, D.C. and Prager, W., Soil-mechanics and plastic analysis or limit design, J. Appl. Mech., 1952, **10**, 157-165.

7. Sinnema, S., Aspects of design and stress calculations in engineering ceramics. Euro-ceramics, 1989, **3**, p. 3.106-3.110. de With, G., Terpstra, R.A. and Metselaar, R., editors, Elsevier, London.

CHARACTERISTICS OF CERAMIC COMPOSITES

G. ZIEGLER
University of Bayreuth,
Institute for Materials Research (IMA),
PO Box 101251, D-8580 Bayreuth, Germany.

ABSTRACT

In this contribution the state-of-the-art of ceramic-matrix composites is briefly reviewed. Additionally, a critical review of the different types of ceramic composites is given. Finally, two important aspects regarding the stress-strain behaviour of the most promising type of ceramic composites (the continuous fibre-reinforced materials) are discussed: some low-temperature processing techniques; and the significance of the interface for the reinforcing mechanisms and the resulting properties.

INTRODUCTION

Characteristic features of the ceramic materials are the solely elastic behaviour without plastic deformation during loading, and catastrophic fracture on overloading. So, monolithic ceramics exhibit fracture strain values of only a few tenths of a percent and fracture toughness values of maximum 8 MN $m^{-3/2}$ (for materials without additional toughening mechanisms). In order to control crack propagation of ceramic materials, development of ceramic-matrix composites incorporating continuous fibres, short fibres, whiskers, particles and platelets, was started. Because of the promising aspects of this class of material, development and characterization of ceramic-matrix composites are carried out widely.

STATE-OF-THE-ART OF CERAMIC-MATRIX COMPOSITES

Composite Systems

In the literature various toughening mechanisms are described (1-9): microcracking; phase transformation; crack deflection; crack bowing; plastic deformation by incorporating metal dispersoids; and particularly whisker- and fibre-reinforcement initiating various mechanisms like load transfer by crack bridging or matrix pre-stressing, interfacial friction and fibre-pull-out. In many cases a combination of various mechanisms take place. During recent years whisker-, particle-, platelet- and fibre-reinforcement have become more and more important.

In the case of whisker reinforcement, crack deflection and crack bridging, and to a minor extent fibre pull-out are effective. Toughness of continuous-fibre-reinforced materials is mainly caused by the pull-out of fibres in the crack wake which absorbs energy. The pull-out is strongly influenced by the sliding resistance of the interface and the properties of the fibres, particularly the statistical distribution of strength. Thus, the characteristics of the interface and the stresses at the interface are of decisive importance. Moreover, fracture energy increases if the Weibull modulus of the fibres is small (4).

Up to now improvement in toughness has been achieved essentially by fracture toughening/phase transformation and by the increase of crack surfaces due to crack deflection and crack branching. Regarding the latter mechanisms, toughness increase is the higher the more the incorporated dispersoids differ from spherical morphology. On a theoretical basis (8) rod-shaped particles are predicted to be more effective toughening agents than disc-shaped particles, which are more effective than spheres. However, the highest toughness values have been observed by incorporating continuous fibres which results in the high energy absorbing pull-out effect.

Many composite combinations have been investigated recently based on different oxide, carbide and nitride matrices. The compounds investigated up to now have mostly been selected using the following criteria: properties of fibres or whiskers, availability of the fibres and whiskers, chemical and physical compatibility of the reinforcements with the matrix material, mechanical and physical characteristics of the matrix as well as the technological possibilities. As matrix material, mainly glass, glass-ceramics, alumina, mullite, zirconia, silicon carbide and silicon nitride have been selected. The processing techniques used are mainly the powder method (slurry infiltration), gas phase reaction-bonding, melt infiltration, chemical vapour infiltration (CVI), the sol-gel route, the polymer pyrolysis technique, as well as directional oxidation (Lanxide process). Based on the selection criteria mentioned above, SiC-whiskers, SiC-fibres as well as uncoated and coated C-fibres have in many cases been used (unidirectional and cross-weaved bidirectional layers of continuous fibres or textile-like multidirectional pre-woven structures). Besides the technical requirements, the availability of reinforcements plays a decisive role. Typical examples are the problems in supply of fibres of high thermal stability as well as platelets and whiskers of the desired morphology.

In many composite systems an improvement in toughness and in the stress-strain behaviour has been obtained (Table 1 and Figure 1), whereby the extent of property improvement is dependent on the type of composite, the morphology of the reinforcement and on a variety of parameters like raw

TABLE 1
Fracture toughness values achieved up to now utilizing various toughening mechanisms (1,11,14)

Toughening Mechanism	Highest Toughness Values [MN $m^{-3/2}$]	Typical System Matrix/Reinforcement
Microcracking	10	Al_2O_3/ZrO_2
Transformation	20	ZrO_2 (MgO)
Particle	8	Si_3N_4/SiC
Whisker/Platelet	8.5	Al_2O_3/SiC
	11	Si_3N_4/SiC
Fibres	> 30	Glass/SiC
	> 25	Glass-Ceramic/SiC
	> 30	SiC/SiC
	16	Si_3N_4/SiC
Metal Dispersion	25	Al_2O_3/Al
		Al_2O_3/Ni

materials, processing techniques and processing conditions. The toughening effect increases from particle- to whisker- to continuous fibre-reinforcement (see Table 2 for oxide-based composites). Combinations of toughening mechanisms seem to be promising. For example, in the case of a combination of ZrO_2-transformation toughening and whisker reinforcement very high toughness values of about 13.5 MN $m^{-3/2}$ were found accompanied by strength values of ∿ 700 MPa (12). Positive results are reported by a combination of toughening mechanisms in mullite (SiC-whiskers and tetragonal ZrO_2) (12) and Si_3N_4 (SiC-whisker and SiC-particles) (Figures 2 and 3). The

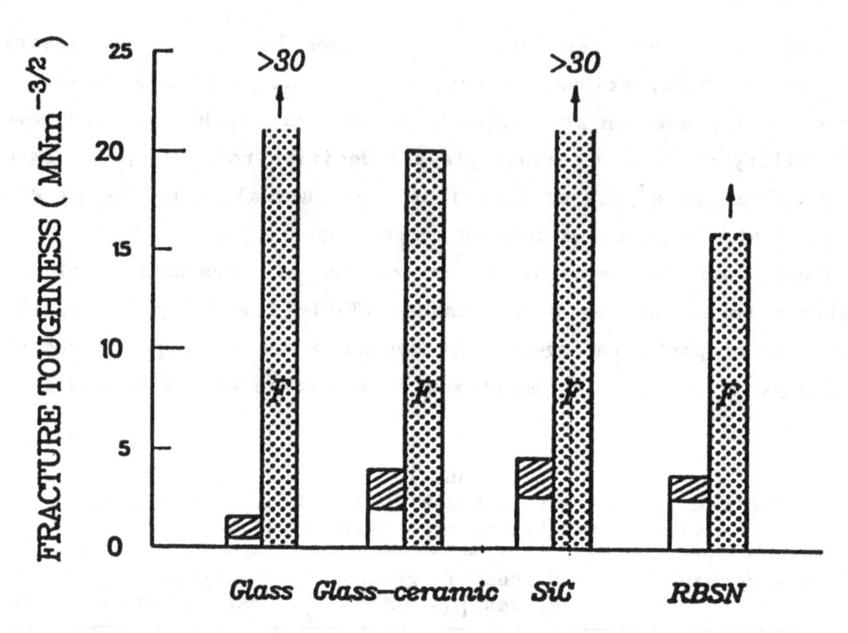

Figure 1. Fracture toughness of fibre-reinforced (F) glass-, glass-ceramic-, SiC- and RBSN- (reaction-sintered Si_3N_4) composites (9-14).

TABLE 2
Examples of oxide-based composites with various reinforcements (12)

Matrix	Reinforcing Component Material	Shape	Strength [MPa]	Toughness [MN m$^{-3/2}$]
Al_2O_3	SiC	Particle	530	-
	SiC	Whisker	800	9
	SiC	Fibre	-	10.5
	ZrO_2	Particle	1080	12
	ZrO_2/SiC	Particle + whisker	1000	13.5
	TiC	Particle	690	4.3
	TiN	Particle	430	4.7
Mullite	SiC	Whisker	450	5.0
	SiC	Fibre	650-850	-
	ZrO_2/SiC	Particle + whisker	580-720	6.7-11
ZrO_2	Mullite	Whisker	-	15
	SiC	Whisker	600	11
	Al_2O_3	Platelet	735	9.5
Spinel	SiC	Whisker	415	-

most promising systems at present are the fibre-reinforced glasses, glass-ceramics (11,13,14) and SiC-fibre-reinforced SiC-matrices produced by the CVI-technique (9). In all these materials high fracture toughness values of 20 (for glass-ceramic) or of higher than 30 MN $m^{-3/2}$ (see Figure 1) (these values are only a qualitative indicator for the toughness), and a significant improvement in stress-strain behaviour have been obtained. Continuous fibre-reinforced reaction-bonded Si_3N_4 also has potential. Fracture stress values have also been increased. Depending on the type of reinforcement, improvements in the scatter of strength data, impact strength, thermal shock and thermal cycling as well as in wear resistance could be achieved. In some cases improvements in high temperature strength and creep behaviour have been reported but the experimental data are partly contradictory.

Critical Review

Depending on the type of reinforcement the various classes of ceramic composite exhibit different values of toughness, fracture strain, fracture stress, residual strength after unloading and damage tolerance; there are thus significant differences in future potential. Furthermore, the various classes of composite present large variations in processing manpower, technological problems and the fabrication complexity and cost.

Continuous-fibre reinforcement has the highest potential for improving stress-strain behaviour and damage tolerance, but on the other hand, the greatest fabrication complexity and cost. One serious problem in processing continuous-fibre reinforced ceramic composites is the limited temperature stability of the fibres. Those fibres having good chemical stability which are of interest for most matrix materials such as SiC fibres, lose their strength during processing at temperatures higher than about 1100°C due to crystallization effects. Only C fibres keep their strength up to very high temperatures, but only in oxygen-free atmosphere. Additionally, the chemical attack of the fibres due to reactions with the matrix material and/ or with gaseous media of the sintering atmosphere has to be considered. A typical example is the infiltration of C or SiC fibre pre-forms with liquid silicon. Mechanical damage of reinforcements, particularly of coated fibres (e.g. coating as protection layer for C fibres or for optimizing the interfacial characteristics), has to be avoided during processing. Moreover, the interfacial characteristics have to be optimized in such a way that at lower stresses load transfer from the matrix to the fibres is

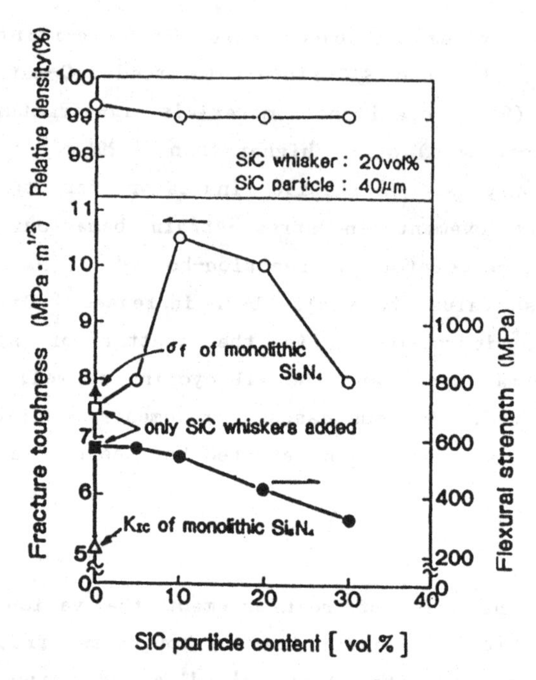

Figure 2. Effect of SiC-particle content on fracture toughness, flexural strength and relative density in Si_3N_4-matrix composites reinforced by SiC-whiskers and SiC-particles (15).

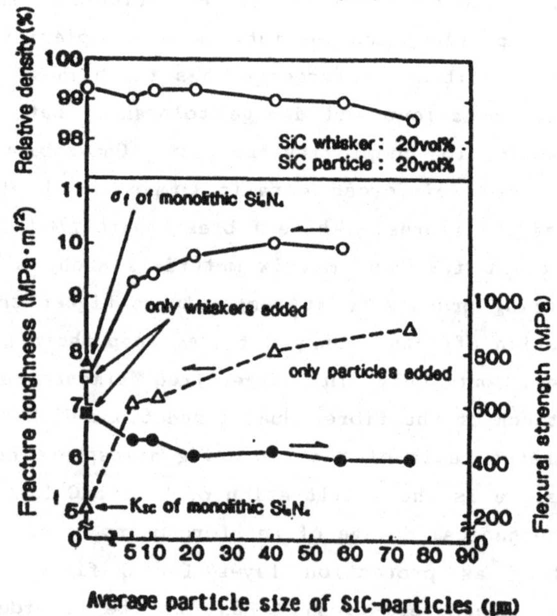

Figure 3. Effect of average particle size of added SiC-particles on fracture toughness, flexural strength and relative density in Si_3N_4-matrix composites reinforced by SiC-whiskers and SiC-particles (15).

possible, and at higher stresses debonding and post-debond friction with subsequent fibre-pull-out can be initiated. To meet these requirements numerous initiatives are in progress to avoid the high processing temperatures of the powder route. As a consequence, some chemical techniques become more and more important, e.g. chemical vapour infiltration (CVI), the sol-gel technique, the pyrolysis of polymer precursors, liquid phase infiltration and directional oxidation. Additional promising routes are reaction sintering of silicon nitride at low temperatures (with minimal shrinkage) and the infiltration of porous ceramic pre-forms by liquid metals. Typical examples are the infiltration of reaction-bonded Si_3N_4 and Al_2O_3.

In many cases the matrix materials prepared by chemical techniques still exhibit a certain amount of residual porosity; this however, is not as critical as for monolithic ceramics due to the high toughness of the composites. Nevertheless, serious problems may arise from oxidation during high-temperature loading.

In the case of whisker reinforcement a major problem is the carcinogenicity of the very fine whiskers ($\phi \sim 0.1$ to 2 μm); in Germany technological work with whiskers was stopped. Furthermore, for increasing the effectiveness of the toughening mechanisms the whisker morphology has to be optimized. From the technological point of view some problems have to be solved: to improve dispersion of whiskers in order to avoid hard agglomerates which may result in strength reduction; to avoid mechanical damage of whiskers during processing; to retain the stability of whiskers during sintering; and to sinter the reinforced materials without applying pressure. For example mechanical dispersion has to be replaced by chemical processing to avoid surface or coating damage, and the reduction of the aspect ratio of the reinforcement.

The alternative to whisker reinforcement is the incorporation of short fibres, or small particles or platelets. For particle and platelet reinforcement the powder route is well suited. In general, property improvement for these types of reinforcement is less effective due to the weaker energy absorbing mechanisms acting in these composites.

In the case of chopped-fibre reinforced composites the main problems are to incorporate a level of chopped fibres, sufficient to initiate reinforcing mechanisms, and to achieve proper dispersion of the fibres without causing additional artificial flaws. Positive effects may be expected if the fibres are oriented perpendicular to the direction of crack propagation. Additionally, depending on the composite system, problems of thermal and chemical instability of the fibres during processing may arise, e.g. with nitrogen during sintering of Si_3N_4.

Regarding hardness and wear resistance, and to a minor extent fracture toughness, promising results have been obtained for various particle-reinforced composites. The main advantage of this type of composite is the ease of processing. Here, for property improvements, the size of the particles, their volume content and the thermodynamic particle/matrix relationship are particularly important. For example, toughness increase is only observed in various systems for coarse dispersoids. In this connection, the size (and morphology) of the dispersoids in relation to these characteristics of the matrix material have to be considered. Based on these facts, promising toughness results have been observed recently by a combination of whisker and particle reinforcement (see Figures 2 and 3) (15). The results are interpreted by consideration of the different reinforcing mechanisms caused by the particles and the whiskers.

Up to now, the introduction of platelets has not been very successful. The slight toughness increase is frequently accompanied with a strength reduction (16). The problem is that only large-sized platelets of relatively high impurity content are available.

Nanocomposites in which very fine second-phase particulates ($\phi < 0.1$ μm) are dispersed within the matrix grains have obvious potential to improve fracture strength significantly even at high temperatures of about 1400°C (17).

The incorporation of ductile metal phases into ceramic materials has been shown to enhance toughness effectively (1,6). An example of recent investigations is the toughness increase by incorporating nickel (13 vol.%) in alumina (by a factor of two compared to monolithic alumina) suggesting a toughening mechanism involving plastic deformation of the metal particles. Maybe the new techniques for fabricating ceramic matrix/metal composites, may further increase crack propagation resistance of ceramics (18): directional metal oxidation; metal infiltration by gas pressure of reaction bonded Si_3N_4 (RBSN) and Al_2O_3 (RBAO); squeeze casting to fill the pores of RBSN and RBAO with various metals (such as Al, Si, intermetallics or super-alloys). However, for all these novel techniques more systematic basic and technological work is necessary.

Based on the state-of-the-art it can be concluded that, notwithstanding critical aspects, ceramic matrix composites have a high future potential.

By selecting the type of reinforcement it will be possible to tailor the property profile of the material. Besides improvement in the stress-strain behaviour, damage tolerance and other properties mentioned above, two perspectives are of future interest. First, to reduce the scatter of mechanical property data, and second to improve the high-temperature strength, utilizing for example the mechanisms of crack bridging and fibre-pull-out in a glass-free matrix material. However, a number of problems have to be solved before these materials can be produced on the large scale and for complex-shaped components with the reproducibility required. In this context only some problems are mentioned: specific technological problems depending on the processing route, further development of low-temperature processing techniques, the lack of thermal stability of fibres, the optimization and control of the interface by coating, changes in microstructure and microchemistry at the interface after thermal exposure, fracture mechanics of interfaces, efforts to deepen the understanding of toughening mechanisms including modelling, the characterization of mechanical and thermo-mechanical properties, the design of components and the development of non-destructive techniques for improving quality control.

LOW-TO-MODERATE TEMPERATURE PROCESSING

The problems of limited thermal and chemical stability of fibres and interfaces are ameliorated at present by two approaches: by coating C-fibres with suitable materials, and by using techniques which allow processing in a temperature range which does not cause fibre degradation. In the first case the damage of coating has to be avoided and the problem of different thermal expansion coefficients has to be considered. In the second case, as mentioned above, special chemical techniques have to be developed.

Promising examples are the CVI-technique and the Lanxide process. For Lanxide processed Al_2O_3/SiC- fibre composites for example strength values of up to 1000 MPa and fracture toughness of up to 29 MN $m^{-3/2}$ are reported (12). The disadvantage may be the residual metal content (between 5 and 30 vol.%) which limits the high-temperature properties. Other examples for moderate-temperature processing are the glass-forming technique, the organo-silicon route and the processing of Si_3N_4-based composites by reaction sintering.

In the case of the glass-forming technique the processing temperature has to be chosen in such a way (Figure 4) that on the one hand fibre strength degradation due to crystallization processes is avoided (for glass composites T < 1300°C), and on the other hand mechanical damage of the fibres during processing, e.g. during hot-pressing, is avoided or limited (T > 1150°C). After optimizing processing parameters the maximum values for mechanical properties which have been obtained for glass and glass-ceramic composites (40 vol.% fibres) are (13,14):

- Glass (optimum processing conditions for Duran-glass/SiC: 1250°C/5 min/10 MPa): $\sigma \sim 1000$ MPa $K_{IC} \sim 34$ MN $m^{-3/2}$

- Glass-ceramic (optimum processing conditions: 1370°C/3 MPa): $\sigma \sim 750$ MPa $K_{IC} \sim 17\text{-}20$ MN $m^{-3/2}$

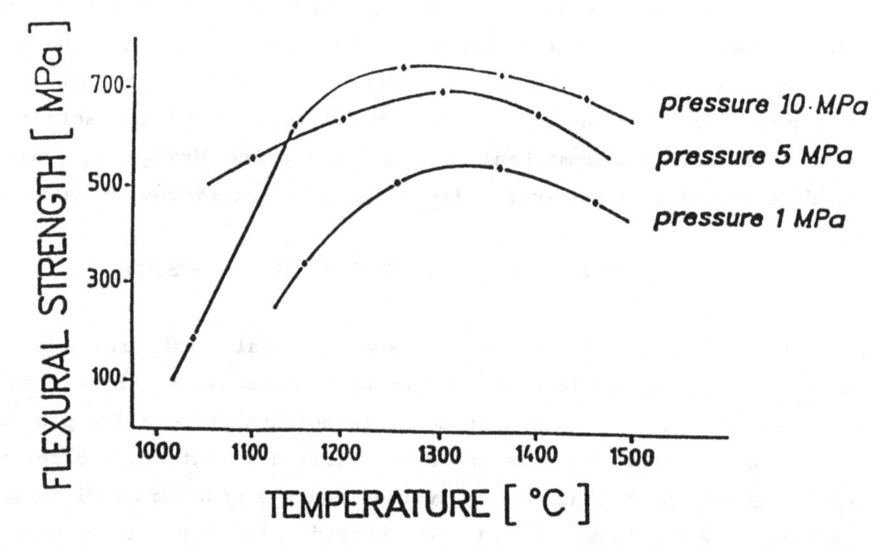

Figure 4. Flexural strength of DURAN glass/SiC (Nicalon)-fibre composites (fibre content 40% ± 2 vol.%) as a function of pressing temperature at various pressures (13).

By using C-fibres to overcome the thermal stability problem of SiC-fibres, the problems of oxidation at the interface arises at application temperatures higher than 600°C. Attempts to use coated C-fibres as a protection layer are up to now not as successful as expected.

Another example of moderate-temperature processing is the development and production of fibre-reinforced ceramic composites by the metal-organic route. A typical example is the processing of SiC- (or Si_3N_4)-based composites. In this case the starting materials may be polysilanes, polycarbosilanes, polysilazanes or polysiloxane. The processing steps for the production of fibre-reinforced SiC based on the polymer-pyrolysis route are indicated in Figure 5. The main advantage of this technique is that temperature does not exceed 1000°C, which causes little degradation in fibre strength. Up to now, promising results have been obtained: pseudo-plastic behaviour and relatively high strength values (Figure 6). The stress-strain behaviour of this type of composites seems to be strongly dependent on the type of fibres and on the type of coating.

In conclusion, it can be stated that there is a potential for this type of composite, but some basic, technological problems have to be solved in the near future. In this context one essential problem is to optimize the chemical parameters controlling the microstructure of the matrix.

Another promising moderate-temperature processing route is the development of continuous fibre-reinforced Si_3N_4 by the reaction-sintering technique. The main problem of reinforcing Si_3N_4 with continuous fibres, following the Si_3N_4-powder route, is the thermal and/or chemical instability of the fibres. Additionally, as in most ceramic composites, the interfacial bonding and the high shrinkage are serious problems. However, using the RBSN-route for reinforcement offers several advantages: the lower processing temperatures (1160-1400°C), the minimized shrinkage, and for high-temperature applications the absence of glassy phases. In contrast, well-known disadvantages are the residual porosity and the problem of chemical stability of fibres during nitridation.

Various types of fibre, such as uncoated and coated C-fibres as well as SiC-fibres (polymer-derived yarns and CVD-monofilaments) have been incorporated in reaction-bonded Si_3N_4 (20). Fibre damage during nitridation was evaluated by tensile testing of the thin single fibres which partially were removed from the compound after processing. The results showed that fibre strength was only slightly affected by heat-treatments under nitridation conditions (Figure 7 (21)). From our experiments it may be concluded that C-fibres, SiC-fibres (polymer-derived) and CVD-SiC monofilaments are potential candidates for reinforcing RBSN. For high-temperature applications this is particularly true for C-fibres. The best results regarding the improvement in stress-strain behaviour are obtained with C- and CVD-SiC (AVCO) fibres (Figures 8 right, 9 and 10). The results with SiC-monofilaments are in good agreement with literature data (22), where high flexural strength of 900 MPa and pseudo-plastic behaviour have been reported. The extent of toughness increase is strongly dependent on

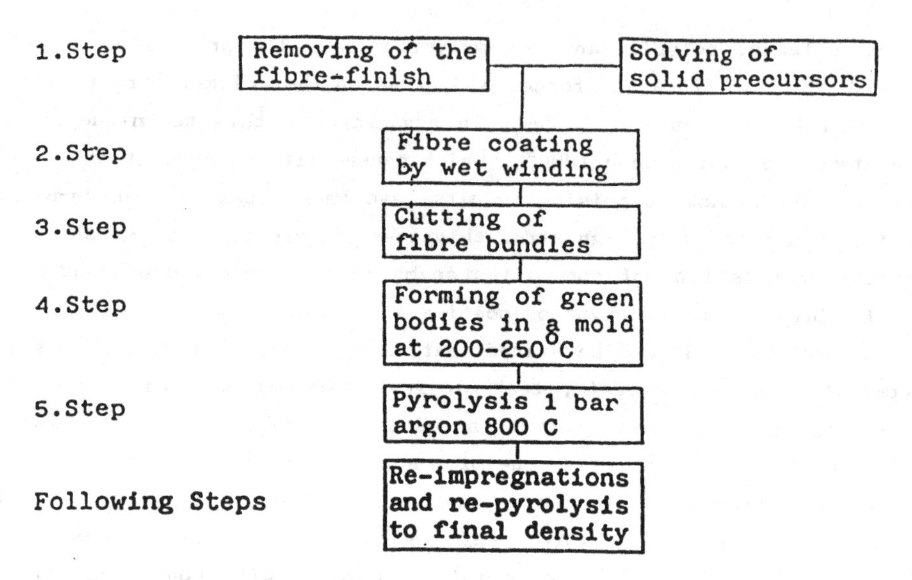

Figure 5. Processing steps during production of fibre-reinforced SiC using the polymer-pyrolysis route (19).

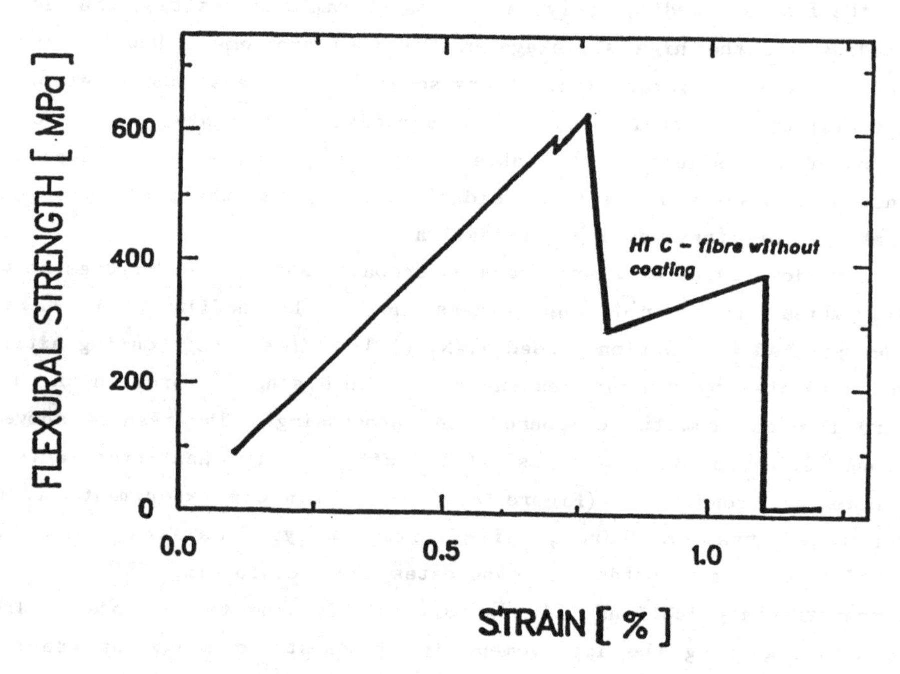

Figure 6. Stress-strain behaviour of a C-fibre reinforced SiC composite produced by the pyrolysis of organo-silicon polymers (fibre content 60 vol.%) (19).

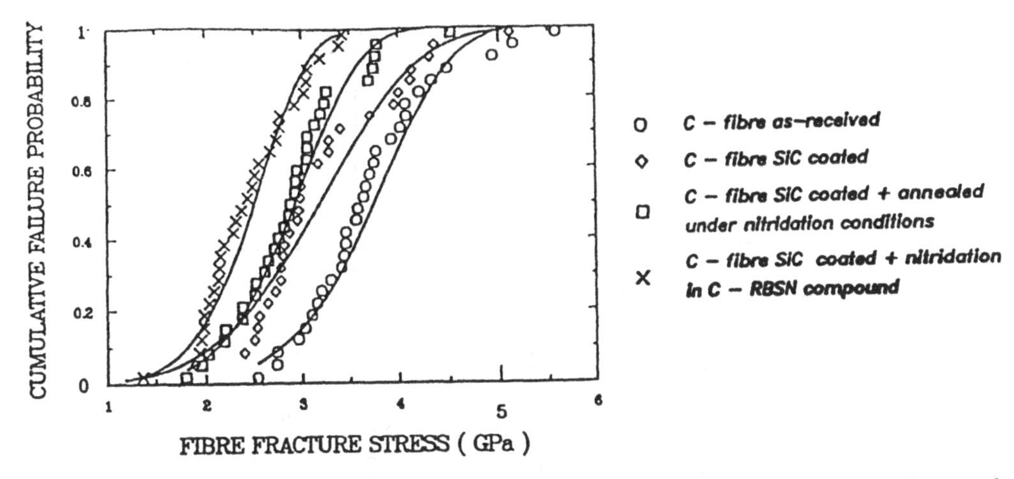

Figure 7. Strength distribution of single C-fibres (M40B) after coating and after different heat-treatment/nitridation conditions (21).

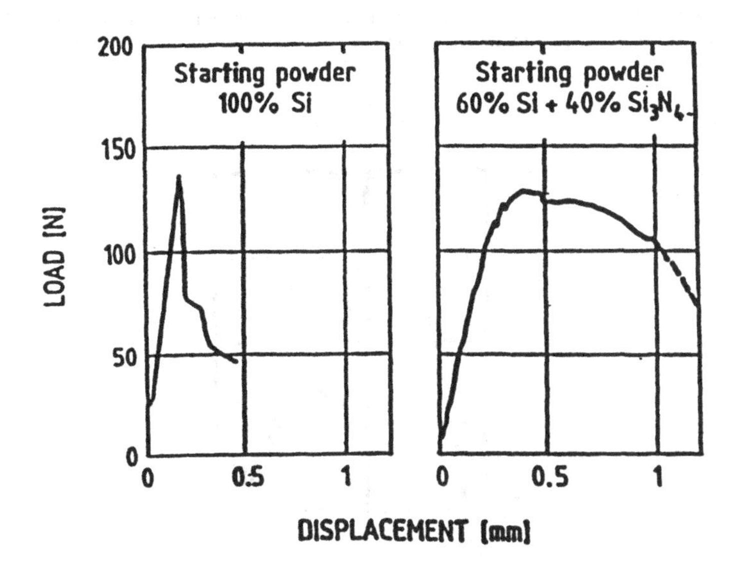

Figure 8. Influence of the starting powder on the stress-strain behaviour of C-fibre reinforced RBSN (20). Fibre: uncoated T800. Nitridation cycle: 1200°C/1 h + 1350°C/10 h.

the starting powder/powder mixture, the nitridation cycle and the type of C- or SiC-fibres. For example, by using a mixture of Si- and Si_3N_4-powder as starting material the stress-strain behaviour could be significantly improved. This is demonstrated in Figure 8 which indicates the change from nearly brittle fracture to controlled fracture behaviour by optimizing the starting powder mixture.

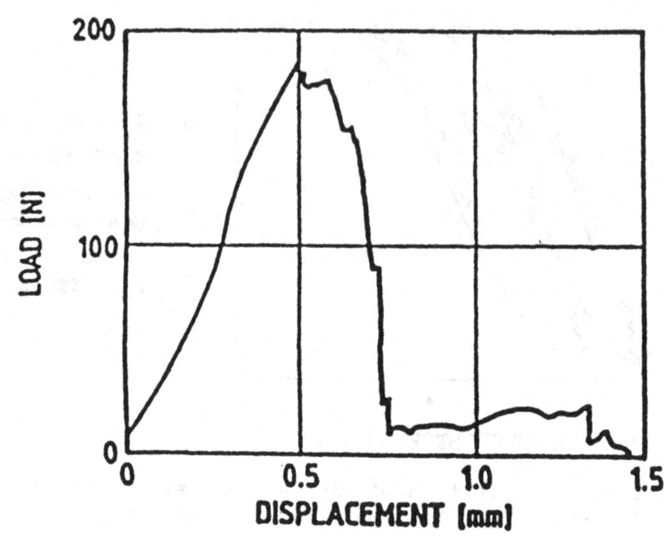

Figure 9. Load-displacement curve for C-fibre reinforced RBSN (20). Fibre: M40B coated with TiN. Starting powder: Si + Si_3N_4 + additives (Y_2O_3 + Al_2O_3).

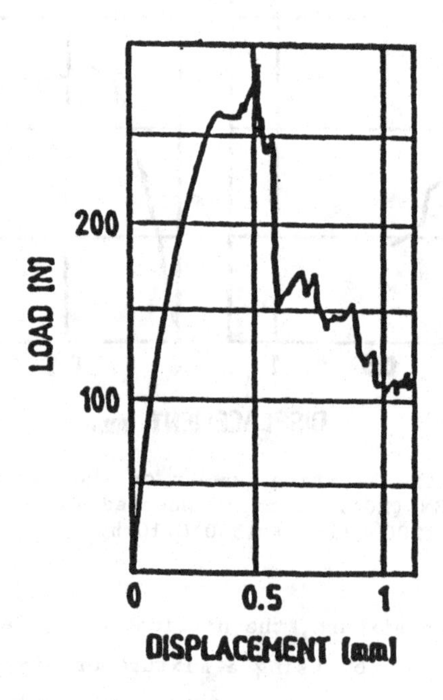

Figure 10. Load-displacement curve for CVD-SiC monofilament reinforced RBSN (20). Fibre: AVCO-SCS-6. Nitridation cycle: 1200°C/1 h + 1350°C/10 h.

One main problem which has still to be solved for this type of material is the optimization of the matrix material, particularly regarding the degree of transformation, total porosity and pore structure.

Summarizing this section: there are a number of techniques, e.g. glass and glass-ceramic forming, the CVI-technique, the organo-silicon route and reaction sintering (in the case of Si_3N_4) which allow processing of ceramic composites at relatively low temperatures without damaging fibres by mechanical or thermal loading and/or chemical reactions. However, in all cases processing has to be carried out very carefully, and basic investigations are still necessary in order to optimize various processing steps.

IMPORTANCE OF THE INTERFACE

From the reinforcing mechanisms it is quite clear that the interface of the reinforcement/matrix plays a decisive role in improving and controlling the mechanical behaviour of ceramic composites, particularly for continuous-fibre reinforced composites. Mechanical properties of these materials are strongly influenced by the debonding and sliding resistance of the interfacial region. This effect of debonding and sliding is demonstrated by two examples from fibre reinforced glasses.

The first example concerns the influence of residual stresses, which are built up at the interface due to differences in the coefficient of thermal expansion (α) between fibres and matrix, on mechanical properties and fracture behaviour. In Figure 11 (see also Table 3) flexural strength data of various SiC (Nicalon) fibre reinforced glass composites as a function of the difference in α are presented. Additionally, the residual stresses are plotted in this graph.

Strength increases to a maximum, and decreases when α of the glass matrix is much higher than that of the SiC- fibre. With increasing α of the matrix, strength increases initially due to the reduction of the tensile stress or the change from tensile to compressive stress. However, for larger differences in α the residual stresses cause a strength reduction. A relationship between fracture type and the difference in α coefficients and hence residual stresses has been found (Figure 12). For low values of α of the matrix, controlled fracture is observed. With increasing α of the matrix strength first increases, however quasi-plastic behaviour is reduced. Very high α-values of the matrix result in low strength. Additionally, the

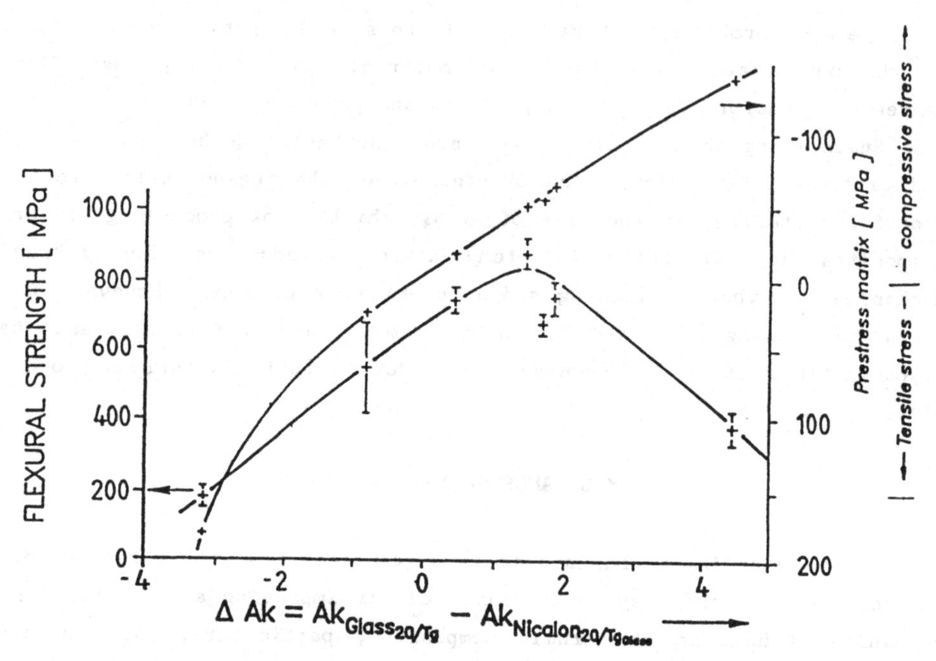

Figure 11. Flexural strength and stress at the interface of SiC (Nicalon) fibre-reinforced-glass composites as a function of the difference in thermal expansion coefficients of various glasses and the SiC-fibres (see Table 3) (14).

TABLE 3
Properties of the Glasses used as Matrix Materials (14)

Glass	E-Modulus (GPa)	Density (g/cm^3)	$\alpha(10^{-6}\ K^{-1})$ 20–300°C	Tg (°C)	V_A(°C) lg η = 4 dPas
SiO_2 + 3% Duran	72	2.20	0.70	1220	>1700
Glass A	73	2.44	2.09	624	1324
Duran	63	2.23	3.25	530	1260
Supremax	90	2.56	4.10	730	1230
Glass C	85	2.54	4.10	643	1180
Glass H	81	2.63	4.60	720	1254
Glass B	85	2.75	6.60	620	1095

α – Thermal expansion coefficient. V_A – Processing temperature.
Tg – Transformation temperature. η – Viscosity.

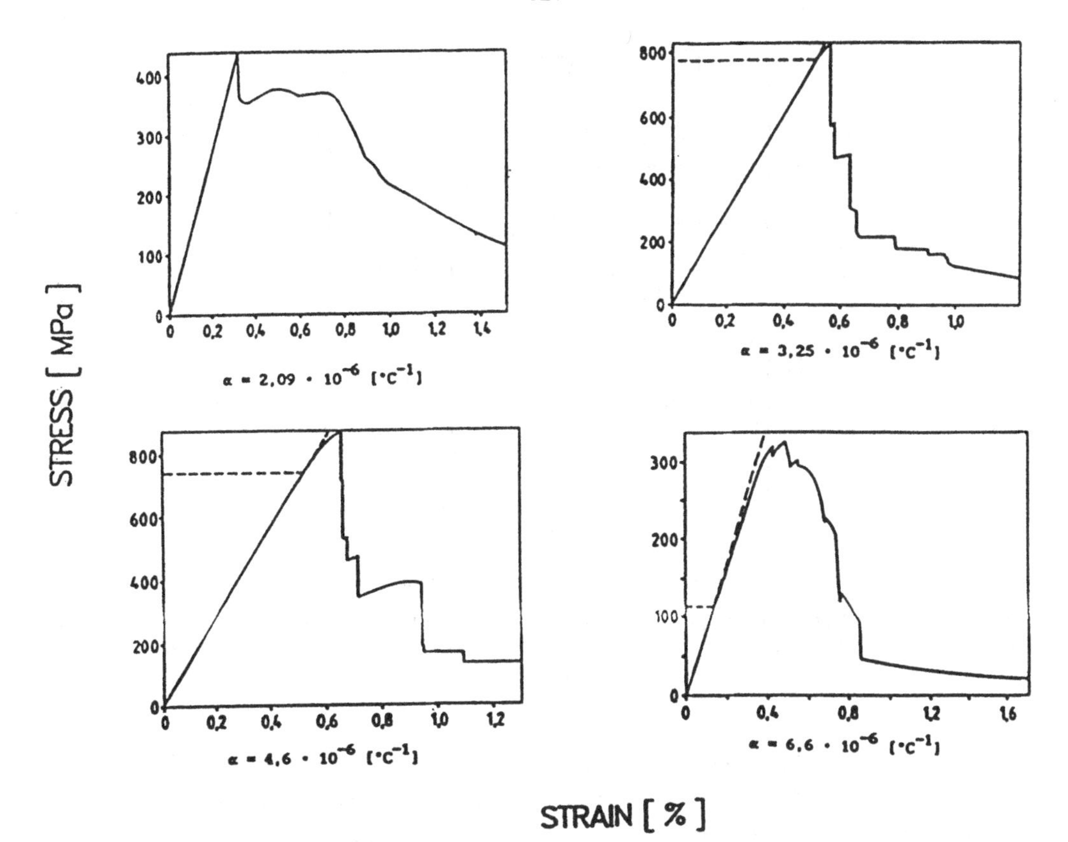

Figure 12. Stress-strain behaviour of SiC (Nicalon)-fibre reinforced glass composites produced with glasses of different coefficients of thermal expansion (see Table 3) (14).

formation of thin layers formed during processing may affect the strength and the stress- strain behaviour. These layers may act as ductile layers on the fibre surface and may favour the pull-out effect.

The second example concerns the influence of microstructural changes at the interface and associated effects on the fibre failure and pull-out. This is demonstrated by experiments with a SiC (Nicalon-fibre reinforced glass-ceramic matrix (lithium aluminium silicate), in which the composites were heated in air to 800°C for 2, 4, 8, 16 and 100 h (23). Heat-treatments under these conditions (Figure 13) caused systematic differences in fracture strain (even after brief heat-treatment periods) and a transition from steady-state matrix cracking in the as-received composite to more brittle failure after heat-treatment. These differences in mechanical behaviour can be explained by microstrucutral changes at the interface. At the interface,

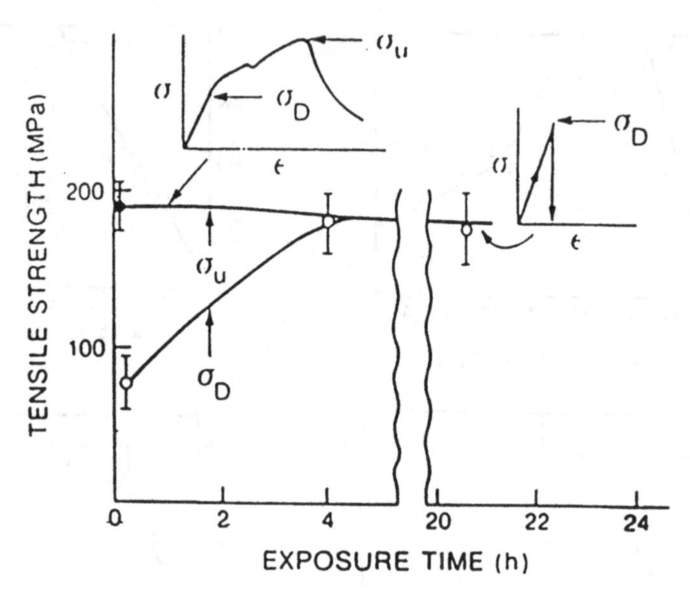

Figure 13. Mechanical property results on heat-treated SiC-fibre reinforced lithium aluminium silicate glass-ceramic. Tensile tests (temperature 800°C) (23).

two distinct interface zones are apparent between the fibre and the matrix in the as-processed state: an amorphous C-layer adjacent to the fibre, and NbC of thickness ranging between 20 and 200 nm (Nb is added as a crystallization aid to the glass). After heat-treatment in air at 800°C, carbon is found to be replaced by amorphous SiO_2, and the carbides of nitrogen replaced by oxides. The SiO_2 thickens with exposure time.

These microstructural changes can now be related to the changes in the stress-strain behaviour which is affected by the interface sliding resistance. The debonding and sliding is governed by the properties of the fibre coating. The microstructural change from C- to SiO_2-layer increases the sliding resistance, resulting in a fracture strain reduction and in a more brittle failure. This explanation is proved by the decrease in the pull-out length after heat treatment (Figure 14 (4)).

From these results it may be concluded that the optimum interphase for high-temperature toughness is a material that behaves essentially as a solid lubricant: either a layered structure or a soft metal thermodynamically compatible with fibre and matrix (4).

These examples demonstrate that the relationship between the characteristics of the interface and the mechanical behaviour of the

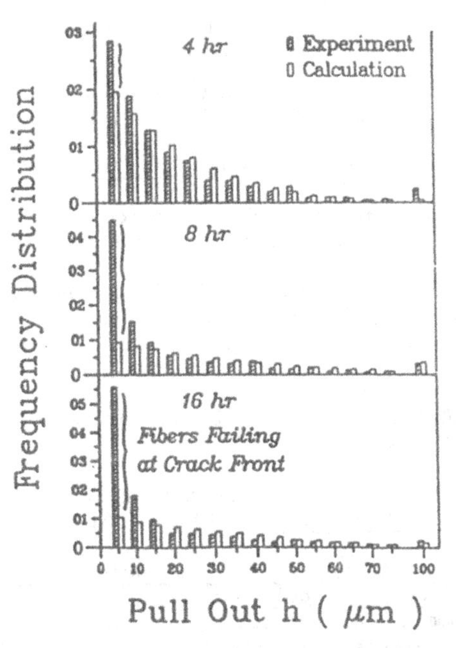

Figure 14. Frequency distribution of measured pull-out lengths and predicted values for fibres that fail at the crack front (23).

composites should be known in more detail for all important composite combinations. This problem area includes:

- The micro-analytical and microstructural characterization of interfaces by means of high resolution and analytical transmission electron microscopy in order to understand the microchemistry and microstructures.

- The analytical approach from a fracture mechanics point of view.

- The measurement of mechanical and physical characteristics of the interface.

Based on these results, experimental efforts have to be made in order to optimize the interface by coating. In some cases not only one layer, but a double layer should be developed in order to protect the fibres and to control the interface.

REFERENCES

1. Rühle, M. and Evans, A.G., High toughness ceramics and ceramic composites, Progr. in Mat. Sci., 1989, **33**, 85-167.

2. Rouby, D., Ceramic-matrix composites, cfi/Ber. DKG, 1989, **66**, 216.

3. Buljan, S.T. and Sarin, V.K., Silicon nitride-based composites, Composites, 1987, **18** (2), 99-106.

4. Thouless, M.D., Sbaizero, O., Sigl, L.S. and Evans, A.G., Effect of interface mechanical properties on pull-out in a SiC-fiber-reinforced lithium aluminium silicate glass-ceramic, J. Am. Ceram. Soc., 1989, **72** (4), 525-532.

5. Freiman, S.W., Brittle fracture behaviour of ceramics, Ceram. Bull., 1988, **67** (2), 392-402.

6. Tuan, W.H. and Brook, R.J., The toughening of alumina with nickel inclusions, J. Eur. Ceram. Soc., 1990, **6**, 31-37.

7. Phillips, D.C., in: Survey of the Technological Requirements for High Temperature Engineering Applications, 1985, 48-73, Office Official Publ. Europ. Comm., Luxembourg, and P.R. Naslain, in: A.R. Bunsell et al. (eds), Development in the Science and Technology of Composite Materials, 1st. ECCM, 24-27th September 1985 in Bordeaux/France, 1985, 34-35.

8. Faber, K.T., Toughening of ceramic materials by crack deflection processes, Ph.D. Thesis, (University of California, Berkely, USA, 1982).

9. Naslain, R., Inorganic matrix composite materials for medium and high temperature applications: A challenge to materials science, in: Development in the Science and Technology of Composite Materials, A.R. Bunsell et al. (eds.), 1st ECCM, 24-27th September 1985, Bordeaux/France, 1985, 34-35.

10. Lange, F.F., Effect of microstructure on strength of Si_3N_4-SiC composite system, J. Am. Ceram. Soc., 1973, **56** (9), 445-450.

11. Ziegler, G., Fibre-reinforced ceramic-matrix composites, Proc. Materialforschung 1988 des BFMT, 12-14th September 1988 in Hamm, Projektleitung Material und Rohstofforschung (PLR) der KFA Jülich GmbH, 1988, 765-786.

12. Lehmann, J. and Ziegler, G., Oxide-based ceramic composites, Proc. Fourth European Conference on Composite Materials ECCM-4, 25-28th September 1990, Suttgart, (In press).

13. Hegeler, H. and Brückner, R., Fibre-reinforced glasses, J. Mat. Sci., 1989, **24**, 1191-1194.

14. Brückner, R., Final Report of the BMFT-Project, Fibre-reinforced glasses and glass-ceramics as composites, March 1989.

15. Kodama, H. and Miyoshi, T., Fabrication and properties of Si_3N_4 composites reinforced by SiC whiskers and particles, Ceram. Eng. Sci. Proc., 1989, **10** (9-10), 1072-1082; and J. Am Ceram. Soc., 1990, **73** (3), 678-683.

16. Heussner, K.-H. and Claussen, N., Yttria- and ceria-stabilized tetragonal zirconia polycrystals (Y-TZP, Ce-TZP) reinforced with Al_2O_3 platelets, J. Eur. Ceram. Soc., 1989, **5**, 193-200.

17. Niihara, K., Izaki, K. and Nakahira, A., The Si_3N_4-SiC nano-composites with high strength at elevated temperatures, J. Jpn. Soc. of Powder and Powdermetallurgy, 1990, **37** (2), 352; and J. Mat. Sci., 1990, **9**, 598-599.

18. Claussen, N., Novel ceramic/metal composites, Proc. Seventh Cimtec World Ceramics Congress 1990, Montecatini Terme, Italy, (In press).

19. Fitzer, E. and Keuthen, M., Reinforced ceramics made of organosilicon compounds, J. Eur. Ceram. Soc., (In press).

20. Kanka, B. and Ziegler, G., Aspects of processing RBSN composites, Proc. Seventh Cimtec World Ceramics Congress, Montecatini Terme, Italy, (In press).

21. Göring, J. and Ziegler, G., Statistical evaluation of the strength behaviour of thin fibre monofilaments for the characterization of fibre-reinforced ceramic composites, Proc. Fourth European Conference on Composite Materials ECCM-4, 25-28th September 1990, Stuttgart, (In press).

22. Bhatt, R.T., Mechanical properties of SiC-fiber-reinforced reaction bonded Si_3N_4 composites, NASA Technical Report 85-C-14, (1985).

23. Bischoff, E., Rühle, M., Sbaizero, O. and Evans, A.G., Microstructural studies of the interfacial zone of a SiC-fiber reinforced lithium aluminium silicate glass-ceramic, J. Am. Ceram. Soc., 1989, **72** (5), 741-745.

WHISKER-REINFORCED COMPOSITES

J.L. BAPTISTA, R.N. CORREIA and J.M. VIEIRA
Department of Ceramics and Glass Engineering,
Centre of Ceramics and Glass (INIC),
University of Aveiro, 3800 Aveiro, Portugal.

ABSTRACT

Models that account for whisker toughening are compared and their consequences discussed in relation to possible composite design for toughness optimization. Sintering of composites with concurrent mechanisms of back stress development by rigid inclusions is analysed and explanations of slow sintering critically evaluated. Colloidal processing of nearly-random whisker composites is discussed with emphasis on the surface state of the particles and its influence on the packing of the green body.

INTRODUCTION

During the last decade a considerable effort has been committed to the development of SiC whisker-reinforced ceramic composites, once it appeared that strong, high-modulus single crystal whiskers with high aspect ratios would be ideal toughening reinforcements. Their toughening potential was expected to be similar to that of fibres, with the advantages of greater thermal and chemical stability, easier processing and greater variety of component design. Although successful composites, like SiC-reinforced Al_2O_3, are already in the market, showing the feasibility of the concept, the toughening increments experienced by other matrices have been less important, bearing in mind the stringent applications they were expected to fulfil. Further improvements are thus necessary.

SiC whisker-reinforced alumina composites have been prepared mainly by hot pressing (1-8), although pressureless sintering has also been tried (9). For hot-pressed material with up to 30 vol.% whiskers, densities greater than 98% theoretical have been attained, with flexural strengths increasing from 150 MPa for the matrix to 700 MPa in 30% whisker composites, while K_{IC} values increased accordingly from 2.5-4.5 MPa.m$^{\frac{1}{2}}$ to 9 MPa.m$^{\frac{1}{2}}$. The actual values of the mechanical parameters vary somewhat with the type of test and the geometry of the testing arrangement with respect to the anisotropic whisker distribution due to sintering under pressure, but no significantly higher values seem to have been achieved. Problems arise with whisker concentrations greater than 30%, namely densification difficulties and whisker agglomeration, allegedly due to matrix grain growth, when higher hot-pressing temperatures are tried in order to avoid residual porosity. Most of the work reports no chemical reaction between SiC and Al_2O_3, but more detailed investigations seem to reveal the presence of an amorphous SiO_2-containing layer at the interface (< 5 nm), even when ultraclean materials are used (3).

In pressureless-sintered composites, complications occur due to hindering of the sintering processes for significant whisker concentrations, greater than say 5%, and possible weight losses at the higher sintering temperatures (6). Sinterability and whisker skeleton formation bear a relationship with the aspect ratio of the whiskers (10). Whisker milling down to an average aspect ratio of 25 and the use of fine, less than 0.5 μm, alumina powders may increase green densities to about 60% for whisker contents of 10 vol.% (9), but difficulties remain, especially if a higher whisker content is envisaged.

In brief, the present paper discusses toughening models for whisker-reinforced composites, analyses the issue of constrained sintering and describes work on colloidal processing of green bodies. Some conclusions are drawn with respect to possible ways of optimizing toughness, decreasing the sintering damage and producing high density green compacts.

TOUGHENING

Discussion of toughening mechanisms in SiC-reinforced alumina, when compared with SiC-reinforced silicon nitride, has frequently been based on the thermoelastic interfacial misfit strains, which are radially compressive in the first system, as opposed to radially tensile in the second. In a recent investigation (3) it was found, however, that the bridging zone and the debond lengths were greater in the former case, which points towards the possibility that interfacial characteristics and whisker roughness effects may overwhelm the influence of misfit strains, as will be discussed later.

Whisker-reinforced mullite has also been investigated (1,11). The toughening reported is about 1 $MPa.m^{\frac{1}{2}}$ (for 10 vol.% whiskers) to 2.5 $MPa.m^{\frac{1}{2}}$ (20% whiskers), in a matrix which shows typically 2 $MPa.m^{\frac{1}{2}}$. The relatively modest improvement, when compared to alumina, has been attributed (1) to a relatively high interfacial fracture energy, which would give a weaker bridging effect, in spite of the fact that the expansion mismatch in the case of mullite should favour debonding. The influence of different chemical bonding strength would similarly apply to soda-lime (Corning 0080) and aluminosilicate (Corning 1723) glasses (1) which, having very different mismatch strains, do show the same toughening in composite form with 20% SiC whiskers. Toughening values reported for SiC-reinforced aluminoborosilicate glasses vary substantially with matrix composition. Increments of about 3 to 4.5 $MPa.m^{\frac{1}{2}}$, over the matrix value of about 1 $MPa.m^{\frac{1}{2}}$, were reported for Corning 7740 (Pyrex) with 20% whiskers of different producers (12), whereas Corning 1723 glass with 20-30% SiC whiskers shows overall toughness of 2 to 3.4 $MPa.m^{\frac{1}{2}}$ (1,13). The mechanisms claimed to be responsible for whisker toughening are (i) crack deflection, possibly associated with crack front bowing, and (ii) bridging, which includes contributions from debonding and pullout.

Crack deflection has been modelled for random volume dispersions of second phase particles of variable shape and aspect ratio (14), and for planar random distributions of whiskers (15) as occurs in hot-pressed composites. For a given particle shape, say rod (whisker), deflection toughening is determined solely by the aspect ratio and volume fraction, increasing with both. The thermoelastic mismatch between particle and matrix plays no role except in inducing the deflection path, namely by tilting the crack front, towards the radially tensile stressed regions surrounding a whisker embedded in a matrix of lower expansion coefficient. Fractographic observations (16), led to the conclusion that deflection was responsible for a 30% increase in K_{IC} in Si_3N_4(m)-20 vol.% SiC(w) composites; however, it was also considered (1) that, for most systems, the deflection distances, angles and frequency are of the same order as found in non-reinforced matrices, whereas the planar random whisker distribution model (15) could not be fit to experimental toughness values. Thus, the existence of an operative deflection mechanism, capable of appreciable toughening, and/or the validity of the model remain to be confirmed.

Bridging models for fibrous reinforcements are based on the situation of aligned particles lying normal to the crack plane. No random arrays have been analyzed so far. Thus whisker reinforcement is discussed assuming the unidirectional problem. Toughening by bridging implies whisker-matrix debonding and progression of the crack front past the intact whisker, which acts as a ligament in the crack wake, where debonding proceeds, with concomitant pullout. Positive contributions to toughening (2) come from the fracture surface energy supplied to the debonded interface, the energy dissipated during pullout as a thermal effect and the elastic energy stored in the whiskers up to the point of rupture and dissipated as acoustic waves; a negative contribution corresponds to the relief of residual thermoelastic energy by the passage of the crack. The asymptotic toughening relative to the matrix level (i.e. in the stationary regime of the composite R-curve) is approximated by Evans (2) as:

$$\Delta G_c \approx \frac{v_f l_d \sigma^2{}_f}{E_f} - v_f l_d E \varepsilon^2{}_i + \frac{4 v_f l_d G_i}{r(1-v_f)} + \frac{2 v_f l_p{}^2 \tau_i}{r} \tag{1}$$

where v_f is the volume fraction of whiskers, σ_f their tensile strength, E the Young's modulus of the composite, E_f the Young's modulus of the whiskers, G_i the matrix/whisker interface toughness, l_d the debond length (in each side of the crack), ε_i the misfit strain, r the whisker radius, τ the sliding stress of whisker against matrix and l_p half the pullout length. This expression is obtained by integration of the surface tractions on the crack surfaces along the bridging zone length, with the added term of interfacial fracture energy. A functional dependence is further assumed

$$\frac{l_d}{r} = F(G_i, \varepsilon_i, \sigma_f, \mu) \tag{2}$$

where F is an unknown function and μ is the whisker/matrix friction coefficient. Equation (1) can be further simplified if it is considered that the negative contribution from the residual strain term is small ($0.3\ J.m^{-2}$ for Si_3N_4/SiC and $2\ J.m^{-2}$ for Al_2O_3/SiC (3)).

An alternative model by Becher et al. (1) using either (i) a fracture mechanics approach in which the bridging zone is considered analogous to a Dugdale-Barenblatt zone, with uniform closure stress (tensile stress on the whiskers) and stationary length and (ii) an energy balance approach, where the excess strain energy release rate is the sum of the elastic energy stored in the whiskers and the pullout work under a shear stress (constant over the debond length) and assuming that the debond length is much greater than the crack opening displacement, produces a fracture toughness dependence of the form

$$\Delta K_c = [K_{cm}{}^2 + \frac{k\sigma_f{}^2 v_f r E G_{cm}}{(1-\nu^2)E_f G_i}]^{\frac{1}{2}} - K_{cm} \qquad (3)$$

ν being the Poisson's ratio of the composite and G_{cm} the matrix toughness, which leads to

$$\Delta G_c = \frac{k' v_f l_d \sigma_f{}^2}{E_f} \qquad (4)$$

which is a functional relationship analogous to the first term of (1), apart from the multiplying factor. In (3) l_d is assumed to be given by Kendall's equation

$$\frac{l_d}{r} = \frac{G_{cm}}{6G_i} \qquad (5)$$

Comparing equations (1) and (4) it is noted that neither the fracture surface energy term nor the pullout term are now explicit. However, the interfacial energy concept in Becher's analysis does include the contributions of chemical bonding, misfit strains and sliding work, whereas in Evans' analysis it is considered merely as chemical bonding contribution. The same considerations would apply to equations (2) and (5). One important difference between the two models is that the matrix fracture energy G_{cm} does not enter in Evans' (2); actually this model makes the debonding contribution depend rather on the fibre/interface fracture energy ratio.

Both models predict that a high volume fraction of high-strength whiskers is desirable for high degrees of toughening; these requirements are, however, limited by the sinterability of the composites and the quality of available whiskers. Improvements in processing parameters of composites and in whisker fabrication are worth considering and, in this respect, the presence of whisker defects, like voids (3,16) and surface damage, which can act as weak spots, is to be avoided.

The other common toughening parameter is the debond length. It is difficult to analyze the dependence of l_d on manageable processing parameters, but it is natural to assume that l_d will scale with r (according to both analyses) and that it will increase with decreasing G_i, increasing

σ_f, decreasing μ and increasing positive ε_i. In this way it could be argued that the maximization of the debond term, for a given whisker concentration, should be attained mainly by maximizing the whisker strength (3); the role of the residual strains and the friction coefficient, namely with respect to whisker roughness, should, however, deserve some attention as well. For whiskers at the point of rupture, the Poisson strain $\nu\sigma_f/E_w$ may be much larger than the residual strain, so that debonding is dominated by the surface roughness of the whisker and residual strains are unimportant; these become important when much larger than the Poisson strain, so that radial interfacial tension promotes debonding, whereas compressive interfaces inhibit it.

The maximization of the elastic bridging contribution, formally expressed by (4) and by the first term of (1), can be achieved by decreasing G_i, increasing σ_f and increasing l_d; since l_d scales with r, increasing whisker diameter will promote toughening, and this was experimentally confirmed in Al_2O_3/SiC (1). It is noted that it is the whisker diameter, not the aspect ratio, which influences toughening; actually, even the residual strains, which can influence the value of l_d, are practically invariant for aspect ratios greater than about 10 (4). It is probably worth performing a simple calculation in this respect: according to the toughening predictions of Becher et al. (1), taking $\sigma_f \approx 10$ GPa, $v_f = 0.2$ and assuming equation (5) with $G_{cm}/G_i \approx 30$, an increase of 10 $MPa.m^{\frac{1}{2}}$ would be obtained for $r \approx 1.2$ µm, with $2l_d \approx 12$ µm, that is an aspect ratio greater than only 10. The implications of the use of milled, relatively coarse whiskers to produce significant toughening are also obvious as far as processing is concerned; on the other hand, the use of shorter whiskers would increase σ_f, which is a distribution-controlled variable. Another important effect on toughening is that of G_i, as demonstrated (1) for SiC whiskers with different surface oxygen contents in an alumina matrix, whereby increasing silica content at the interface decreases toughening by promoting stronger chemical bonding, thus inhibiting the debonding process. Whisker coatings may play a significant role in this respect: besides being able to modify the residual stresses (5) they can act as reaction barriers.

The pullout contribution is still controversial. The pullout length l_p contributes strongly to the pullout term, but direct measures of this parameter in the same system (Al_2O_3/SiC) by different authors vary from a few r to less than r. It is argued (2,3) that low flexural strength whiskers, like SiC, when inclined to the plane of the crack would tend to fail by bending, thus producing little pullout; in this way, the pullout term could be discarded in many randomly-reinforced composites.

Bridging can also take place in the matrix, whereby large grains may act as ligaments in the crack wake, resulting in further toughening. A coarse-grained matrix would then be desirable, and a toughening effect has indeed been found in alumina, from 2.5 to 4.5 MPa.m$^{\frac{1}{2}}$, by increasing the grain size from 1-2 to 15-20 μm; this increase in "baseline" toughness is not altered by alloying with up to 30 vol.% SiC whiskers (1).

The above discussion leads to some conclusions regarding toughness optimization by the use of whiskers. It is apparent that the use of short, relatively thick, high-strength whiskers is desired. On the other hand, whiskers should be non-reactive towards the matrix, which is not always the case. For example, SiC whiskers have an external silica layer that acts as a bonding agent, not only with the matrix but also with the glassy phase used as sintering aid; apart from this, silicon carbide itself shows some solubility in those secondary phases. Two routes seem possible for avoiding these problems: the first one relies on the coating of whiskers by a non-reactive insoluble layer like carbon, and the second envisages the use of SiC-saturated second phases as sintering additions. Another factor of importance is the attainable compromise between residual strains (which are imposed by the thermoelastic properties of matrix and whisker - with or without coating), predicted Poisson's effect and whisker rugosity and strength, in order to simultaneously profit from debonding and attrition effects during pullout. Given the relationship between G and K, for the same G_c, a greater K_c, and thus a greater composite strength is achieved for a high Young's modulus of the composite. High values of E/E_f would thus be of advantage, and this can be achieved by a relatively rigid, though tough matrix.

Toughness optimization is just one of the factors that determine composite strengthening; critical defect size is the other. Processing routes play a key role in this respect and results appearing in the literature frequently show (17-19) that, in similar systems, although increasingly higher whisker concentrations may lead to higher fracture toughness they can actually produce lower fracture strengths.

DENSIFICATION OF CERAMIC-MATRIX COMPOSITES

Inclusions, either particles, whiskers or platelets, that have been used as strengthening agents for ceramic materials have the drawback (20,21) of demanding very high sintering temperatures for full densification of the composites. Even then, this can seldom be realized without applied pressure (22,23). Inclusions have the potential to reduce the densification rate of the ceramic matrix in comparison to the inclusion-free matrix (21,22-34). Stresses develop at the onset of sintering (24-31) in association with the differential shrinkage of the matrix in relation to rigid inclusion. As in compacts with internal agglomeration (32,33) the stresses decrease the sintering driving force and may cause sintering damage. Two types of damaged microstructures with quite different mechanical quality are observed: (i) arrays of voids and large pores randomly distributed in the matrix (21,31,34), (ii) crack-like flaws that extend from the rigid inclusion (23,31,34-36).

"Back Stress" Relaxation

The stress that results from compact inhomogeneity, the "back stress", its magnitude and transiency have been investigated as the basis for interpreting some important features of the hindered densification and aspects of microstructure development during sintering of composites (21,31,34). Emphasis has been given to the mechanisms of stress relaxation through viscous creep of the porous matrix (21,24-30,34). Fracture strain and void growth compensate for the matrix/inclusion strain misfit. They are concurrent mechanisms for stress relaxation, but these mechanisms are poorly understood (21,31,35,37).

The available models of the densification kinetics with the inclusion back stress (24-26,28,29) assume that the inclusions and matrix are continuous media, that there is no sintering damage so that the composite material will have the same microstructural evolution as the inclusion-free matrix, in spite of the back stress field (21,24-30). The models differ in relation to the contribution of viscoelasticity of the porous body, in the fundamental constitutive laws of sintering, creep and elastic deformation of the porous matrix (30). The analysis has been recently reviewed (30). The main achievements are in summary (30,38): (i) Due to strain rate misfit, a compressive radial stress, σ_r and a tensile hoop stress, σ_θ, are generated in the matrix by the inclusion. Both stresses add together so that the mean hydrostatic stress, σ_H, is not a function of position, being a function of inclusion volume fraction, v_f, of intrinsic sintering stress of the inclusion free matrix, Σ, and of the viscous Poisson's ratio, ν_p (30,38);

(ii) In the timescale of the characteristic time constants for creep and densification (30,38), the time dependent Poisson's ratio becomes the viscous Poisson's ratio, ν_p, that being only a function of the viscosity ratio: $\nu_p = (3-2\eta_c/\eta_d)/[2+(3+\eta_c/\eta_d)]$, where η_c and η_d are the uniaxial viscosities for creep and densification respectively (30,39).

There has been argument on the physical meaning of ν_p (30,38-40). ν_p sets a limit on the maximum values of stress that should develop at the inclusion-matrix interface, which are $|\sigma(a)_r| \leq |2\Sigma|$ and $|\sigma(a)_\theta| \leq |\Sigma|$. The values of ν_p also set the maximum value of the viscosity ratio η_c/η_d, which must be below 3/2 for ν_p to be positive (30,38), negative values of ν_p being thermodynamically allowed but not observed (30,38). The viscosity ratio, η_c/η_d of ceramic matrices at high temperatures, as determined from loading dilatometry results, was recently reviewed by one of us (39). The viscosity ratio for the ceramic materials must be very close to unity at the onset of sintering, having a slow decrease as the internal load bearing area increases with densification and being almost independent of temperature. This is against previous analyses that have foreseen high values of the relative back stress based on large differences of the creep relaxation time in comparison to the densification relaxation time (21,24,26,34), the values of η_c/η_d for solid state sintering of ceramics supposedly being in the wrong side ($\eta_c >> \eta_d$) (21,24,41,42). Also, the ratio $\beta = (G_c/K_c)(\eta_d/\eta_c)$ for TiO_2-Al_2O_3 (34) and amorphous SiO_2-SiC_p (31) composites was determined as $\beta = 0.01$ and $\beta = 0.04$ respectively, by using Raj et al. viscoelastic analysis (24). Support was claimed for the viscoelastic argument in the first case, but the value of β in the second case is unexpectedly low. Both the TiO_2 and SiO_2 matrices showed crack growth and sintering damage (31,34).

Constraining Network Effects

In back stress analysis the multiple inclusions are assumed to interact only by the linear combination of the back stress of each inclusion (43), full density being always potentially achieved (25,27,44). Analysis of the inclusion spacial distribution in relation to the limiting particle volume fraction of the inclusion network, s, the inclusion volume fraction, v_f, the inclusion size, D_i, and matrix grain size, G, brings in an end-point density to the composite (43,45). Figure 1a gives a summary of the main relationships between inclusions and matrix according to the constraining network model (45), where α is the ratio between inclusion-pair centre distance, l_n, and the shortest inclusion centre separation distance of the composite, l_o. The values of s and α for a network of spherical inclusions with the regular tetrakaidecahedron as the primitive cell are also given in Figure 1a. The values of s decrease for angular inclusions or whiskers (43,45). Experimental results in the Al_2O_3-SiC_w system set (45) s = 0.1. Good dispersion of the whiskers will lead to values of α approaching 1, at high volume fractions of whiskers (45). Accounting for the contribution of the inclusion diameter to the length of the line that joins the centres of neighbouring inclusions (45), compressive and tensile stresses are expected to arise in the matrix between network pairs with the shorter and larger than average separation distances respectively. The constraining network stress will be locally combined with any stress due to strain rate misfit around the inclusion (43,45). The constraining network model has not been developed to calculate the stress field, the stress relaxation and composite densification rates (30,45). The values of α and s that were necessary to force coincidence of calculated densification data to the experimental data of the ZnO-SiC_p composites (28) are within the lattice pair limitations of the terakaidecahedron cell, the effects of the constraining network on composite densification being presumed important (45). But, similar calculations for the Al_2O_3-Al_2O_{3p} composites failed to yield values of α with clear physical meaning (21). Assuming, as before, that composite densification occurs without changes in the microstructure-densification relationships of the matrix, such as grain growth and sintering damage, is a true limitation of the model (21,30).

In Figure 1b, we distinguish two different regions within the matrix, according to the constraining network model (45). When the composite end-point density is reached, region (I), that has been submitted to compressive stress, must be close to full density, while region (II) remains porous. In real systems the porosity must continuously change from region (I) to region (II). Relative density of region (I) is simply assumed to be ρ = 1 at this stage. From this point onwards, the dense region (I), which has finite viscosity, is only allowed to creep, as do the liquid bridges between solid particles in liquid phase sintering (46). Re-arrangement kinetics permitting, the full stability of the inclusion network will only be achieved if s is above 0.5 (28,46). Region (II) has a non null sintering potential that comes from the pore with negative average surface curvature inside the region (26), but it is constrained from densifying. The shrinkage coming from inclusion re-arrangement controlled by creep in region (I) will be slow. It is mostly dependent on matrix viscosity and

A) GEOMETRICAL RELATIONSHIPS

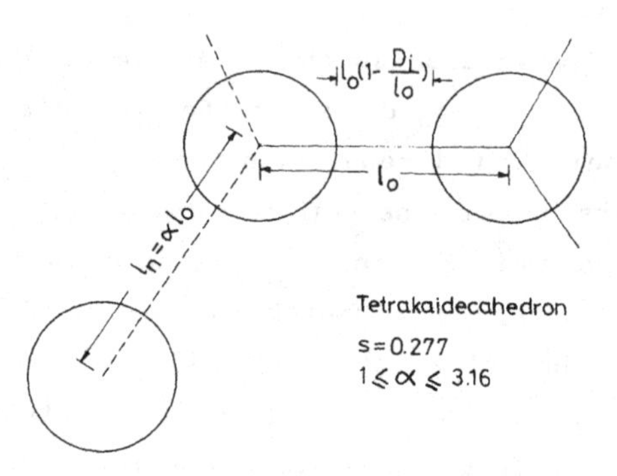

B) END-POINT DENSITY

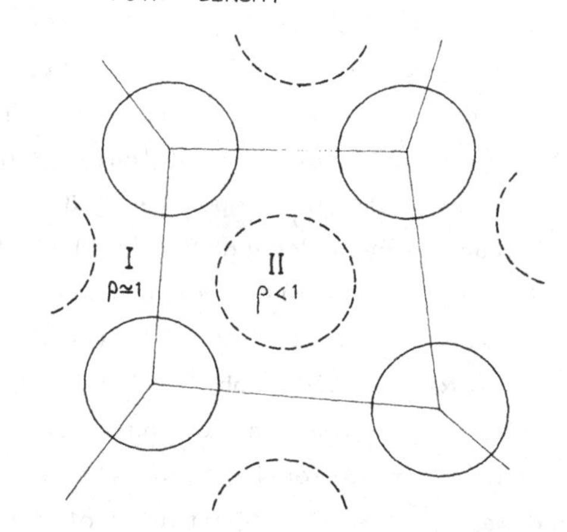

Figure 1. Constrained network model: (a) - Geometrical relationships for the regular tetrakaidecahedron primitive cell; (b) - Local density in relation to the inclusion centre distances.

inclusion size (30,34). Mechanisms of matter transport that were concurrent with densification (30,45,47,48) will remain in full in region (II). Their rates are proportional to a given power of the reciprocal of matrix grain size (48,49), so they can be faster than network re-arrangement. These mechanisms can be listed as follows: (a) diffusion controlled pore coalescence, even bloating of closed pores (50); (b) opening of necks between particles of low co-ordination number that stay in the inter-agglomerate space (33,37,47,51); (c) particle coarsening, grain growth with the concomitant coalescence of mobile pores at constant pore volume fraction (30,49,50); and (d) (only in solid state sintering) microdensification at nearly constant pore volume fraction with transition of porosity from a mixture of pores of low and high co-ordination number to stable porosity (33,34,51,52).

Pore growth in the constrained region (II) lowers the sintering potential (39), so it reduces the stresses in the composite. The average curvature of the liquid meniscus at pore surfaces for small and large pores in amorphous systems and in liquid phase sintering is always negative (26,46,50,53), the sintering potential being preserved at lower values after the microstructure transitions (a) to (c) above. The sintering potential of region (II) will come to zero in solid state sintering if microstructure transition (d) becomes completed in the constrained region (II) (34). Grain growth reduces pore co-ordination number, so that stable pores eventually become unstable, the densification driving force being locally restored (33,52).

Pores contribute to local stress enhancement. Expressions for the stress intensification factor of porosity have been established for all the stages of sintering (54-56). Tensile stress intensification is seriously high on loosely sintered particles. Depending on the magnitude and signal of the local stress, these contacts can undergo desintering (negative neck-growth) or fracture (51,52). Pores with negative surface curvature of their equilibrium shapes have increased stability against desintering (52). Large, closed pores can either shrink or expand under small local stresses in solid state sintering (52). They are the less effective in respect to stress intensification in the matrix (54). Processing must be controlled in order to minimize the sources of defects of the type (b) and (d) above, as these transitions contribute in different ways and severity to rapid decrease of the sintering potential.

Grain growth in nearly dense areas, like region (I), is often faster than in more porous regions (26,28,49). The creep and densification viscosities being a function of grain size (26,57), the creep viscosity of region (I) is increased in comparison to creep and densification viscosity of region (II). In Figure 2, the dependence of linear densification strain rates of three different types of inclusion-free matrices are represented as a function of relative density, ρ. The slope of the curves in Figure 2 includes the contributions of the dependencies of sintering pressure, and of densification viscosity on density (57). The creep rate of many ceramic matrices being almost proportional to the densification rate (41), Figure 2 can also be seen as representing the decrease of matrix creep rate with density. Type (A) kinetics is representative of amorphous (glass) matrices although type (B) kinetics has also been observed in glass matrices (53). Type (B) densification kinetics has been observed in solid state sintering (57), the porosity being uniformly distributed. Densification kinetics of type (C) are often observed in solid state and in liquid phase sintering of systems with ill defined microstructures (21,44,58), the sharp decrease of densification rate at the end of curve (C) being due to rapid grain growth. The points ρ_I and ρ_{II} in Figure 2 represent the densities of region (I) and (II) of Figure 1b. They have large differences in creep rate for matrices of type (C). From the condition of homogeneous average shrinkage, the stress gradients between region (I) and (II) will increase (21) in the same way as the slopes of the curves in Figure 2. Type (A) matrices (glasses) are the only ones where the constraining stresses will remain low and sintering damage of any type (a or b above) is expected to be at the minimum. So, matrices with densification kinetics of type (A) would be preferable. It would imply a move towards amorphous matrix composite processing. Two phase, polycrystalline ceramics, like some grades of t-ZrO_2 (44), where grain growth is hindered by the second phase, develop into dense polycrystalline matrices with the finest grain sizes and show superplasticity. These matrices are also expected to overcome, as the glasses, the constraining network effects.

Inclusion and whisker size effects on composite densification have been reported (20-22,34,36,59,60). For inclusion sizes, $D_i < 10G$, the effect is strong. For fixed values of v_f and G, the number of matrix grains fitting the smallest gap between spherical inclusions is:

$$n = [(s/f)^{1/3}-1](D_i/G) \quad (6)$$

For the tetrakaidecahedron primitive cell and $(D_i/G) \approx 10$, the maximum value of v_f, which fulfils the condition that there is at least one matrix particle in between every inclusion is typically lower than s = 0.27, $v_f^* = 0.20$. There will be from 1 to 3 matrix particles in between first order neighbour inclusions, this being a limit to the models that treat the inclusion-matrix as continuous systems. The single particle in between inclusion pairs will oppose network shrinkage and densification, unless it deforms or the inclusion-to-particle contact shrinks. Contact shrinkage is allowed in the following conditions: (i) the particle deforms by viscous flow (53), (ii) there is a reactive liquid at the contact (46),

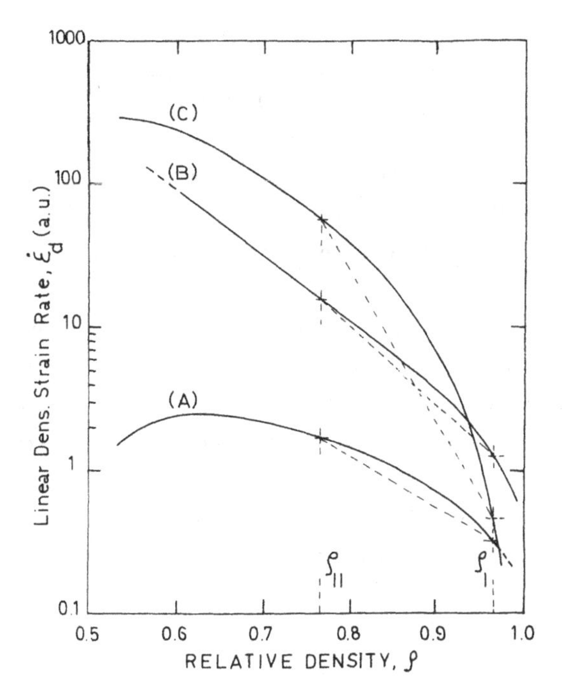

Figure 2. Linear densification strain rates of glass and ceramic inclusion-free matrices.

(iii) chemically identical inclusions and particles in solid state sintering (21,48). The solid state sintering shrinkage kinetics of contacts of different chemical composition can be hindered by vacancy condensation at the neck or by volumetric expansion of the binary reaction products (61,62). Coalescence of the small particle to the inclusion, if of identical composition, may also yield abnormal grain growth, the annihilation of the bridging particle bringing neighbour inclusions into direct contact (21).

When $(D_i/G) >> 10$, the critical density ρ_{cm} of the porous region in between inclusions with $l_n > l_o$, region (II) in Figure 1b, can be calculated from the inclusion size-to-inclusion centre distance ratio, from the maximum matrix shrinkage, that is made equal to the shrinkage corresponding to ρ = 1, region (I) in Figure 1b, and from the corresponding values of α, and of the initial density, ρ_o, being:

$$\rho_{cm} = \frac{\rho_o}{[1-(1-\rho^{1/3})F]^3} \tag{7a}$$

where

$$F = \frac{[1-(v_f/s)^{1/3}]}{[1-(1-(1/\alpha)(v_f/s)^{1/3}]} \tag{7b}$$

Values of ρ_{cm} are plotted in Figure 3a and 3b. The value of $\alpha = 3.16$ that is chosen for Figure 3a corresponds to the largest distance between network sites of the tetrakaidecahedron primitive cell. ρ_{cm} is just slightly above the initial density, ρ_o, when $v_f = 0.20$. Accounting for improved inclusion dispersion, α was adjusted from 3 to 1.2 in Figure 3b, and v_f is made constant. Figure 3b shows that by improving initial matrix density and by preserving good inclusion dispersion, the final stage of sintering might become accessible without much disturbance from the constraining network. This result is supported by high final densities of pressureless sintered Al_2O_3-SiC_w composites, prepared by colloidal processing techniques, with relative green densities close to $\rho_o \approx 0.70$ (20).

Sintering Damage

The tensile stress at the inclusion-matrix interface, where it has the maximum value, $\sigma_\theta(a) \leq \Sigma$, can rise above the breaking strength of the porous matrix (23,24,34,35). Radial cracks emanating from the inclusions were observed in sintering of TiO_2-$Al_2O_{3_p}$ composites (23,34) and in model experiments (35). The sintering of the TiO_2-$Al_2O_{3_p}$ composite (23,34) shows that there is a minimum inclusion size for the net observation of radial flaws in the composite. Fragile fracture criteria (23,24) and creep crack growth mechanisms of the composite flaws (35) are compared in Figure 4, a summary of the main relationships that are used to calculate critical inclusion size being given in the figure. The crack opening displacement rate in the composite matches the linear densification rate of the inclusion-free matrix (35), $\dot{u}/a \approx \dot{\varepsilon}_d$, where a is the flaw size. The fracture stress intensity factor for creep crack growth is lower than for brittle fracture (35), $K^* \approx 0.1\ K_{Ic}$. The bound for the breaking stress was set equal to the maximum hoop stress (30), $\theta_\theta^{max} \approx -\Sigma$. For initial densities close to $\rho_o = 0.50$, the sintering stress of several ceramic powder compacts is a known function of the initial particle size, G_o, $\Sigma_o = AG_o^{-1}$, where $A = 0.60$ for solid state sintering and $A = 0.12$ for amorphous and liquid phase sintering (39). As the radial cracks develop from the full perimeter of the inclusion (23,24,34,35), the size of the critical flaws a_c is made equal to

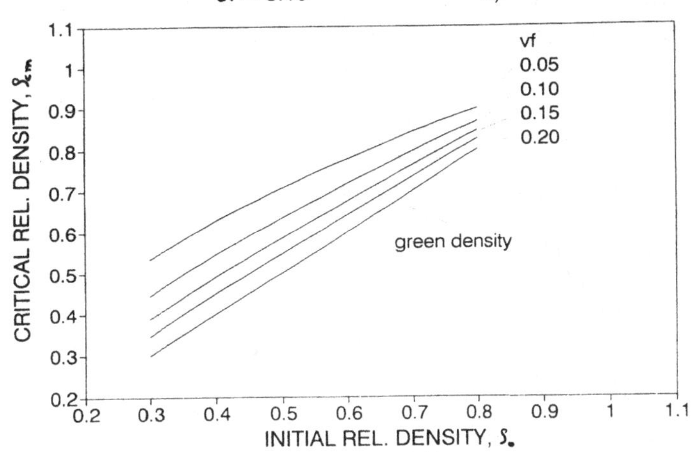

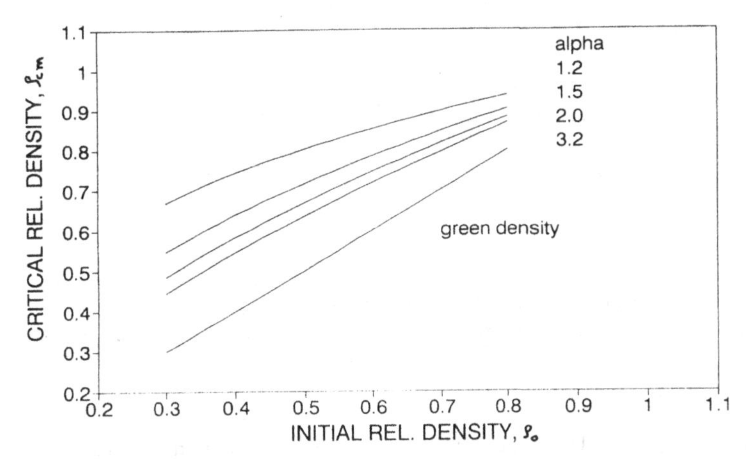

Figure 3. Critical density between neighbour spherical inclusions with centre distances above the average: (a) - Inclusion volume fraction effect; (b) - Inclusion dispersion effect.

D_i, the inclusion size. The critical inclusion size for growth of radial flaws becomes proportional to G_o^2, Figure 4, the porportionality being supported by experimental data (23,34,35). This figure also shows that sol-gel derived powder matrices (either amorphous or crystalline) may be cracked by micron-sized inclusions through creep crack growth. The critical value of the relative inclusion size $(D_i/G_o)_c$ is itself proportional to G_o. There is experimental evidence that the matrix flaws can initiate from weak points that result from drying, debonding and intradomain differential microdensification before the isothermal condition is reached, once there is a change of the linear dimensions of the composite (23,34,37).

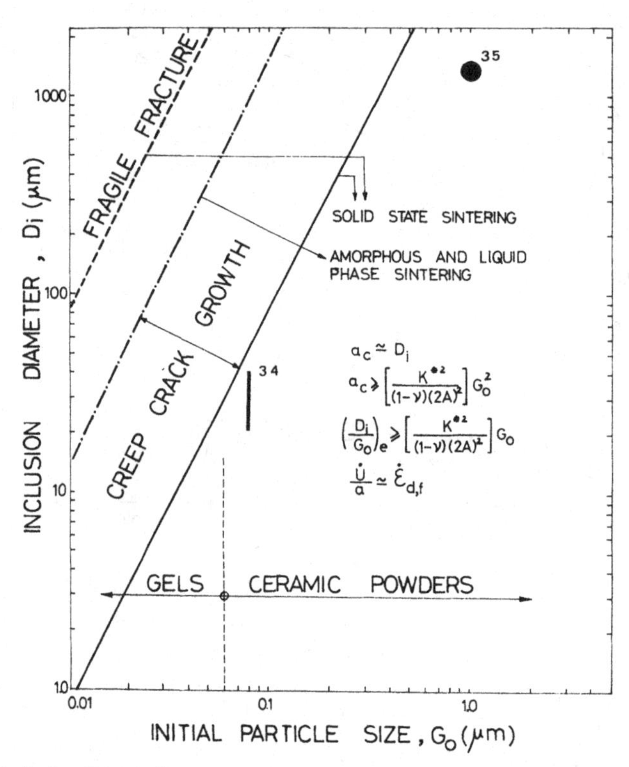

Figure 4. Critical inclusion size for fragile fracture or creep crack growth resulting from the tensile hoop stress at the inclusion-matrix interface.

As cracking results from the balance of strain misfit versus matrix strength, the maximum linear dimension of the inclusion is the more important to define the greatest misfit stress (31,37). Cracks in whisker and in fibre reinforced composites develop normal to the inclusion length with nearly constant spacing along the length (31). Bidimensional networks of flaws are also observed to develop during sintering of particulate films on rigid substrates (37). Cracking can be reduced by pre-heat treatments that strengthen the matrix particle contact (e.g. controlled surface diffusion) (37).

There is evidence of increased microstructural damage effects (abnormal pores, flaws, poor particle packing) with higher inclusion volume fractions.

Difficulties in obtaining equal compact green densities by cold isostatic pressing with increasing whisker loadings are reported. The composite densification rates of Si_3N_4-SiC_w (22) and of Al_2O_3-Al_2O_{3p} (21) composites are shifted along the density axis according to their differences in green density or in the initial densities at the onset of isothermal sintering respectively. The densification rates of the composites corrected for this shift are close to the densification rate of the inclusion-free matrices. Typical microstructural defects of these composites are regions of low relative density, 5-10 µm wide, around inclusion clusters or regions of low relative density alternating with other regions of high density with similar dimensions, the last regions being uniformly distributed in the matrix (21).

COLLOIDAL FORMING

Processing at the green body stage is recognized as one of the factors able to control the design of the microstructure and the densification of ceramic materials. This is even more so in the case of composites containing whiskers since the whiskers have a high surface area and therefore a strong tendency to form agglomerates and entangle, due to their fibrous nature, increasing the difficulties of obtaining good dispersion and good packing with the matrix ceramic powders by the usual techniques of dry mixing. Whisker agglomeration will lead to inhomogeneous density in the compacts, increasing therefore the probability of developing sintering flaws which will compromise the strength of the composites.

Colloidal processing (63,64) is a consistent approach to minimize the size of packing defects in ceramic materials and it can conceivably be used to produce homogeneous dispersions of the two components in a ceramic-ceramic composite green body. The colloidal approach also allows direct forming of complex shape compacts by various techniques like slip-casting, pressure casting, pressure filtration and extrusion.

As for any kind of solid particles in a wetting liquid environment, the dispersion conditions depend on physical and chemical interactions at the solid-liquid interface. The surface state of the particles determines if dispersion or agglomeration with flocculation will take place. The pH of aqueous suspensions (water being the most largely used suspension medium) is used to control the sign and magnitude of the surface charge of the suspended particles. Surface adsorbed species will dissociate according to the pH of the medium, either by accepting or donating a proton from or to the liquid surrounding the particles. Usually it is possible to define two pH regions, one where the surface charge of the suspended particles is positive and another where it is negative. At pH values near the point where the surface charge changes sign (isoelectric point, i.e.p.) the magnitude of the electrical charge in the surface is low and agglomeration of the particles with flocculation will take place, whereas at pH values far from the i.e.p. the magnitude of the charge is usually high and electrostatic repulsion between the particles promotes good dispersion and stability of the suspension against coagulation.

Let us take for example silicon carbide reinforced silicon nitride, a system where a large number of conflicting experimental results are available (17-19) which point towards the importance of processing conditions. Several surface characterization techniques like SMS, ESCA and XPS (65-68) have been used to characterize the surface of either silicon nitride or silicon carbide. The colloidal behaviour of water suspensions has also deserved attention (65,67,69-73). It has been shown that the surface of commercially available powders or whiskers contains, besides other minor impurities, substantial amounts of oxygen, forming an oxide-like layer identified in both cases by the detection of Si-O bonds. In both cases also, surface oxidation/reduction could be used to change the colloidal behaviour of the water-suspended particles.

The surface oxygen content of the silicon carbide can be decreased by heating in a reducing gas atmosphere or by acid leaching (66,67). Increase of the oxygen content can also be easily achieved by heating in air (1). It was found that the isoelectric point of silicon carbide powders or whiskers moved to lower pH values after the oxidative treatment, resembling the behaviour of silica suspensions where acidic silanol groups (Si-OH) determine the solid liquid interaction. More basic surface groups, arising from hydrolysis of silicon carbide surface bonds to form carboxilate (COO^-) groups, were assumed to change the isoelectric point of less oxidized whiskers towards high pH values (67).

Studies of aqueous suspensions of Si_3N_4 powders have shown that their colloidal behaviour can be interpreted as being due to the presence of acidic silanol groups and basic amine groups (Si_2NH, $SiNH_2$) adsorbed at the surface (65,73). Surface titration studies (70,71) have shown that the apparent silanol site densities were similar in silicon nitride and silicon carbide. The amine groups in silicon nitride originated from hydrolysis of Si-N bonds. Like in silicon carbide, the relative concentrations of the adsorbed species will determine the sign and magnitude of the net surface charge of the silicon nitride particles at different pH values, which is reflected in a large variety of isoelectric points found in silicon nitride powders from different manufacturers (66,71).

Electrophoretic measurements have shown that either by oxidation of the silicon nitride particle surfaces or by acid leaching of the more basic amine groups (73), the isoelectric point of the silicon nitride suspensions can be moved to lower pH values with concomitant increases in the absolute value of the negative zeta potential of the particle surfaces at higher pH.

Two approaches are possible in colloidal processing of mixtures of silicon nitride particles with silicon carbide whiskers. The use of high pH values allows a good dispersion of the particles in suspension due to their electrostatic repulsion. At high pH, as can been in Figure 5 both kinds of particles have negatively charged surfaces and high zeta potential which can even be increased by the use of surfactants (73,74). This region permits the use of concentrated stabilized suspensions which, if well mixed, could maintain homogeneity during forming by any of the colloidal forming techniques (74).

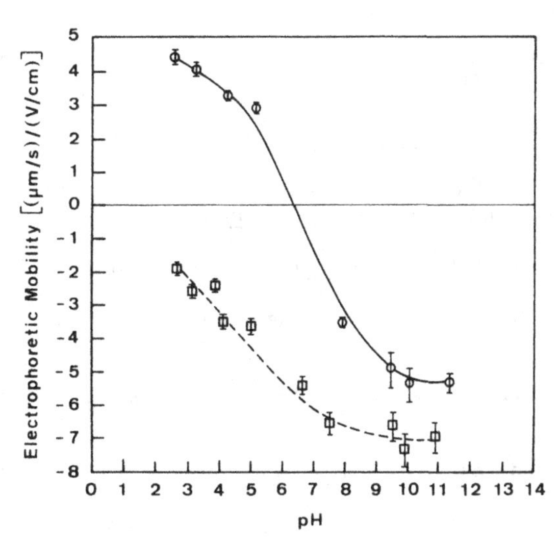

Figure 5. Electrophoretic mobility of Si_3N_4 particles (o) and SiC whiskers (□) (74).

A different approach can be used to obtain homogeneous mixed dispersions of the two kinds of particles. It consists in separately obtaining good dispersions of each kind at high pH and mixing the two suspensions with simultaneous adjustment of the pH value to the acidic region (pH 3-4) (74,75). In this region, as can be seen in Figure 5 the surface charge of the silicon carbide whiskers continues to be negative whereas the surface charge of the silicon nitride powder is already positive. It is expected that on mixing, the electrostatic attraction between the oppositely charged particles, will promote an homogeneous dispersion. It was indeed observed (74,75) that the surface of the SiC whiskers was covered by the Si_3N_4 smaller particles. Good homogeneity was obtained either by pressure filtration (74) or by freeze drying (75) of the suspensions. Since sintering in this system requires the presence of sintering additives, their colloidal behaviour has to be taken into account. Also the change of the whisker surfaces by coatings which prevent the development of a strong interface between the matrix and the whiskers, would result in a completely different colloidal behaviour which will need to be investigated in every case in order that the colloidal methodology for forming defect-free green ceramic bodies can be used.

REFERENCES

1. Becher, P.F, Hsueh, C.-H., Angelini, P. and Tiegs, T.N., J. Am. Ceram. Soc., 1988, **71**, 1050-61.

2. Evans, A.G., J. Am. Ceram. Soc., 1990, **73**, 187-206.

3. Campbell, G.H., Rühle, M., Dalgleish, B.J. and Evans, A.G., Paper 57-SI-89, 91st Annual Meeting Am. Ceram. Soc., 1989.

4. Li, Z. and Bradt, R.C., Mater. Sci. Forum, 1988, **34-36**, 511.

5. Hsueh, C.-H., Becher, P.F. and Angelini, P., J. Am. Ceram. Soc., 1988, **71**, 929-33.

6. Homeny, J., Vaughn, W.L. and Ferber, M.K., Am. Ceram. Soc. Bull., 1987, **67**, 333-8.

7. Becher, P.F. and Wei, G.C., J. Am. Ceram. Soc., 1984, **67**, C267-9.

8. Iio, S., Watanabe, M., Matsubara, M. and Matsuo, Y., J. Am. Ceram. Soc., 1989, **72**, 1880-4.

9. Tiegs, T.N. and Becher, P.F., Am. Ceram. Soc. Bull., 1987, **66**, 339-42.

10. Holm, E.A. and Cima, M.J., J. Am. Ceram. Soc., 1989, **72**, 303-5.

11. Osendi, M.I., Bender, B.A. and Lewis III, D., J. Am. Ceram. Soc., 1989, **72**, 1049-54.

12. Gac, F.D., Petrovic, J.J., Milewski, J.V. and Shalek, P.D., Ceram. Eng. Sci. Proc., 1986, **7**, 978-82.

13. Gadkaree, K.P. and Chyung, K., Am. Ceram. Soc. Bull., 1986, **65**, 370-6.

14. Faber, K.T. and Evans, A.G., Acta Metall., 1983, **31**, 565-76.

15. Liu, H., Weisskopf, K.-L. and Petzow, G., J. Am. Ceram. Soc., 1989, **72**, 559-63.

16. Akimune, Y., Katano, Y. and Matoba, K., J. Am. Ceram. Soc., 1989, **72**, 791-8.

17. Bellosi, A., and De Portu, G., Mater. Sci. Eng., 1989, **A109**, 357-62.

18. Shalek, P., et al., Am. Ceram. Soc. Bull., 1986, **65**, 351-6.

19. Lundberg, R., et al., Am. Ceram. Soc. Bull., 1987, **66**, 330-3.

20. Sacks, M.D., Lee, W.H. and Rojas, O.E., Ceram. Eng. Sci. Proc., 1988, **9** (7-8), 741-754.

21. Tuan, W.H., Gilbart, E.R. and Brook, J., J. Mater. Sci., 1989, **24**, 1062-1068.

22. Hoffmann, M.J., Greil, P. and Petzow, G., Sci. Ceram., 1988, **14**, 825-830.

23. Bordia, R.K. and Raj, R., Adv. Ceram., 1988, **3** (2), 122-126.

24. Raj, R. and Bordia, R.K., Acta Metall., 1984, **32** (7), 1003-1019.

25. Hsueh, C.H., J. Mater. Sci., 1986, **21**, 2067-2072.

26. Hsueh, C.H., Evans, A.G., Cannon, R.M. and Brook, R.J., Acta Metall., 1986, **34** (5), 927-936.

27. Hsueh, C.H., Evans, A.G. and McMeeking, R.M., J. Am. Ceram. Soc., 1986, **69** (4), C-64-C-66.

28. De Jonghe, L.C., Rahaman, M.N. and Hsueh, C.H., Acta Metall., 1986, **34** (2), 1467-1671.

29. Scherer, G.W., J. Am. Ceram. Soc., 1987, **70** (10), 719-725.

30. Bordia, R.K. and Scherer, G.W., Acta Metall., 1988, **36** (9), 2393-2416.

31. Clegg, W.J., Alford, N.McN. and Birchall, J.D., Br. Ceram. Proc., 1987, **39**, 247-254.

32. Lange, F.F. and Metcalf, M., J. Am. Ceram. Soc., 1983, **66** (6), 398-406.

33. Lange, F.F., J. Am. Ceram. Soc., 1984, **67** (2), 83-89.

34. Bordia, R.K. and Raj, R., J. Am. Ceram. Soc., 1988, **71** (4), 302-310.

35. Ostertag, C.P., Charalambides, P.G. and Evans, A.G., Acta Metall., 1989, **37** (7), 2077-2084.

36. Lnage, F.F., Lam, D.C.C. and Sudre, O., in The Processing and Mechanical Properties of High Temperature/High Performance Composites, A.G. Evans and R. Mehrabian, Eds., University of California, Vol. 6, Section 4, Part 2, 1989.

37. Garino, T.J. and Bowen, H.K., J. Am. Ceram. Soc., 1987, **70** (11), C-315-C-317.

38. Scherer, G.W., J. Am. Ceram. Soc., 1988, **71** (6), C-315-C-316.

39. Vieira, J.M. and Brook, R.J., "Scale Effects in the Use of Loading Dilatormetry for the Study of Sintering", For publication, 1991.

40. Hsueh, C.H., J. Am. Ceram. Soc., 1988, **71** (6), C-314-C-315.

41. Chu, M.Y., De Jonghe, L.C. and Rahaman, M.N., Acta Metall., 1989, **37** (5), 1415-1420.

42. Hsueh, C.H., J. Am. Ceram. Soc., 1988, **71** (10), C-442-C-444.

43. Rahaman, M.N. and De Jonghe, L.C., J. Am. Ceram. Soc., 1987, **70** (12), C-348-C-351.

44. Tuan, W.H. and Brook, R.J., J. Mater. Sci., 1989, **24**, 1953-1958.

45. Lange, F.F., J. Mater. Res., 1987, **2** (1), 59-65.

46. Huppmann, W., J. Mater. Sci. Res., 1975, **10**, 359-378.

47. Cannon, R.M. and Carter, W.C., J. Am. Ceram. Soc., 1989, **72** (8), 1550-1555.

48. Ashby, M.F., Acta Metall., 1974, **22** (3), 275-289.

49. Greskovich, C. and Lay, K.W., J. Am. Ceram. Soc., 1972, **55** (3), 142-146.

50. Oh, U.C., Chung, Y.S., Kim, D.Y. and Yoon, D.N., J. Am. Ceram. Soc., 1988, **71** (10), 854-857.

51. Weiser, M. and De Jonghe, L.C., J. Am. Ceram. Soc., 1986, **69** (11), 822-826.

52. Evans, A.G. and Hsueh, C.H., J. Am. Ceram. Soc., 1986, **69** (6), 444-448.

53. Rahaman, M.N., De Jonghe, L.C., Scherer, G.W. and Brook, R.J., J. Am. Ceram. Soc., 1987, **70** (10), 766-774.

54. Coble, R.L, J. Appl. Phys., 1970, **41** (12), 4798-4807.

55. Beere, W., J. Mater. Sci., 1975, **10**, 1434-1440.

56. Vieira, J.M. and Brook, R.J., J. Am. Ceram. Soc., 1984, **67** (4), 245-249.

57. Rahaman, M.N., De Jonghe, L.C. and Brook, R.J., J. Am. Ceram. Soc., 1986, **69** (1), 53-58.

58. Rahaman, M.N., De Jonghe, L.C. and Chu, M.Y., Adv. Ceram. Mater., 1988, **3** (4), 393-397.

59. Weiser, M.W. and De Jonghe, L.C., J. Am. Ceram. Soc., 1988, **71** (3), C-125-C-127.

60. Kimura, T., Kajiyama, H., Kim, J. and Yamaguchi, T., J. Am. Ceram. Soc., 1989, **72** (1), 140-141.

61. Kuczynski, G., Sci. Sintering, 1977, **9** (3), 243-264.

62. Cambier, F., Leblud, C. and Anseau, M.R., Ceram. Int., 1982, **8** (2), 77-78.

63. Bleier, A., Ultrastructure Processing of Ceramics, Glasses and Composites, Ed. L.L. Hench and D.R. Ulrich, John Wiley, N.Y., 1984, pp. 391-403.

64. Lange, F.F., J. Am. Ceram. Soc., 1989, **72**, 3-15.

65. Bergstrom, L. and Pugh, R.J., J. Am. Ceram. Soc., 1989, **72**, 103-09.

66. Rahaman, M.N., Boiteux, Y. and De Jonghe, L.C., Am. Ceram. Soc. Bull., 1988, **71**, 1086-93.

67. Adair, J.H., Mutsuddy, B.C. and Drauglis, E.J., Adv. Ceram. Mater., 1988, **3**, 231-34.

68. Busca, G., Lorencelli, V., Porceli, G., Baraton, M.I.,Quintard, P. and Marchand, R., Mat. Chem. Phys., 1986, **14**, 123-40.

69. Crimp, M.J., Johnson, R.E. Jr., Halloran, J.W. and Feke, D.L., in Science of Ceramic Chemical Processing, Ed. L.L. Hench and R.D. Ulrich, Wiley, New York, 1986, pp. 539-49.

70. Whitman, P.K. and Feke, D.L., J. Am. Ceram. Soc., 1988, **71**, 1086-93.

71. Whitman, P.K. and Feke, D.L., Adv. Ceram. Mater., 1986, **1**, 366-70.

72. Hartman, M.J.A.M., Van Dijen, F.K., Metselaar, R. and Siskens, C.A.M., J. Phys., 1986, **47**, C1-79-83.

73. Stadelmann, H., Petzow, G. and Greil, P., Europ. Ceram. Soc., 1989, **5**, 155-163.

74. Almeida, J.C.M., Sacramento, J.M.G., Correia, R.N., Fonseca, A.T. and Baptista, J.L., in Fabrication Technology, Ed. R.W. Davidge and D.P. Thompson, The Institute of Ceramics, Shelton, Stoke-on-Trent, U.K., 1990, pp. 179-86.

75. Crimp, M.J. and Piller, R.C., in Fabrication Technology, Ed. R.W. Davidge and D.P. Thompson, The Institute of Ceramics, Shelton, Stoke-on-Trent, U.K., 1990, pp. 199-204.

CERAMIC MATRIX COMPOSITES - HIGH PERFORMANCE DAMAGE TOLERANT MATERIALS

M. PARLIER and J.F. STOHR
ONERA, B.P. 72,
F 92322 Chatillon Cedex,
France.

ABSTRACT

Ceramic matrix composites (CMCs) are non-brittle, tough and damage tolerant materials as compared with monolithic ceramics. The non-brittle dissipative behaviour of CMCs results from the achievement, during their processing, of a fibre-matrix bond sufficiently weak to allow, during loading, multi-cracking of the matrix to develop without breaking the fibre. This unique feature, together with the good high temperature environmental behaviour of these materials resulting from the nature of the matrix, either silicon carbide or oxide, accounts for their rapid development during the last two decades. Therefore, the applications of these materials, restricted in a first step to military use, are now entering civil programmes such as hypersonic vehicles or turbojet engines.

INTRODUCTION

Considerable attention has been paid, for the last two decades, to the main routes leading to a substantial reduction of the weight of the aircraft or missile structures and that of propulsion systems. This has resulted in new structural design, together with the development of low density, damage tolerant materials able to operate at high temperature under high stresses. For new materials development the prime objectives that must be considered, beside specific mechanical properties, are cost effectiveness related to the component manufacture on the one hand and life duration in the severe environmental working conditions on the other.

The advent of high temperature CMC components in the aerospace industry mainly originates from the ability of this class of material to undergo a non-brittle damage-tolerant behaviour, contrary to monolithic ceramics which always exhibit a brittle behaviour related to a statistical flaw distribution. This particular property of CMCs, rather unexpected from a material made out of two brittle constituents, is due to debonding of fibres from the matrix, allowing cracking of the matrix to develop without it leading to failure of the material.

Moreover, since the first carbon-carbon composites appeared in the early 60's, additional matrices have been used including both covalent, mainly silicon carbide, and oxide constituents, therefore leading to an overall excellent environment behaviour.

These combined properties presented by CMCs resulted in a large endeavour for industrial development together with comprehensive research studies. These two aspects will be illustrated for both silicon carbide and glass ceramic matrix composites reinforced either by carbon or silicon carbide continuous fibres.

MONOLITHIC CERAMICS

Mechanical Behaviour of Monolithic Ceramics

Contrary to metallic materials, for which it is generally recognized that the failure has a deterministic nature leading to a unique value of the rupture stress, the rupture stress of monolithic ceramics is strongly volume dependent since it is controlled by the flaw distribution. This volume dependence is generally well described by a Weibull law which gives the cumulative probability of rupture P* as a function of the applied stress:

$$P(\sigma) = 1 - \exp\left[- V \frac{\sigma - \sigma_u}{\sigma_o} \right]^m \qquad (1)$$

where P is the fraction of specimens breaking below a given stress, m is Weibull modulus, V the volume, σ the applied stress, σ_u and σ_o the threshold stress and a scaling parameter respectively. However a direct application of this statistical approach cannot be used to derive values of the stress conservative enough for the design of components. Moreover, pre-existing flaws may grow within the components and reach the critical size leading to a brittle failure, which is unacceptable from the designer point of view.

Although not included in the Weibull statistical approach, a time dependence life prediction taking into account the flaw growth rate as a function of the applied stress can be introduced by expressing K_{Ii} in terms of a failure probability. The corresponding relationship giving the failure time as a function of the cumulative probability of rupture P can be derived from the following two equations. Firstly, for an applied stress σ_a, the corresponding stress intensity factor K_{Ii} is related to the critical stress intensity factor K_{Ic} by

$$K_{Ii} = K_{Ic} \, (\sigma_a/\sigma_{Ic}) \tag{2}$$

where σ_{Ic} is the stress leading to the immediate rupture of a specimen containing a defect of size a (where $K_{Ic} = \sigma_{Ic} \, Y \sqrt{a}$).

Secondly, substituting the stress derived from a cumulative Weibull law with σ_u = o into equation (2) leads to:

$$K_{Ii} = K_{Ic} \, (\sigma_a/\sigma_o) \, (\log \, [1-P])^{-1/m} \tag{3}$$

The time of rupture can be subsequently obtained from the defect growth rate V with:

$$V = \frac{da}{dt} = \frac{2K_I}{\sigma_a{}^2 Y^2} \frac{dK_I}{dt} \tag{4}$$

By an integration between the two limits K_{Ii} and K_{Ic}, the rupture time t_r is,

$$t_r = \frac{2}{\sigma_a{}^2 Y^2} \int_{K_{Ii}}^{K_{Ic}} \frac{K_1}{V} \, dK_1 \tag{5}$$

which leads to

$$t_r = \sigma_a{}^{-n} \, f(P) \tag{6}$$

due to the fact that for most ceramic materials the crack velocity is related to the stress intensity factor by a power law.

Using this simplified approach, the rupture time t_r as a power function of the applied stress is illustrated in Figure 1 for an ultra low expansion glass (1).

It can further be seen from this diagram that for the same probability of rupture, a two-fold increase in the applied stress results in a reduction of the failure time by about 10^8. This clearly demonstrates that designing with monolithic ceramics will be a difficult exercise. From a materials point of view, reliability improvement of ceramic components has resulted in two different actions; the first one relies on the reduction of the flaw size while the second is related to structural control.

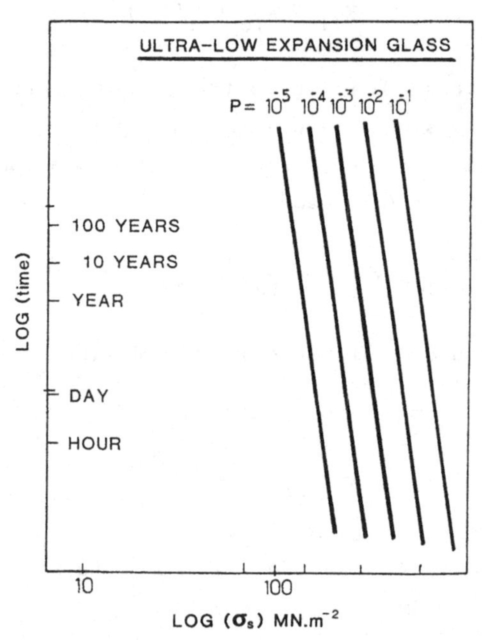

Figure 1. Failure time as a function of applied stress for ultra low expansion glass. The numbers (P) indicate the failure probability (1).

Improvement of Monolithic Ceramics Reliability

A significant endeavour has been devoted to reducing the flaw size and to evaluate the relative seriousness of a variety of flaws. As an example, it has been shown in sintered silicon nitride that amongst WC, Fe, C or Si inclusions, voids, and surface cracks, the most deleterious flaw appeared to be surface cracks (2). The three main ways of improving monolithic ceramics reliability are through the processing route, non destructive examination (NDE) and toughening.

Significant improvements of Si_3N_4 strength has been reported by changing the processing route, Table 1. It should be noted that for the hipped specimens, a large increase of the rupture strength together with a reduction in scatter have been achieved.

TABLE 1
Dependence of Si_3N_4 average modulus of rupture (MOR) and Weibull modulus with processing route

Processing Route	MOR (MPa)	Weibull Modulus
Isopressing and sintering	697	9
Injection moulding and sintering	834	20
Slip casting and sintering	897	14
Injection moulding and hipping	986	20

The second means used to improve monolithic ceramic reliability is NDE which truncates the distribution of failure stress by rejecting specimens or components containing large flaws. It is worth mentioning the recent results obtained using microfocus radiography; voids with a size as small as 22 µm are currently detected, thus permitting an average MOR of 766 MPa to be ascertained for Si_3N_4 parts (4). However, even if reliability of monolithic ceramics can be improved, the ability to detect critical flaws at highly stressed locations in thick cross sections components will always remain limited.

The third way to improve reliability of monolithic ceramics is toughening. It is generally recognized that four different toughening means (Figure 2) may be used (2):

- Crack deflection toughening using either a two phase material resulting from either precipitation, allotropic transformation, or the grain boundaries. The crack can be diverted away from its usual propagation plane (i.e. normal to the applied stress) either by tilting or twisting, the twisting contribution being more efficient as far as the released energy rate is concerned (5).
- Microcrack toughening where local microcracking originates from residual stresses due to either thermal expansion anisotropy or phase transformation with a change in volume. The two phenomena generally put forward to explain blunting of the main crack are a reduced modulus of the cracked zone on the one hand and expansion of the microcracked zone on the other, the latter giving the larger effect.
- Phase transformation toughening where the transformation involved is induced during the crack growth, within the stress concentration zone near the crack tip, in small untransformed particles which undergo a

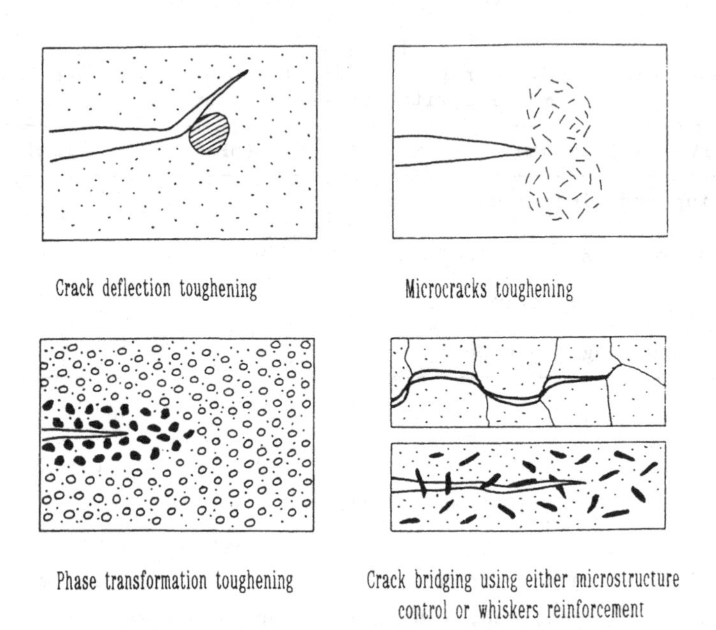

Figure 2. The four means of improving ceramic reliability.

martensitic transform with a change in volume. This concept relies mainly on the use of a dispersion of fine grain zirconia in oxide matrices: the zirconia particles will retain, at room temperature, their high temperature tetragonal structure due to constraining of the matrix, provided their size is small enough (6). During crack growth, extension of the transformed zone along the crack surfaces will give rise to a toughening. This toughening process has been applied with success to alumina, mullite, and stabilized zirconia. As an example, very high rupture stresses above 2 GPa and toughness ranging between 5 and 8 MPa $\sqrt{m}$ have been reported in both Al_2O_3/ZrO_2 and tetragonal zirconia (7).

- Crack bridging where toughening relies on an increase of the energy release rate during crack growth due to bridging of the crack faces either by ceramic grains or by whiskers. In both cases the increase of the released energy rate during crack growth arises from dissipative phenomena occurring either at the fibre-matrix interface or at the grain boundary during pullout. Large increases in toughness have been reported for silicon carbide reinforced alumina or silicon nitride with Kc values as high as 12 MPa $\sqrt{m}$ (8,9).

Table 2 summarizes improvements achieved by the different means for a Si_3N_4 ceramic (3).

TABLE 2
Property improvement achieved in Si_3N_4 monolithic ceramics

Property	Current Capability	Improvement	
		Processing	Toughening
Modulus of Rupture (MPa)	766	897	1069
Weibull Modulus	10	14	10
Toughness (MPa $\sqrt{m}$)	4.3	4.3	6
NDE - Detectability relative to current matrix	1	1	1.95*

* Toughening has enhanced both the MOR and the critical flaw size and therefore the defect detectability.

However, despite the significant progress achieved concerning the reliability of monolithic ceramics, it should be kept in mind that, in intricately shaped components, stress concentrations generally located within the thick cross sections will always exist which may lead to premature failure. This brittle behaviour of monolithic ceramics cannot be overcome, even if the rupture stress is raised, which limits component design to a safe lifetime approach. For aerospace use, this condition is too restrictive for structural components due to the wide use of damage tolerant design. A good way to overcome this monolithic ceramic limitation is to move to CMCs which exhibit damage tolerant behaviour.

MECHANICAL BEHAVIOUR OF CERAMIC MATRIX COMPOSITES

CMCs reinforced by continuous ceramic fibres may, under certain conditions, exhibit a non-brittle failure. The origin of this particular behaviour (rather unexpected from a material where the two phases - the matrix and the fibres - are brittle) is due to the ability of the matrix to develop multi-cracking, without breaking the fibres, therefore allowing the applied load to be borne. The first analysis was proposed by (10,11) some twenty years ago for materials with a unidirectional (UD) reinforcement.

Dissipative Failure of Unidirectional Materials: the ACK Model

The linear mechanics approach: Aveston, Cooper and Kelly (11) demonstrate, through a simple linear mechanics analysis, that UD ceramic-ceramic composites may present two different rupture modes depending upon the fibre volume fraction V_f. For low volume fraction of fibres, the composite failure is brittle with the propagation of a single crack, which breaks both the fibres and the matrix. In contrast, for high fibre volume fraction, multicracking of the matrix develops, provided that:

- The fibre volume fraction V_f exceeds a critical value V_f^c determined by the relationship $\varepsilon_{fu} \geq \varepsilon_{mu}$ (1+a) where ε_{fu} and ε_{mu} are the elongation to rupture of the fibres and the matrix respectively, $a = E_m V_m / E_f V_f$ being the ratio of the elastic moduli of the matrix and fibres weighted by their respective matrix and fibre volume fractions. This relationship simply expresses that the load borne by the matrix prior to its failure can be sustained by the fibres once the matrix has failed.
- A weak fibre-matrix bond can be achieved allowing the matrix crack which propagates in mode I to be diverted to the fibre-matrix interface into mode II, thus allowing the fibres to be circumvented by the crack without failing. Behind the crack tip and within the crack plane, the load is therefore fully sustained by the fibres alone; reloading of the matrix on both sides of the crack is then due to load transfer from the fibres within a radius r of the matrix owing to the interfacial shear which occurs at a constant frictional shear stress τ_i over the debonded length l_c such that:

$$l_c = \frac{V_m}{V_f} \frac{\sigma_{mu} r}{2\tau_i} \tag{7}$$

The variation of the longitudinal stresses within the fibre and the matrix is shown in Figure 3. Within the limits of the deterministic ACK approach where the stress to rupture of the matrix is supposed to be unique, the ultimate width of the matrix blocks at crack saturation lies between l_c and $2l_c$ (Figure 4), due to the fact that the matrix elongation to rupture can be reached within a matrix block of width $2l_c+\varepsilon$, whereas it cannot within a block of length $2l_c-\varepsilon$.

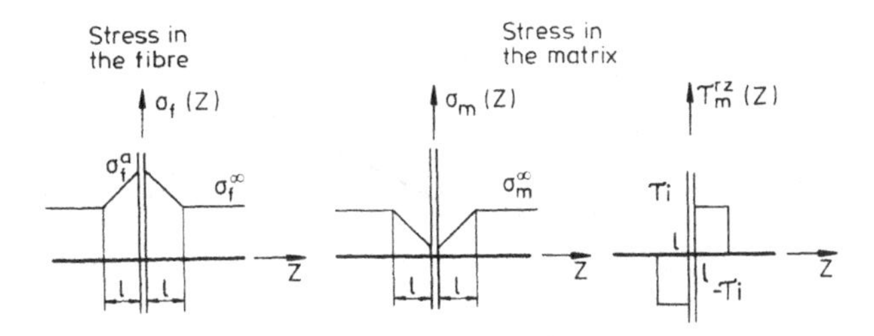

Figure 3. Stress distribution in the fibre and the matrix in front of a matrix crack for a tensile loaded CMC.

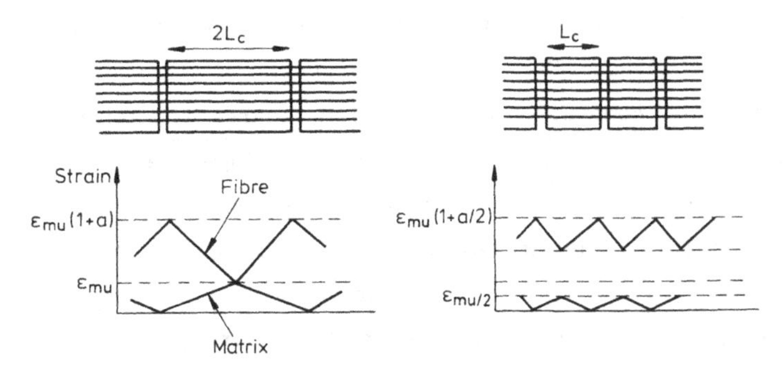

Figure 4. Multifissuration of the matrix: the strain distribution within matrix blocks of width l_c and $2l_c$.

The tensile curve of a UD composite will thus comprise three domains (Figure 5):

- A first elastic domain which ends with the first matrix crack.
- A plateau region corresponding to multicracking of the matrix occurring at the constant rupture stress of the matrix.
- Finally, a linear stage corresponding to incremental loading of the fibres alone, leading to an apparent modulus of E_fV_f; this third stage ends at the failure stress of the fibres within the composite.

Energetic approach of the matrix fissuration: In the above simple linear mechanics approach, the elongation to rupture of the matrix within the composite has been supposed to be unique and equal to that of the monolithic matrix, which has been shown not to be the case for a number of composites with a brittle matrix (12). To account for this phenomenon of an

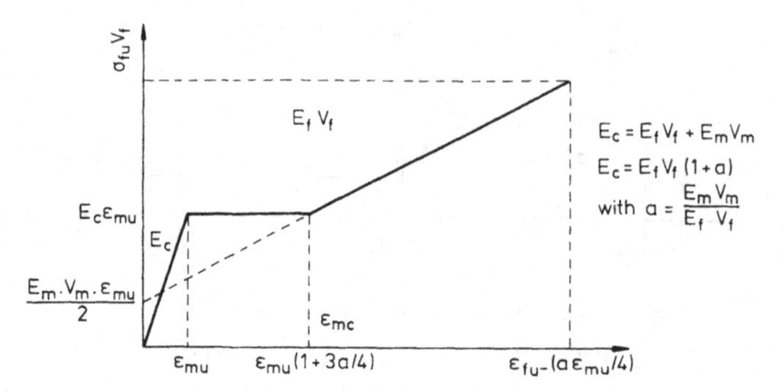

Figure 5. Theoretical crack curve of a unidirectional ceramic matrix composite according to ACK.

elongation to rupture of the matrix within the composite larger than that of the matrix taken alone, ACK (11) have proposed a rupture criterion of the matrix within the composite based on an energetic approach which only considers the initial stage of the uncracked composite and the final stage where the matrix crack runs throughout the whole cross-section of the specimen. The first matrix crack running throughout the whole specimen cross-section appears when the energy released by the composite due to the matrix cracking ΔU becomes larger than the energy necessary to make the matrix fail within a whole cross-section $2\gamma_m V_m + \gamma_i$, where γ_m is the energy necessary for the crack to grow and γ_i that required to ensure debonding of the fibres from the matrix over the length l_c. Considering in their energy balance a simple solid friction of the fibres within their matrix sheaths along the debonded length and neglecting the energy γ_i necessary for this debonding, ACK demonstrated that the elongation to rupture of the matrix within the composite ε^* was given by:

$$\varepsilon^*_{mu} = \left[\frac{12\tau_i \gamma_m E_f V_f^2}{E_c E_m^2 r V_m}\right]^{1/3} \qquad (8)$$

This relationship accounts for the reported dependence of the matrix elongation to rupture within the composite upon both the radius and volume fraction of fibres as reported by Cooper and Sillwood (12).

From this analysis, ACK concluded that if the elongation to rupture of the matrix within the composite ε^*_{mu} is larger than that of the matrix taken alone ε_{mu}, then the first cracking of the composite will occur at ε^*_{mu}. On the contrary, if ε^*_{mu} remains smaller than ε_{mu}, cracking of the composite occurs at ε_{mu}; ACK were obliged to include this assumption due to the fact that the relationship giving ε^*_{mu} does not approach the matrix value ε_{mu} as V_m tends towards zero.

However, the ACK analysis which present similarities with fracture mechanics does not take into account the growth stage of the matrix crack within the composite. Some attempts made in that area, using the conventional homogenisation technique (13,14) currently used to derive constitutive laws for crack growth, also fail to take into account the local stiffening effects of the fibres. This is of prime importance to improve understanding of the parameters controlling the matrix crack growth in a CMC.

Modelling of the Matrix Crack Growth in UD Composites

The rupture mechanics local approach: As the local approach developed by Peres (15) to consider the validity of the previous models was too intricate to refer to an analytical calculation, a numerical simulation has been used. The composite is schematized by a matrix plate onto which are bonded the aligned fibres (Figure 6). The load transfer between the fibres and the matrix occurs as in the ACK model due to interfacial shear. The crack growth is studied using the energy release rate at the crack tip $G_I(a)$ according to the following relationship:

$$G_I(a) = \lim_{\Delta a \to 0} \int_0^{\Delta a} (R.\Delta u)dx \qquad (9)$$

where R represents the crack closing forces and Δu the displacement field resulting from a growth of the crack of an incremental step Δa.

With a static loading at a constant stress and using an energy balance similar to that used by ACK, but considered here for each step of the crack growth, Peres derived the energy released during the crack growth Δu as being half the value of the sum of the work done by external forces and that resulting from the friction at the fibre-matrix interface during the gliding of the fibres in their matrix sheaths. The variation of this work calculated for each step of the crack growth Δa is plotted as a function of the crack length for both a fully rigid fibre-matrix bond and a weak fibre-matrix bond in Figure 7. The energy released at the crack tip is a pseudo-periodic function of the crack length with a series of minima, the lower each minimum, the higher the fibre-matrix cohesion. For a fully rigid bond, it appears that in front of each fibre, the crack can no longer grow whatever the matrix critical energy release rate $G_{I\,m}^{c}$. Therefore, for a non fully rigid bond, the crack will grow only if the energy of the minimum is higher than the matrix critical energy release rate $G_{I\,m}^{c}$, so that the most efficient obstacle will be the first fibre.

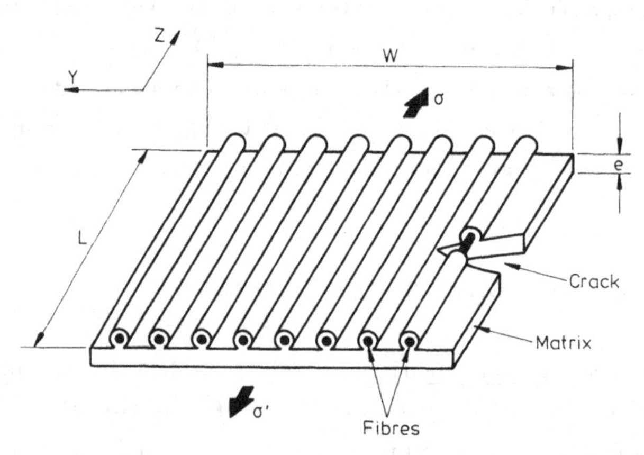

Figure 6. The single layer unidirectional composite used for the crack growth modelling.

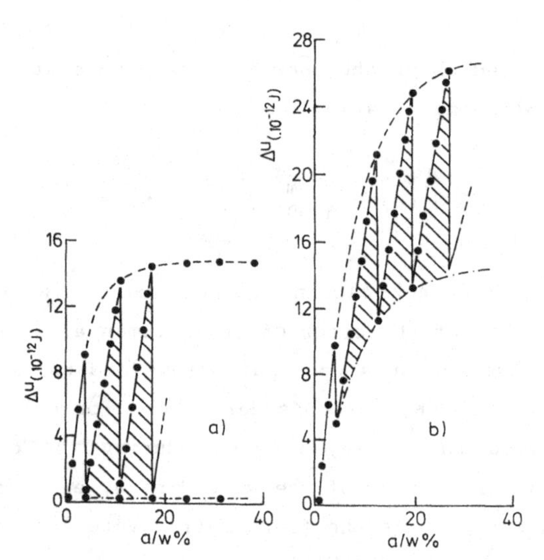

Figure 7. Energy released at the crack tip during the crack growth: (a) with no relative fibre-matrix displacement, (b) with decohesion and glide at the fibre-matrix interface (τ_i = 10 MPa).

The minimum value of the energy released at the crack tip increases during the crack growth as the interfacial shear strength decreases. The total energy released during the crack growth is well depicted by the area subtended by the pseudo-periodic curve. Therefore a weak fibre-matrix shear strength will lead to a high value of the energy released during the crack growth, and then to a non-brittle dissipative failure.

However, these conclusions derived in the case of a simple single layer ideal composite will be modified in the case of a real unidirectional composite due to averaging effects as pointed out by Anquez et al. (16). These come from either the third dimension which exists in real unidirectional composites or from statistical scatter of the interfacial shear stress and the local fibre volume fraction which both have a strong influence on the value of the minimum of the released energy. Nevertheless, these fluctuations will mainly prevail during the first stages of the crack growth, i.e. for interaction with the very first fibres. During further propagation of the crack, the released energy will therefore lie between the lower and the upper limits of the Δu curve of Figure 7. Moreover, it is worth noting that this mean value of the energy necessary for the crack to propagate throughout the whole cross-section of the composite is similar to that of the ACK model.

In conclusion of this analysis of the multifissuration of UD composites, it may be recalled that:

- The ACK model allows a good interpretation of the matrix multi-fissuration despite its incorrectness relative to fracture mechanics.
- The more accurate Peres model which takes into account the propagation of the crack at a local scale clearly demonstrates: firstly, the stiffening role played by the fibres in a way similar to what is observed in a stiffened metallic panel; secondly, the primary importance of the value of the fibre-matrix interfacial friction shear-stress relative to the dissipative failure energy of the composite; the lower the friction stress, the higher the energy released at failure.

Consequences on CMC processing: The production of a CMC exhibiting a non-brittle failure requires:

- A weak fibre-matrix bond to allow the crack to be diverted at the fibre-matrix interface from mode I to mode II; this easy fibre decohesion from the matrix in front of the matrix crack can be obtained by avoiding chemical reactions between the fibre and the matrix.

- An interfacial friction shear stress low enough to provide a dissipative failure, but above a threshold value τ_0 to allow the composite to function (Figure 8); as the value of the fibre-matrix interfacial shear stress mainly results from both the friction coefficient and the radial stresses developed during cooling, the elastic moduli and the thermal expansion coefficients of the fibres and the matrix have to be fitted. These constraints reduce the choice of constituents, fibre and matrix, that can be put together to realize a composite. An additional degree of liberty is provided by the interphase inserted between the fibre and the matrix which plays the double role of a diffusion barrier and a mechanical fuse. In such a complex system, it is obvious that an accurate control of the friction stress obtained is absolutely necessary for the development of new systems. This control can be achieved by measurements using micro-indentation techniques.

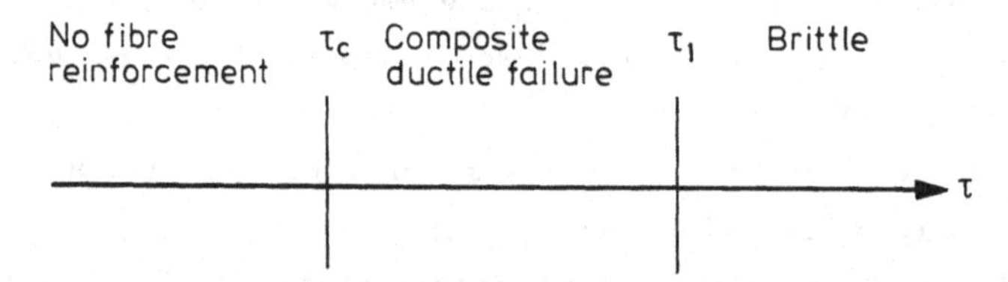

Figure 8. Schematic diagram highlighting the friction shear stress domain of interest to obtain a dissipative failure.

Instrumented Micro-indentation and Fibre-matrix Interface Control

The instrumented micro-indentation technique presents, relative to the micro-indentation test first used by Marshall (17), the advantage of a precise determination of the fibre-matrix friction stress due to a simultaneous recording of the applied load and the fibre displacement.

A typical load-displacement curve exhibits three stages (Figure 9):

- The first stage, which corresponds to the indentation of the fibre, ends with debonding of the fibre from the matrix.
- The second stage, which corresponds to the glide of the debonded length of the fibre in its matrix sheath, ends when the indenter reaches the matrix.
- The third stage corresponding to the indentation of the matrix.

Analysis of the second stage of the load-displacement curve using a simple shear-lag model allows derivation of a constant value of the interfacial friction shear stress.

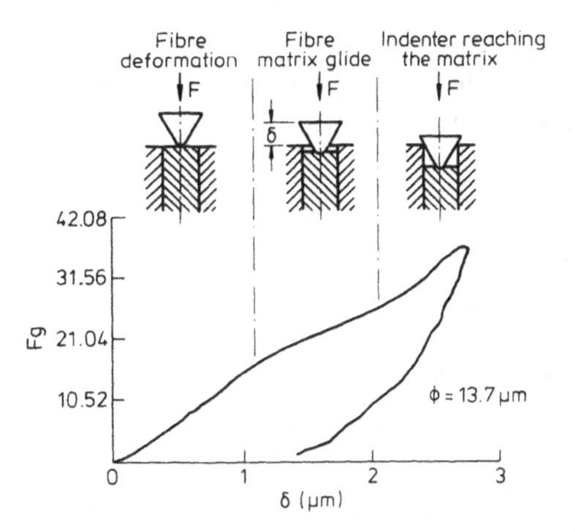

Figure 9. Instrumented Vickers indentation test: a typical load-displacement curve.

This technique allows an accurate discrimination between SiC-LAS composites processed by different routes so that the nature of the interphase present at the fibre-matrix interface can be modified accordingly. Table 3 clearly shows that interface tailoring allows variation of interfacial friction shear-stress by a factor between 2 and 3.

TABLE 3
Mechanical properties of unidirectional SiC/LAS composites in relation to the fibre-matrix interfacial friction shear stress

Composite	τ_i (MPa)	Flexural Strength (MPa)	Feature of the Rupture
SiC/LAS	11	1020	Dissipative (pullout)
SiC + C/LAS	20	600	Rough rupture surface
SiC + Nb_2O_5/LAS	25	580	Rough rupture surface

This highlights the correlation between the friction shear stress, the flexural rupture strength, and the kind of rupture. It can easily be seen that a high rupture stress and a dissipative rupture can only be achieved provided that the friction shear stress remains sufficiently low. This is further evidenced in Figure 10 which shows two fracture surfaces:

- The first one totally flat, characteristic of a brittle failure, corresponds to a SiC/LAS composite with a strong bond between the fibre and the matrix.
- The second one exhibits much pullout, characteristic of a dissipative failure, corresponding to the composite having a low value of τ_i (11 MPa).

Similar conclusions can be drawn for other CMCs. Therefore, for each CMC a critical value of the interfacial friction shear stress below which a

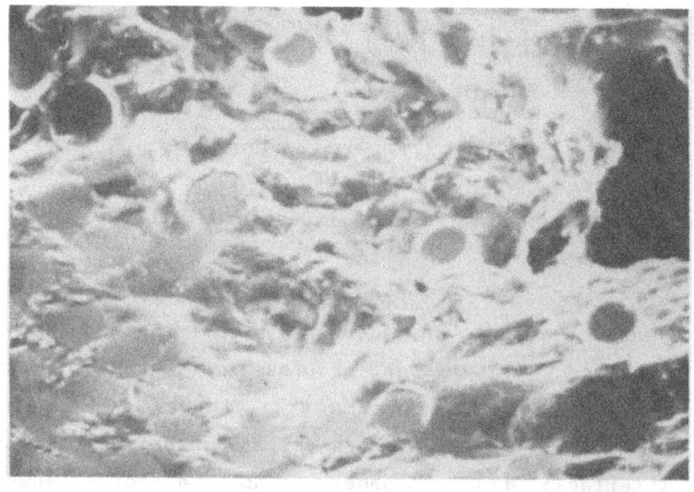

Figure 10. Rupture surfaces observed in SiC/LAS composites (SEM micrographs). Above - flat surface of a brittle composite, x700. Below - pullout of fibres observed in composites exhibiting dissipative failure, x16.

dissipative failure is observed has to be derived, as for example about 100 MPa for a SiC-SiC composite as compared to 10 MPa for the SiC-LAS composite.

In conclusion, instrumented micro-indentation appears as a very efficient test method to determine the onset of the fibre decohesion and glide thus allowing an accurate measurement of the interfacial friction shear stress. Moreover this method may, in the future, develop as a non destructive testing method to control the interfacial characteristics necessary to achieve a dissipative failure.

THE MAIN PROCESSING ROUTES FOR CMCs

Most of the CMCs in use today in the aerospace industry are designed starting from the fibre preform. Therefore, processing of these composite components cannot rely on the sintering technique as used for monolithic ceramics, for the two following reasons:

- The limited thermal stability of oxide or silicon carbide fibres does not allow heating to the temperature level of a sintering cycle.
- The constraining of shrinkage due to the presence of fibres would lead to severely cracked components.

Thus CMC processing relies mainly on techniques based on infiltration of the fibre preform. The matrix is obtained by the decomposition of a gas phase, the pyrolysis of an infiltrated and transformed liquid phase or infiltration by a glassy phase.

The Chemical Vapour Infiltration (CVI) Technique

The CVI technique allows processing of any complex shape starting from the fibre preform with a 2D stacking sequence or a 3D or nD weaving (n = 4.5) generally held in a tooling (Figure 11).

In the CVI reactor, the matrix is obtained by the decomposition of the gaseous species within the open porosity of the preform. Therefore, two different phenomena may control the matrix growth rate in the depth of the preform:

- The mass transfer of the reactants and products of reaction through the porosity occurring only by diffusion in an isothermal process.
- The kinetics of the chemical reactions.

It is now well known that a good in-depth deposition of the matrix will be obtained if CVI is performed under both low pressure and temperature, conditions under which the matrix growth rate is controlled by the chemical reaction kinetics (19,20).

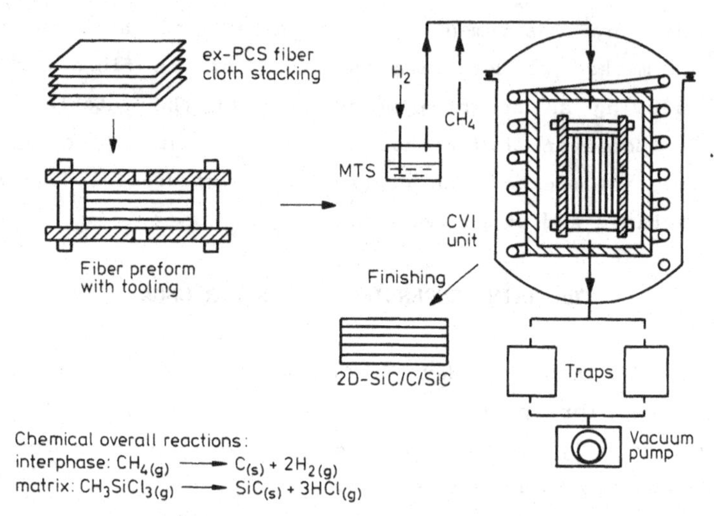

Figure 11. CVI processing steps for a CMC (after 20).

Processing of CMCs using CVI has been demonstrated for numerous matrices either of the covalent type such as C, SiC, Si_3N_4, B_4C, TiC, BN or of the oxide type such as ZrO_2 and Al_2O_3. Nevertheless, composites with a good homogeneity of the matrix and a limited amount of porosity can only be achieved through several infiltration cycles with an intermediate trimming of the component necessary to restore the surface porosity.

This processing route leads to good quality components at the expense of long processing times (several months) which remain however acceptable for the aerospace industry. CVI CMC processing has been developed in France by SEP (21) mainly for the production of SiC/SiC and C/SiC components today in use on rocket or turbojet engines.

However, the long manufacturing time for the CVI route has, up to now, restricted the use of CMC components to the aerospace industry for cost effectiveness reasons. Development of the CVI technique to reduce the processing time has been proved to be successful. It is worth mentioning the reactor-vessel, designed by the ORNL in the United States (22), which allows a fifty-fold increase of the deposition rate through both pressure and temperature gradients; unfortunately, this processing technology is only useful for the manufacture of thin components with not too intricate shapes.

The limitations of the CVI technique regarding the production cycle duration has resulted in a search for more rapid and efficient processing routes.

Processing of Composites using Liquid Phase Infiltration of the Fibre Preform

Three different kinds of matrices are relevant to the liquid phase infiltration technique (23):

- The glass-ceramic matrices, due to the fact that the glassy phase exhibits a viscosity allowing densification at a temperature compatible with the thermal stability of the fibres.
- Covalent matrices such as SiC or Si_3N_4 which can be obtained by the pyrolysis of polymeric precursors at a sufficiently low temperature to avoid any fibre degradation.
- Glass-ceramic or oxide matrices such as cordierite or alumina and zirconia, which can be obtained by an infiltration of the fibre preform by organometallic precursors using the sol-gel route followed by a "sintering" cycle at low temperature.

<u>The glass hot melt route for glass-ceramic composites</u>: As an example SiC Nicalon/LAS composites can be processed using either hot pressing of prepreg tapes at a temperature ensuring a viscosity below 5000 poises, or by injection moulding of the fibre preform at the same temperature level (18). This last process is relevant to components of simple shape and axial symmetry.

<u>The polymeric precursor route for covalent matrix composites</u>: The idea of using silicon based polymeric precursors to mould ceramic composites originates from the processing route used for carbon-carbon composites. However, the production of a SiC or a Si_3N_4 matrix is much more difficult than a carbon matrix due to the density different between the polymeric resin and the ceramic leading to a high shrinkage - 75% as compared to only 35% in the case of carbon-matrix composites.

In order to overcome this difficulty, it is desirable to limit the role of the ceramic derived from the resin to an efficient binder. To achieve this end, a two-step infiltration process has been developed at ONERA (23):

- The first step consists of an infiltration of the fibre preform by fine submicronic ceramic particles achieved using a colloidal filtration technique, which allows filling of 50% of the free volume within the fibre preform.

- The second step consists in an injection of the powder infiltrated preform with a suitable polymeric precursor followed, first by a curing cycle, then by pyrolysis.

This processing route allows the achievement of a sound composite with a residual porosity around 10% in the case of a 2D fibre preform.

However, these good results concerning the porosity can only be obtained provided that the polymeric precursor exhibits a combined set of properties allowing both a good matrix infiltration and a high ceramic yield. For this purpose, a specific polymer precursor of a silicon carbide matrix has been developed by ONERA and IRAP (24,25) while ATOCHEM has developed polymeric precursors of SiC/Si_3N_4 matrix (26,27). As an illustration, the polyvinylsilane (PVS) developed at ONERA will be described in some more detail.

PVS polymeric precursor with the general formulation

$$\left[\begin{array}{cc} CH_2\text{—}CH & R_3 \\ | & | \\ \text{—}Si & \text{—}Si \\ | & | \\ R_1 & R_2 \end{array}\right]_n$$

presents the combined set of properties required for ceramic matrix processing:

- A relatively low molecular weight ranging between 3500 and 4500 in order to achieve a sufficiently low viscosity at room temperature (20 poises) allowing the fibre infiltration at a temperature compatible with the gelling time.
- A thermoset behaviour - a pre-requisite in the case of a ceramic precursor - with a cross-linking temperature sufficiently below that of onset of pyrolysis (Figure 12), to avoid bubble forming during the curing cycle, which would lead to a high level of porosity.

PVS exhibits a high matrix yield of 65% with a final composition of the ceramic product very near that of SiC-Nicalon fibres, as illustrated in Table 4.

Due to the overall good properties of the PVS precursor a flexible liquid processing route has been achieved for the manufacture of silicon carbide matrix composites. In order to check the reliability of the process, different silicon carbide fibre woven fabrics 1D, 2D, 3D have been used for composite processing based on the two-step infiltration previously described. As shown in Table 5, the flexural rupture strength of the

SiC/SiC composites processed by the liquid route compare favourably to that of composites processed by the CVI technology.

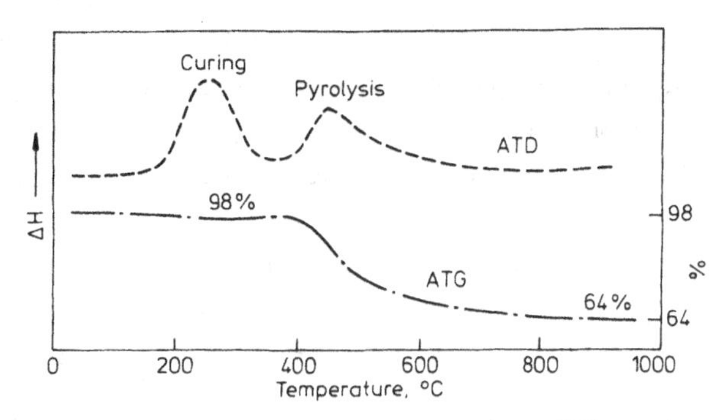

Figure 12. Thermal characteristics of the PVS polymeric precursor.

TABLE 4
Chemical analysis of the ceramic matrix and fibres obtained from polymeric precursors (weight %)

Ceramic	Species		
	SiC	SiO_2	C
SiC (PVS)	68%	8%	24%
SiC (Nicalon)	66%	17%	17%

TABLE 5
Bending rupture strength (MPa) as a function of the processing route for SiC/SiC composites

Fibre Architecture	Procesing Route		
	CVI (SEP)	CVI (Oak Ridge)	Liquid Route (ONERA)
1D	700 (V_f = 0.45)	400 (V_f = 0.45)	500 (V_f = 0.45)
2D	350 (V_f = 0.45)	300 (V_f = 0.40)	200 (V_f = 0.30)
3D	-	-	250 (V_f = 0.45)

The organometallic sol-gel route for oxide or ceramic glass matrix composites (28): The sol-gel route has been known for a long time in the ceramic industry due to its ability to allow the production of both monoliths and powders with a high purity and homogeneity; this process has entered production for the manufacture of ferroelectric ceramics such as PLZT (29,30) and zirconia powders. Moreover, sol-gel technology has been recently extended to processing of oxide coatings for optical applications (31,32).

The sol-gel processing route is, therefore, a versatile technique since it permits the production of solid oxides from a gel starting from a liquid metallic or non-metallic alkoxide by the following hydrolysis-polycondensation reactions:

$$M(OR)_n + nH_2O \rightarrow M(OH)_n + nR\text{-}OH \quad \text{(hydrolysis)}$$

$$M(OH)_n \rightarrow MO_{n/2} + n/2H_2O \quad \text{(polycondensation)}$$

Tailoring of the processing parameters such as pH, water/alkoxide ratio, chemical nature of the alkoxide, allows an accurate control of the gel structure through its pore size and volume fraction.

Application of the sol-gel route to CMC manufacture is however very recent due to the basic necessity to overcome the difficulties arising from the considerable shrinkage so as to take advantage of the low densification temperature of the matrix allowed by this processing route. Therefore, in a way similar to that used for the manufacture of the SiC/SiC composite, a two step infiltration process has also been applied for the sol-gel route. The first step consists in an infiltration of the fibre preform by ceramic powders while the second step includes both the infiltration of the powder pre-infiltrated fibre preform and the subsequent gelification through controlled hydrolysis.

This route has been successively used for the processing of CMCs with SiC-Nicalon or C fibres as a reinforcement and alumina, zirconia or mullite matrices. In comparison with SiC/Al_2O_3 composites processed by the conventional powder infiltration and sintering technique, which lead to highly porous materials owing to the necessity of limiting the densification temperature to 1200°C (33), the sol-gel route allows a reduction of the porosity level from 30 to 10%.

POTENTIAL AND FUTURE DEVELOPMENT OF CMCs

As already state the constraints regarding both the fibre-matrix compatibility and the processing routes have restricted the industrially developed materials to a few systems. Starting from carbon-carbon composites first developed in the early sixties, a number of composites have today reached an industrial stage, as for example SiC-SiC a C-SiC developed by SEP or protected C-C and SiC-glass developed by Aerospatiale.

Today, these composites have demonstrated their capability for use as structural components in the aerospace industry, mainly for ramjet, rocket booster and turbo engine components, due to their good specific strength at high temperature (Figure 13).

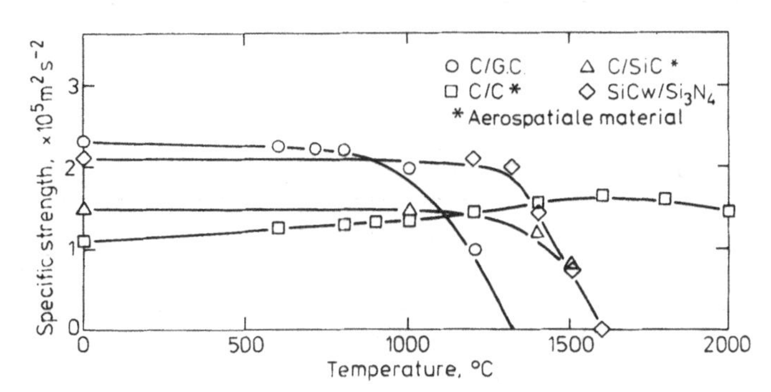

Figure 13. Specific strength versus temperature for some CMCs in use today.

Mechanical Properties

SiC/LAS composites are promising candidates for a use up to 650°C with a potential development to 1000°C, due to both their good mechanical properties (Table 6) and a cost effective processing route by hot pressing or injection moulding.

SiC/SiC and C/SiC composites have potential use temperatures of 1400°C and over 1600°C respectively. Table 7 gives some of the properties of SEP composites with a 2D reinforcement (34). The small strength loss observed at 1400°C on SiC/SiC composites protected against oxidation probably originates from fibre structure changes which are known to occur over 1200°C for Nicalon fibres (35). However, the values of the strength at 1400°C (over 140 MPa) clearly confirm that SiC Nicalon fibres can be used up to that temperature provided that some care is taken regarding their interaction with the environment at high temperature (36).

TABLE 6
Parameters of some of the CMCs in use today

Properties	Composites		
	SiC/LAS	SiC/SiC	C/SiC
Geometry of 2D woven fabrics	Satin weave	Plain weave	Plain weave
Total volume fraction of fibre	40%	40%	45%
Apparent specific gravity (g/cm^3)	2.5	2.5	2.1
Open porosity	2%	10%	10%
Ultimate bending strength (MPa) (Typical values) 20°C	300	300	500
900°C	250	400	700

TABLE 7
Mechanical properties of 2D SiC-SiC and C-SiC composites (after 34)

Properties	SiC-SiC			C-SiC		
	23°C	1000°C	1400°C	23°C	1000°C	1400°C
Fibre volume fraction (%)		40			45	
Density		2.5			2.1	
Open porosity (%)		10			10	
Tensile strength (MPa)	200	200	150	350	350	330
Elongation to rupture (%)	0.3	0.4	0.5	0.9	0.9	
Tensile Young's modulus (GPa)	230	200	170	90	100	100
Flexural strength (MPa)	300	400	280	500	700	700
Compressive strength // (MPa)	580	480	300	580	600	700
Compressive strength ⊥ (MPa)	420	380	250	420	450	500
Interlaminar shear strength (MPa)	40	35	25	35	35	35
"Toughness" K_{IR} ($MPa\sqrt{m}$)	30	30	30	32	32	32

Ageing Behaviour

Another problem that must be addressed here is the ageing behaviour of these composites, mainly due to their interaction with the environment, which may affect one of the three constituents of the composite, i.e. the fibre, the interphase or the matrix.

For every composite except carbon-carbon, the matrix generally offers an intrinsic oxidation resistance, which is the case for glass and SiC matrices. For SiC matrix composites, the oxidation resistance arises from the ability of this SiC matrix to form a protective silica layer by oxidation; however, at high temperatures, i.e. above 1200°C, a protective coating may be used for SiC matrix composites in order to increase their durability.

SiC-LAS composites have been shown to exhibit a rapid degradation of their properties, when tested in air and at high temperatures, probably due to an oxidation of the fibre-matrix interphase leading to a dissipative/ brittle transition around 800°C (37).

SiC-SiC composites and C-SiC composites have been shown to keep their properties up to much higher temperatures, allowing them to be used for short durations up to 1600°C. Moreover, SiC-SiC components have been demonstrated for medium duration use of a few hundred hours in engines (32).

Designing with CMCs: The problem of constitutive laws

Long fibre ceramic composites have been demonstrated to exhibit a non-brittle, damage tolerant behaviour; however, component designing requires, as for metallic alloys, constitutive laws giving precise knowledge of the mechanical properties variation in service.

Based on the same damage approach as used for metallic materials, work has begun for ceramic-ceramic composites, considered here as elastic damageable materials.

From a macroscopic standpoint, a damage parameter D is defined by the following relationship:

$$\tilde{E}(D) = E_o\ (1-D)$$

where $\tilde{E}(D)$ is the stiffness tensor of the damaged material, E_o the stiffness tensor of the virgin material and D the damage parameter; the value of D increases from zero in the virgin material to one at failure.

Macroscopically, the damage parameter D can be determined by a series of experiments performed on composites loaded beyond the elastic limit. From a microscopic standpoint, the damage observed in the composite results from multicracking of the matrix, and delamination. For example, in a UD material such as SiC-LAS for which the unique damage is matrix cracking, it has been shown that the number of matrix macro-cracks grows from the first fissuration of the matrix up to fracture where the saturation is reached.

Let n_S be the number of cracks at rupture and n the number of cracks at any point between the elastic limit and rupture; in that particular case, the damage D can be estimated very simply by the ratio n/n_S, which fulfils the requirements of D = 0 for the virgin material and D = 1 for the ruined material.

For a real component, the composite used will be made using either laminates or tissue piling. Damage within the composite will therefore be more complex than simple matrix cracking of each ply, and a simple relationship between the local damage and the macroscopic damage can no longer be derived. To overcome this difficulty, the approach will rely on the homogenisation technique with the assumption that the composite can be depicted as a medium with a periodic structure in which a unit cell can be defined.

An analysis of the damage occurring at the microscopic level will then provide the data necessary to relate the macroscopic damage D to the microscopic local damage d, through homogenisation. The behaviour of the composite will then be depicted by that of a homogeneous equivalent material, using the usual damage approach.

Applications

Applications for CMCs concern mainly:

- Engine components such as nozzle flaps or after burner flame rigs.
- Rocket booster components such as expandable or rigid nozzle cones.
- Ramjet or scramjet combustors and nozzle cones.
- Thermal protection system and hot structures for the Hermes space shuttle such as leading edges for the wings, winglets and tiles for the fuselage or the wings (intrados).
- In the near future, these materials will be more extensively used for structural parts of hypersonic vehicles and propulsion systems.

For the time being, however, work still remains to be done, both on constitutive laws and environmental behaviour, to provide the designers with the necessary data and tools. Figure 14 gives an idea of the potential of covalent matrix composites with a carbon or a silicon carbide matrix reinforced by carbon fibres and carbon or silicon carbide fibres respectively, as the French SEP Company envisages it today (38); according to this diagram, high use temperatures within the range 2500 to 3000K are restricted to carbon-carbon and only for short durations around a minute, whereas the C-SiC composite exhibits a potential operating domain ranging from 2500K down to 1500K with corresponding working times within the range a few minutes to 10 hours.

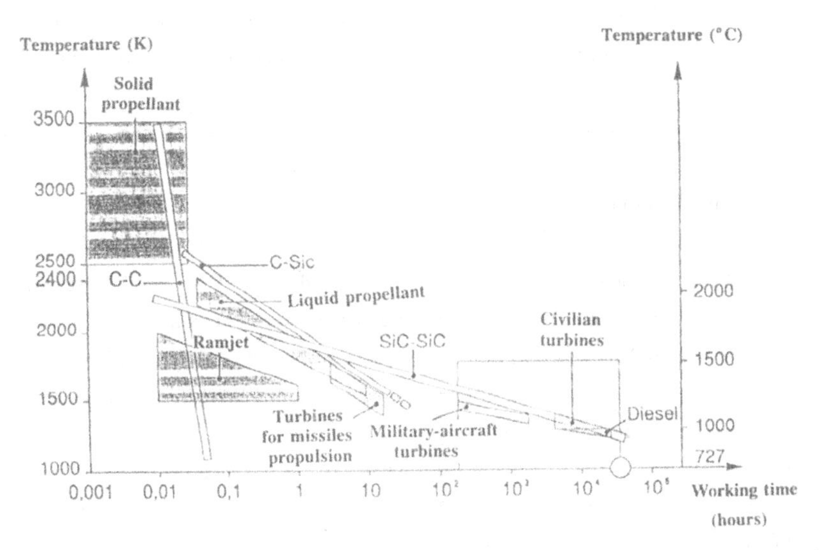

Figure 14. Potential use temperature as a function of working time for some ceramic-ceramic composites able to operate above 1300K.

SiC-SiC composites cover a much wider range from military application with the operating times similar to that of C-SiC but at somewhat lower temperatures (2300K), to civil applications for turbojet components with an operating temperature around 1300K and lives ranging between 5.10^3 and 5.10^4 h. These predictions are further confirmed by the two following components currently under evaluation.

Technological demonstrator components have been realised by SEP for the C/SiC and SiC/SiC composites and by Aerospatiale for the carbon-carbon composite; these components have proceeded successfully through characterization regarding the requirements of the Hermes space shuttle.

For the turbo engine components, rig and engine tests performed at SNECMA show a good behaviour of the SiC-SiC components; moreover the after burner flame rig has experienced flights on a M53 engine. Ceramic-ceramic nozzles are currently used for booster rockets or satellite attitude engines.

CONCLUSIONS

CMCs exhibit a non-brittle dissipative behaviour thus rendering these materials damage tolerant. A local fracture mechanics approach due to Peres further demonstrates that the matrix macrofissuration can be delayed with increasing fibre volume fraction thus expressing differently the increase of the matrix rupture strain of a ceramic matrix within a composite stated by ACK.

This non-brittle dissipation failure of CMCs relies upon the generation of a weak fibre-matrix bond which is achieved during processing by the insertion of an interphase, generally carbon, between the fibre and the matrix.

Glass-ceramic composites are good candidates for applications up to 650°C with a potential development at higher temperatures.

SiC-SiC and C-SiC composites have already reached a production stage and have been used as components for rocket propulsion for some years, with working temperatures up to 1700°C.

Further work is needed in the development of both new thermostructural composites and environmental and mechanical behaviour understanding so as to develop the tools needed for component design.

REFERENCES

1. Weiderhorn, S.M., "Reliability, life prediction and proof testing of ceramics", Ceramics for High Performance Applications, Ed. J.J. Burke et al., 1974, p. 633.

2. Marshall, D.B. and Ritter, J.E., "Reliability of advanced structural ceramics and ceramic matrix composites - A review", Ceramic Bulletin, 1987, **66**.

3. Pasto, A.E., "Current issues in silicon nitride structural ceramics", MRS Bulletin, 1 Oct, 1987, p. 73.

4. Cotter, D. and Koenigsberg, W., "Microfocus Radiography of high performance silicon nitride ceramics", Conference on Non Destructive Testing of High Performance Ceramics, Boston, 1987.

5. Faber, K.T. and Evans, A.G., "Crack deflection Processes", Acta Met., 1983, **31**, p. 565-584.

6. Evans, A.G. and Heuer, A.H., "Review - Transformation toughening in ceramics: Martensitic transformation in crack tip stress fields", J. Am. Ceram. Soc., 1980, **63**, 241-48.

7. Tsukuma, D., Kubota, Y. and Tsukide, T., "Thermal and Mechanical properties of Y_2O_3 stabilized tetragonal Zirconia Polycrystals", Advances in Ceramics, **12**, Science and Technology of Zirconia II. Ed. N. Claussen, M. Ruhle and A.H. Heuer, Am. Cer. Soc. Colombus, OH, 1984.

8. Becher, P.F. and Wei, G.C., "Toughening behaviour of SiC whisker reinforced alumina", J. Am. Ceram. Soc., 1984, **67**, C267-269.

9. Shalek, P.D., Petrovic, J.J., Hurley, G.F. and Gac, F.D., "Hot pressed SiC whiskers/Si_3N_4 matrix composites", Am. Ceram. Soc. Bull., 1986, **65**, p. 351-56.

10. Kelly, A., "Strong Solids", (Second edition), Clarendon Pres, Oxford, 1973.

11. Aveston, J., Cooper, G.A. and Kelly, A., "Single and multiple fracture, in properties of fibre composites", Conference Proceedings NPL, p. 15, IPC, Science and Technology Press, Guildford, 1971.

12. Cooper, G.A. and Sillwood, J., J. Mat. Sci., 1972, **7**, 325.

13. Marshall, D.B., Cox, B.N. and Evans, A.G., Acta Met., 1985, **33**, 11, p. 2013-2021.

14. McCartney, L.N., Proc. Roy. Soc., 1987, **A409**, 329.

15. Peres, P., "Analyse théorique et expérimentale du rôle des paramètres de microstructure sur le compoortement des composites à matrice fragile", INSA Thesis, 6 September, 1988.

16. Anquez, L., Costa, P., Peres, P. and Stohr, J.F., "Déformation et rupture dans les composites céramique-céramique", Colloque de métallurgie du CEN, Saclay, June 1989.

17. Marshall, D.B., "An indentation method for measuring fibre friction stress in ceramic composites", J. Amer. Ceram. Soc., 1984, **67**, 12.

18. Parlier, M., Ritti, M.H., Stohr, J.F. and Vignesoult, S., "Silicon fibre reinforced glass ceramic matrix composites: A high temperature material for high performance application", ICAS 90, Stockholm, 9-14 September, 1990.

19. Naslain, R. and Langlais, F., "CVD processing of ceramic-ceramic composite materials", Mater. Sci. Res., 1986, **20**, 145-164.

20. Fitzer, E., Hegen, D. and Strohmeier, H., "Possibility of gas phase impregnation with silicon carbide", Rev. int. hautes temp. refrac., 1980, **17**, 23-32.

21. French Patent 82.01025, "Structure composites de type réfractaire et son procédé de fabrication", 22 Janvier, 1982.

22. Stinton, D.P., Caputo, A.J. and Lowden, R.A., "Synthesis of fibre-reinforced SiC composites by chemical vapour infiltration", Ceram. Bull., 1986, **165**.

23. Jamet, J.F., Anquez, L., Parlier, M., Ritti, M.H., Peres, P. and Grateau, L., "Composite céramique: Relations entre microstructure et rupture", l'Aéronautique et l'Astronautique, 1987, **123/124**, p. 128-142.

24. French Patent 89.00774, "Polysilanes et leur procédé de préparation", 23 January, 1989.

25. Larche, F., Elissalde, D., Parlier, M. and Noireaux, P., "Effect of the heat treatment temperature on the transformation of a SiC precursor", To be published in the proceedings of "Internatioanl Fine Ceramics Workshop", Nagoya, 1990.

26. French Patent 87.18215, "Polysiloxalanes et leur procédé de préparation, leur utilisation comme précurseurs de céramique et lesdites céramiques", 28 December, 1987.

27. Colombier, C., "Studies of new polysilazanes precursor to SiC-N-O ceramics", Proceedings of the 1st European Ceramic Society Conference, Maastricht, Vol. I, 18-23 June, 1989, p. 143-152.

28. Colomban, P., "Gel technology in Ceramics, Glass Ceramics and Ceramic-Ceramic Composites", Ceramic International, Elsevier Science, 1989, 503-50.

29. Snow, G.S., "Fabrication of Transport Electronic PLZT by atmosphere Sintering", J. Am. Ceram. Soc., 1973, **56**, 91-6.

30. Thompson, J. Jr., "Chemical Preparation of PLZT Powders from Aqueous Nitrate Solution", Am. Ceram. Soc. Bull., 1973, **53**, 421-5.

31. Phalippou, J., Woigner, T. and Praffas, M., Part I: Synthesis of monolithic silica aerogels, J. Mat. Sci., 1990, **25**, 3111.

32. Woigner, T., Phalippou, J. and Praffas, M., Part II: Aerogel-glass transformation, J. Mat. Sci., 1990, **25**, 3118.

33. Jamet, J.F., Abbe, D. and Guyot, M.H., "Interface and matrix optimization in sintered ceramic composites", ICCMV, San Diego, 5 July, 1985.

34. Cavalier, J.C., Lacombe, A. and Rouges, J.M., "Ceramic matrix composites, new materials with very high performances", ECCM3, A.R. Bunsell, P. Lamicq and A. Massian, eds., Elsevier Applied Science, p. 99-110, 1989.

35. Oberlin, H., To be published.

36. Jamet, J.F., 12th Tech., Rimini, October 1987.

37. Brennan, J.J. and Prewo, K.M., J. Mat. Sci., 1989, **17**, 2371-83.

MACHINING ADVANCED CERAMICS - A CHALLENGE IN PRODUCTION TECHNOLOGY

W. KÖNIG and E. VERLEMANN
Fraunhofer Institut für Produktionstechnologie,
Steinbachstrasse 17, 5100 Aachen,
Germany.

ABSTRACT

Demanding requirements in terms of material properties have led to the use of ceramic materials in many mechanical and process engineering applications. In view of the advantages of ceramics, such as superior hardness and compressive strength, it may be asked why these materials are not employed more frequently in industrial applications, particularly in larger-scale production. Apart from their brittleness, one reason may be found in machinability problems. The paper reviews problems related to the machining of advanced ceramics and presents a number of solutions for improved machining to give reliable component quality.

INTRODUCTION

Despite great improvements in sintering processes for ceramic parts, components for most applications still require finishing. Component quality is decisive, so that faults in this final processing step frequently lead to an expensive reject rate.

Problems encountered in finishing high-performance ceramics are caused precisely by their excellent material properties, which are closely associated with corresponding machinability problems (Figure 1).

The wide range of processes applicable to metal working is not available in the case of ceramics. Processes with a geometrically defined cutting edge, like milling and turning, can be used only on a laboratory scale, since the extreme hardness of the material rapidly destroys the diamond tools employed.

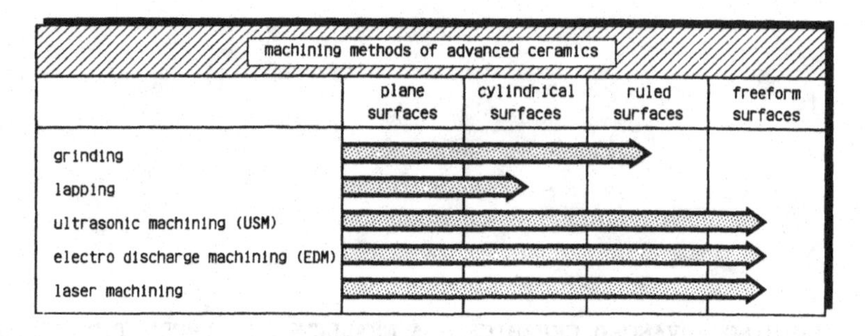

difficulties in machining of advanced ceramics

- potential surface damage — decrease of component strength
- small material removal rates — long machining time
- high tool wear and/or applications of superabrasives necessary — high tool costs
- high machining forces (grinding) — problems in accuracy of shape

Figure 1. Processes for machining high-performance ceramics and the problems encountered in using them.

The majority of industrial production relies on processes with a geometrically undefined cutting edge, where wear is distributed between a large number of cutting edges. The most important conventional processes include grinding, using exclusively diamond abrasives, and lapping, for which hard, yet less expensive materials such as boron carbide and silicon carbide can generally be employed.

Innovative processes include ultrasonic machining, electro-discharge machining and laser machining. Increased activity by research institutes has led to the optimization of these processes in recent years, so that process control strategies for example for ultrasonic machining have been developed, permitting their use in industrial applications.

Nonetheless, all machining processes encounter a common set of problems inherent to the material. Any form of processing entails potential damage to the surface and may lead to a substantial reduction in component strength. As a result, protective machining parameters are often chosen, leading to lengthy and hence costly machining times.

Despite use of the hardest cutting materials, all the processes with the exception of laser beam processing involve high tool wear, making machining difficult.

Analysis of these problems indicates the following set of tasks for the production engineer:

Control of existing processes must be optimized to provide finishing processes for ceramics which are both compatible with the material and economically acceptable. At the same time, new processes need to be developed and tested in order to provide a wider range of available ceramic shaping processes in the future.

The objective of the present contribution is to present the results of research into the conventional grinding and lapping processes which show the possibility of finishing ceramics to reliable quality at economic cost. In addition, the paper presents an innovative laser assisted turning process for high-density silicon nitride, indicating the prospects for hot machining with geometrically defined cutting edges.

GRINDING

Grinding with diamond tools is the process most frequently used to machine ceramic components. The advantage of grinding lies in the large number of processes available and the ability to realise both roughing and finishing processes. Grinding attains the highest removal rates on ceramics, while a properly designed process can also achieve the frequently demanded "mirror finish", i.e. maximum surface quality of the workpiece (1).

All grinding processes share the same elementary working principle of material removal by an individual tool edge. Diamond grain cutting edges projecting from the bonding penetrate the workpiece at high speed on circular paths, removing material from it.

With ceramic materials, various factors, for example cutting edge shape and chip thickness, decide which chip forming mechanism predominates during material removal. This chip forming mechanism can be investigated in single-grit grinding tests, in which a single diamond grain attached to the periphery of a grinding wheel contacts the workpiece under the conditions prevailing in the grinding process (Figure 2).

In silicon infiltrated silicon carbide, differences in chip thicknesses cause substantial differences in the chip forming mechanism. The type of material removal is of significant importance for the subsequent strength of the component (2).

Small chip thicknesses (h_c = 0.06-0.6 μm) are associated with distinct plastic parting phenomena in the material. The single cutting edge cuts through the material, with little accumulation of material, and creates an unmistakable scratch path.

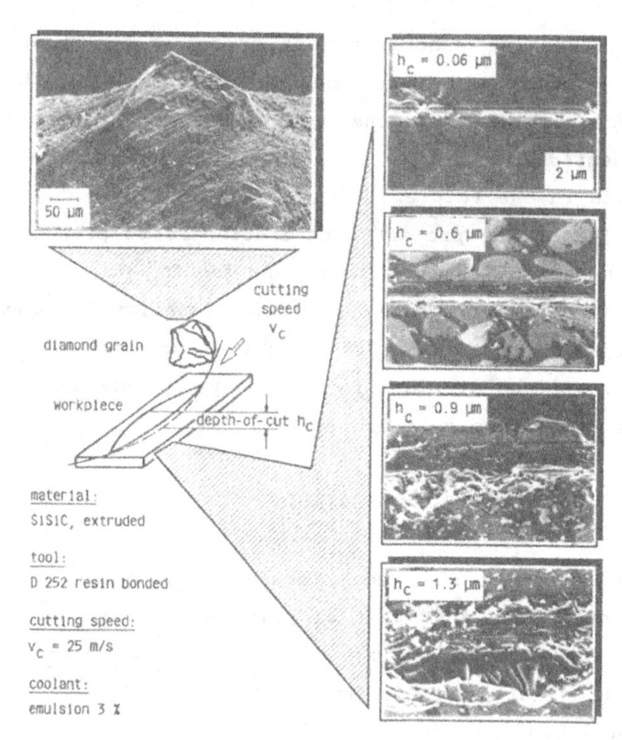

Figure 2. Single-grit grinding of SiSiC.

At greater chip thicknesses (h_c = 0.9-1.3 µm), the removal mechanism tends towards significant chipping at the edges of the scratch path. Cracks leading outwards from the intervening diamond grain are evidently induced far beyond the actual width and depth of cut, causing larger particles of the material to chip away.

The plastic removal phenomena, with their apparent advantages for this material, have been confirmed in other investigations. Both Inasaki (3) and Shore (4) have demonstrated that plastic removal phenomena are key mechanisms for the removal of ceramic.

In the real grinding process, the chip thickness at the single grain can be controlled via a number of factors. Apart from tool choice, the grinding process parameters such as infeed, feed rate and cutting speed all substantially affect chip thickness.

In conventional machining, the specific material removal rate is almost always achieved by using small infeed and high feed rates. High infeed and low feed rate also enable the same amount of material to be removed. This process is called creep-feed grinding. The major differences between these two processes is the size of contact area between the grinding wheel and workpiece (Figure 3).

In the case of creep-feed grinding, the contact length is essentially larger than in other grinding procedures. Consequently, the material removal rate is distributed over a large number of cutting edges, so that the depth of cut of the individual cutting edge is smaller. This procedure will have considerable effect on both the machining process and final results.

A comparison of practical results for alternating grinding (the conventional process strategy) and creep-feed grinding of various high-performance ceramics demonstrates the advantages of creep-feed grinding in terms of component strength and surface quality for a constant removal rate (Figure 4).

Irrespective of the type of ceramic (i.e. carbide, oxide or nitride) creep-feed grinding achieves higher component strength coupled with improved reproducibility and surface quality. The greater number of cutting edge contacts, however, results in an increase in cutting forces.

Since the reduced chip thickness at the single grain apparently has a favourable influence on the working result, logical development of this process strategy involves the use of higher cutting speeds. Higher cutting speeds at otherwise constant machining parameters reduce chip thicknesses at the single grain, as more cutting edges per unit time can machine the same volume of workpiece material.

An increase in cutting speed provides an economically interesting opportunity to increase the removal rate by simultaneously raising both cutting speed and feed rate, without enlarging chip thickness at the single grain. Figure 5 shows the results of this process control strategy.

The most significant result is undoubtedly the ability to double the removal rate while increasing component strength through a choice of parameters matched to the material properties. Improved surface quality constitutes a further advantage of this optimized process control.

Disadvantages of this form of grinding are, however, the increased normal forces which impose severe demands on the rigidity of the machine tool employed. In general, however, the advantages of this process are so evident that their use on suitable grinding machines can lead to a substantially improved performance in ceramic part manufacture.

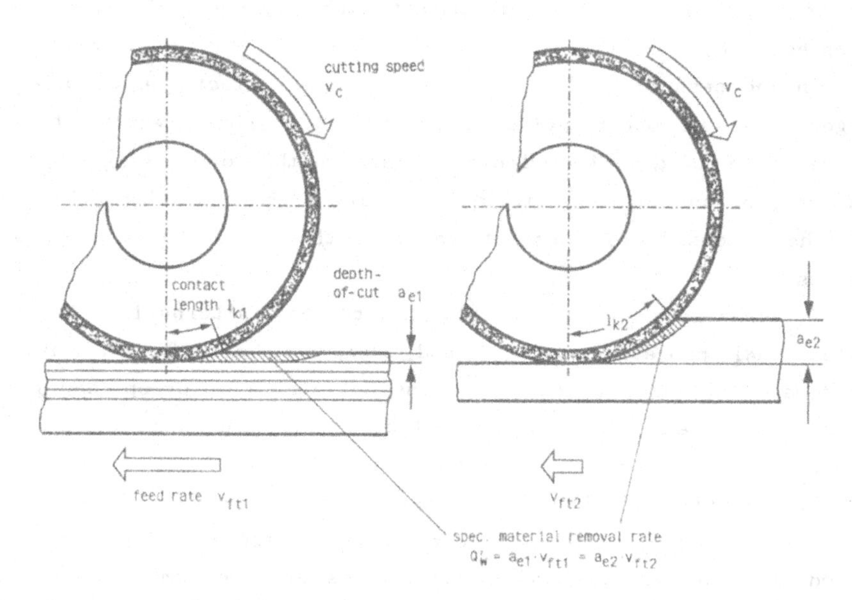

Figure 3. Contact conditions in surface grinding.

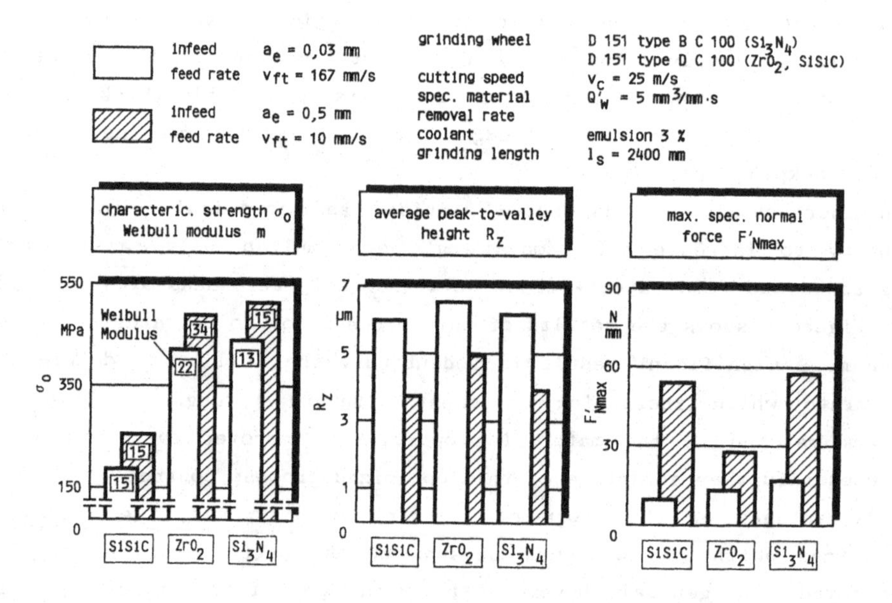

Figure 4. Comparison of alternating and creep-feed grinding of advanced ceramics at a constant removal rate.

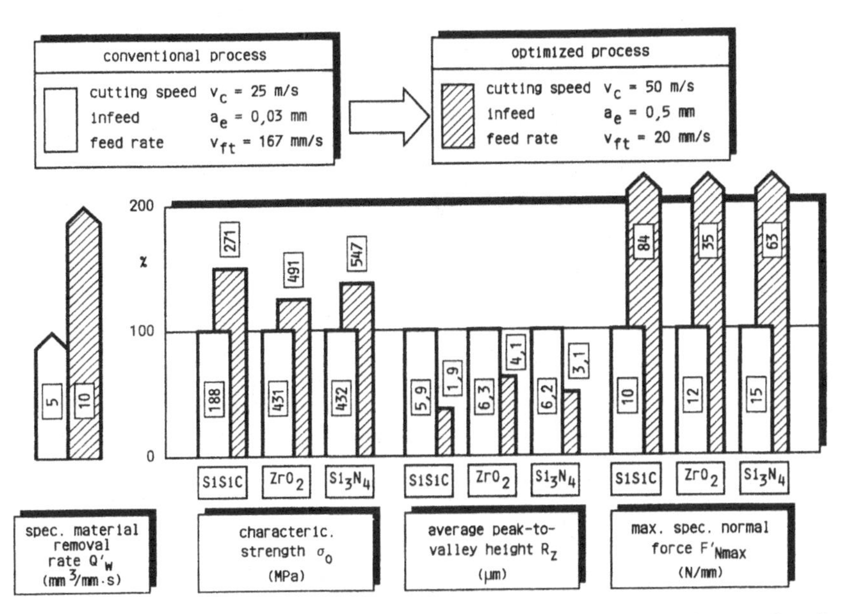

Figure 5. Comparison of conventional and optimized grinding of advanced ceramics, illustrated by SiSiC, ZrO_2, Si_3N_4.

LAPPING

Lapping is frequently used to finish ceramic components with extreme requirements for surface quality and accuracy of shape. Flat lapping has process-specific advantages compared with grinding, especially for face machining of ceramic components such as sealing washers and axial face seals, which deserve closer examination here.

A fundamental advantage is implicit in the lapping principle (Figure 6), which leaves no marks of preferred orientation on the workpiece surface. During the lapping process, driven carrier wheels guide the workpieces between top and bottom lapping wheels rotating in opposite directions. A lapping agent suspension is fed via channels in the top lapping wheel, introducing hard grains to the lapping gap between the workpiece and lapping wheel surfaces, where they remove material.

The process induces working marks of no preferred orientation on the workpiece surface. Variations in strength for components of the same shape and material, which can occur in grinding (5), are impossible with lapping.

An additional advantage of lapping lies in the ability to use grain materials only slightly harder than the workpiece material, since grains

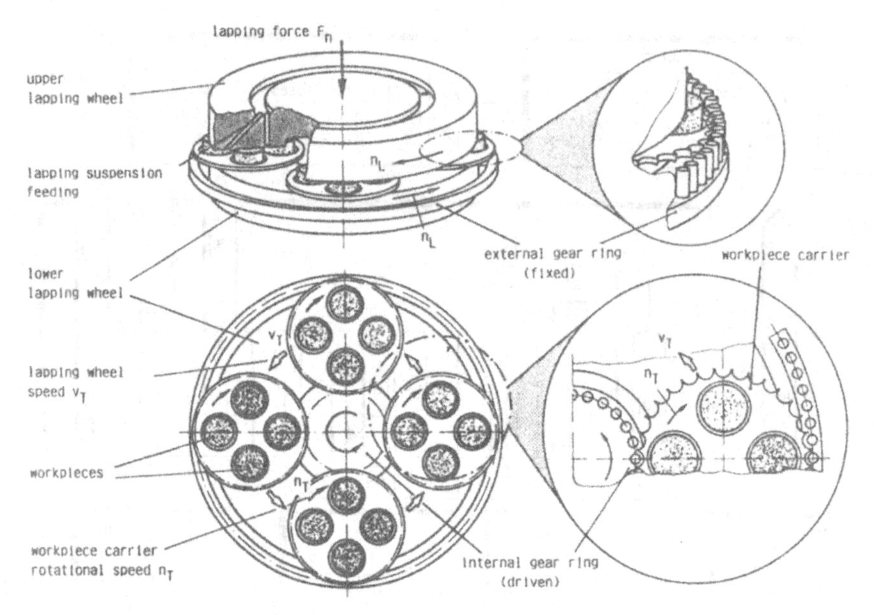

Figure 6. Principle of flat lapping.

which have already passed through the contact zone are not re-used. Instead of diamond, the cheaper boron carbide can frequently be used.

The often delicate procedure of clamping the workpiece in the machine, as required for other processes, is unnecessary in lapping where workpieces are fed loosely to the guide wheels. Since the process also enables a number of workpieces to be processed simultaneously, it appears an obvious choice for removing sintering distortion of ceramic components. A prerequisite is, however, that the removal rate should be economically acceptable.

The determining factors governing the removal rate for flat lapping are the grain size of the lapping abrasive, the normal force between the lapping wheel, lapping grain and workpiece, and the speed of rotation of the lapping wheel.

Given optimum matching of all parameters, it is possible to achieve distinct increases in the removal rate without adversely affecting component quality and, indeed, even improving it (Figure 7).

Apart from the obvious increase in the removal rate achieved by increasing the contact pressure, a further increase can be achieved by using larger lapping grains at higher contact pressure and constant lapping agent concentration. This is accompanied by an improvement in surface quality the causes of which are not immediately evident.

It is necessary to consider the effective mechanism of material removal. When a new grain enters the working gap, it is proportionally loaded by the normal force acting upon it, the larger grains being subjected to the greater forces. This results in larger grains splitting immediately as they enter the gap, with a corresponding reduction in mean particle size. In consequence, the number of contacting grains is considerably increased, and the greater contact force is distributed over a large number of grain contacts. The specific load per grain falls accordingly, i.e. the size of the particles removed is reduced but their number is greatly increased. The sum effect of these phenomena is then manifested as a higher removal rate and improved surface quality.

These effects are not confined to SiSiC, but are also observable with Al_2O_3 and ZrO_2. For an optimum grain size, lapping wheel speed and contact force, relatively high removal rates coupled with high surface quality can therefore be achieved with ceramic materials.

Tests to determine possible damage to components caused by roughing did not reveal any significant sacrifice in terms of component strength. Strength tests with Al_2O_3, for example, indicated that under otherwise

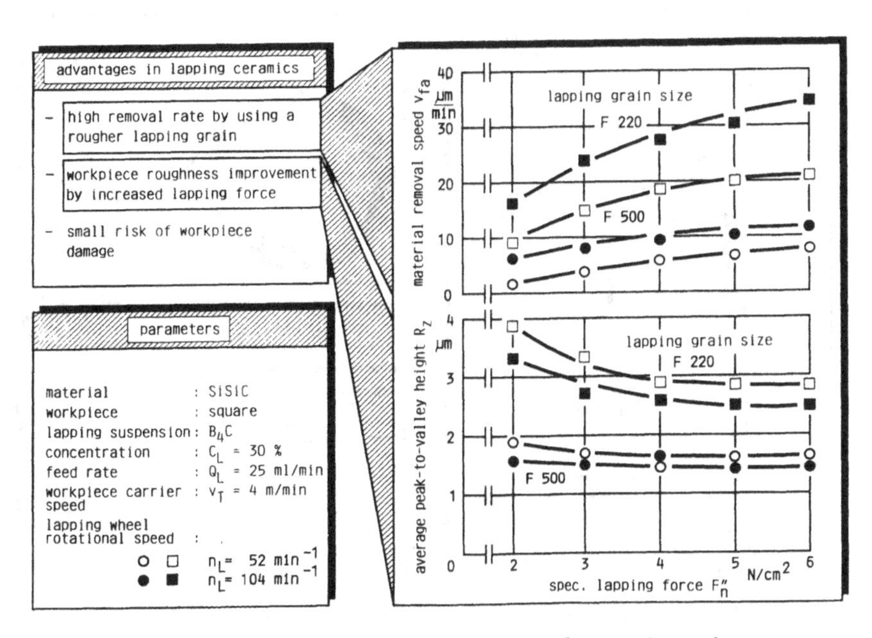

Figure 7. Improved removal rate and surface roughness in a lapping process.

constant conditions, an increase in grain size from F500 to F220 increases the removal rate by 200%, but reduces component strength by less than 10%.

Industrial application of these results could include multi-stage design of the lapping process (6), enabling ceramic components in the sintered state to be processed economically by means of lapping (Figure 8).

Use of this processing strategy on an axial face seal offers substantial advantages compared with conventional production using double face grinding and subsequent finish lapping.

In the case of the axial face seal, a complete machine batch (40 parts) can be finished to polishing standard in a production time of 16 min. Following a pre-lapping step, the roughing allowance is removed by large grain lapping (F220). After a flushing step to remove coarse grains, finishing continues with fine grain (F500). Since a batch of 40 components can be finished per machine load, the processing time per component is approximately 16 s, a value which cannot be attained by the conventional processing sequence.

In general, results show that high lapping removal rates, qualifying lapping for use as a roughing process, can be achieved by well-adapted process control. Rough lapping achieves high removal rate and extremely good surface qualities with no significant loss of component strength.

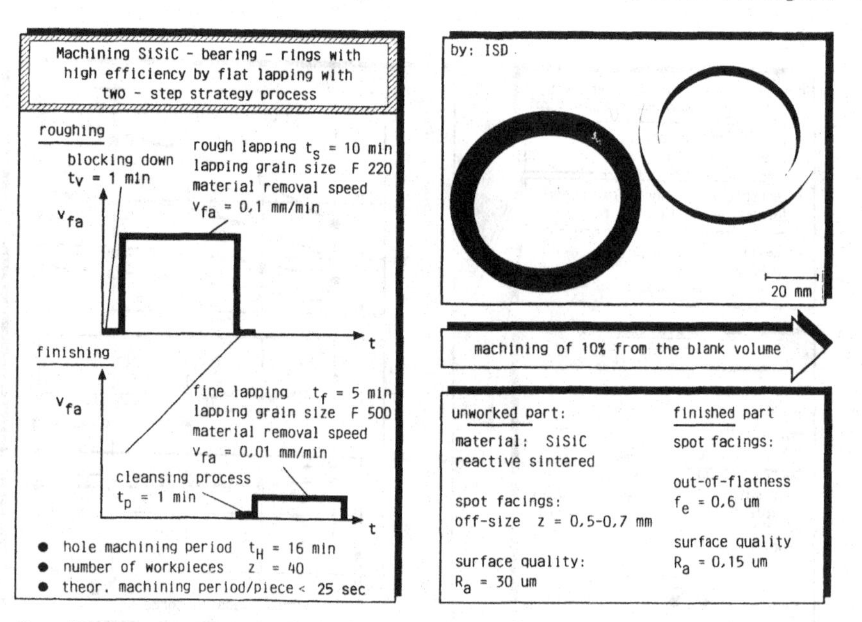

Figure 8. Higher cost-effectiveness through mutli-stage lapping.

LASER ASSISTED TURNING OF HIGH DENSITY SILICON NITRIDE

In addition to optimizing existing ceramic processing methods such as grinding and lapping, production technology also needs to develop and test new alternatives for ceramic processing to provide new shaping processes for ceramic components which may be better matched to the materials and more cost effective.

Especially in the case of hot pressed silicon nitride, the range of processing options can be extended by the use of material-specific hot machining processes. At the grain boundaries of the largely crystalline microstructure of silicon nitride there are amorphous zones which may be expected to undergo limited plasticity at a temperature in excess of 1000°C (7).

A possible approach may therefore be to process surface heated HPSN with geometrically defined cutting edges, i.e. in a turning process. Whereas other studies (8,9) have employed a gas or plasma torch to heat the workpiece surface zone, a laser was available for this purpose in the tests at the Fraunhofer IPT. Owing to its extremely high energy density, good energy input and narrow heat affected zone, the laser represents an excellent energy source for this application (Figure 9).

During processing, the workpiece is preheated by the laser so that the cutting tool can machine pre-softened material. The feed motions of the laser and the cutter are co-ordinated in such a way that the laser focus always slightly precedes the cutting edge, ensuring the appropriate workpiece temperature at the machining point.

The potential of this process was investigated as part of a feasibility study. Apart from the manipulated variables of the turning process, selection of a suitable tool and an optimum workpiece surface zone temperature were points of key interest in the investigations. The most important result to emerge was the ability to form a chip provided that an optimum workpiece surface zone temperature is maintained and CBN cutting materials are used.

Measurements of workpiece temperature using a spectral pyrometer showed that the best working results are achieved at minimum temperatures of approximately 1100°C. As expected, no stable chip flow can be attained at temperatures below 1100°C, as partial softening of the material has not yet commenced; the majority of the material splits away, substantial chipping occurs at the workpiece surface and increased tool fracture is encountered.

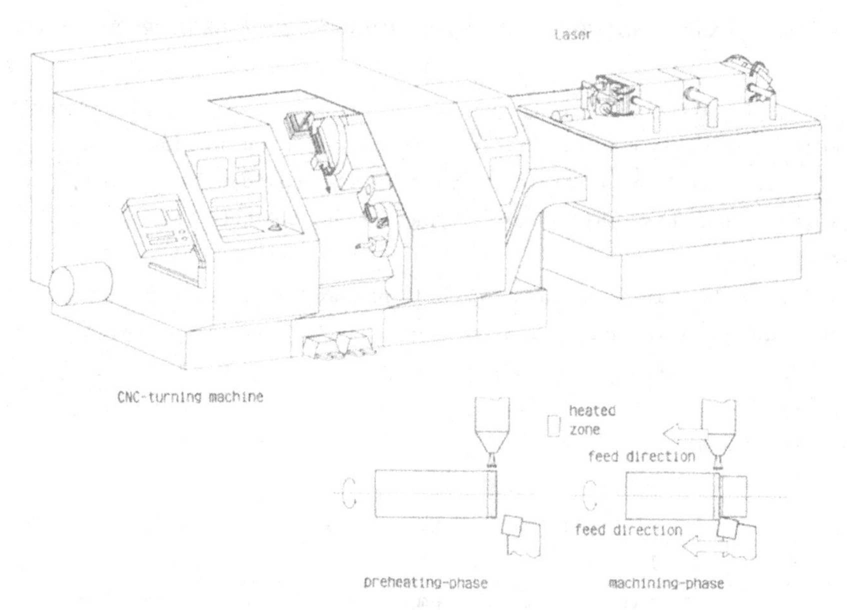

Figure 9. Principle of laser assisted turning.

At excessively high temperature (> 1300°C), the optimum chip formation temperature is apparently exceeded. The material is not machined continuously; instead, overheated particles of the material are ejected from the contact zone. Thermal loading leads to severe flank wear on the tool, resulting in extremely short tool lives.

If the workpiece surface zone temperatures is maintained at approximately 1100°C, good surface qualities can be obtained with laser-aided turning (Figure 10).

Surface qualities approaching finished standard can be achieved with this process. However, high cutting speeds result not only in excellent surface qualities but in increased tool wear. Low cutting speeds in the 20 m/min range evidently favour relatively stable flow chip formation. Surface qualities are slightly poorer, but tool life is increased. The removal rates of about 5 mm^3/s are, however, extremely low compared with conventional processes. Given a constant optimum processing temperature, tool wear manifests itself primarily as flank wear. There are no cutting edge breakouts even in interrupted cuts.

To sum up, it may be stated that the laser assisted turning of HPSN may represent an interesting future alternative to conventional methods. The disadvantages of low tool lives and removal rates currently associated with

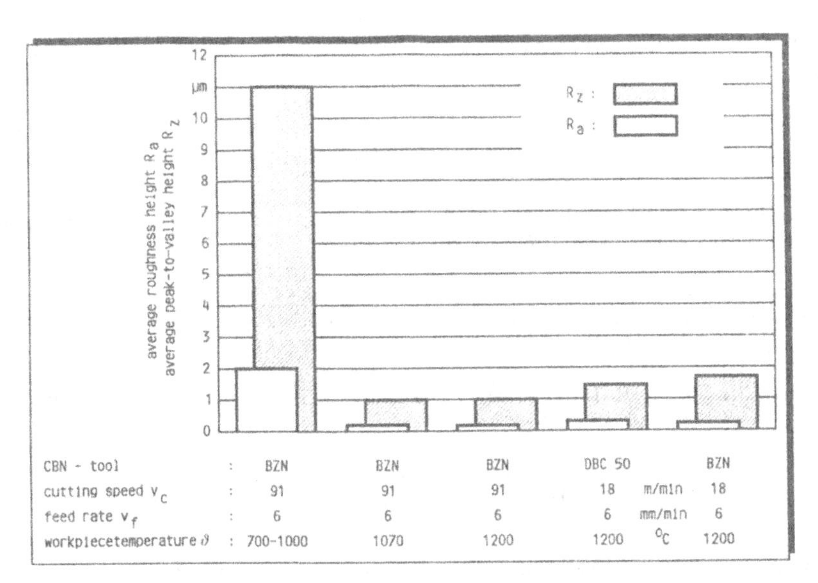

Figure 10. Attainable surface qualities with laser assisted turning.

the process will need to be counteracted by appropriate process optimization.

SUMMARY

Materials with excellent physical properties are unsuitable for technical applications unless they can be manufactured in the form of appropriate components. In the high-performance ceramics sector only a minority of parts will be usable immediately after sintering. Broad industrial application will thus be difficult unless suitable finishing processes which are both compatible with the materials and economically acceptable can be devised.

Production technology is therefore confronted with the task of adapting existing processes for use on ceramics and of developing new processes which can provide a wider range of alternatives.

The paper demonstrates that the conventional grinding and lapping processes definitely possess reserves capable of being exploited for economic production adapted to these materials. In addition, the example of laser assisted turning of hot pressed silicon nitride serves to illustrate the potential of innovative techniques for the future processing of ceramics.

REFERENCES

1. König, W. and Kleinevoss, R., Spiegelglanz an keramischen Bauteilen, Industrieanzeiger, 1987, **109**, 11-13.

2. König, W., Wemhöner, W. and Berweiler, W., Optimierung von Bauteilqualität und Festigkeit beim Schleifen von siliziuminfiltriertem Siliziumkarbid, Keramishe Zeitung, 1988, **40**, 764-768.

3. Inasaki, I., Grinding of hard and brittle materials, Annals of The CIRP, 1987, **36**, 463-471.

4. Shore, P., State of the art in "damage-free" grinding of advanced engineering ceramics, Vortrag Nanotechnology Forum Meeting, 13 September 1989.

5. Kessel, H., Bearbeitung von heissgepresstem Siliziumitrid und anderen keramischen Werkstoffen mit Diamantwerkzeugen, IDR, **10**, 1976, 128-133.

6. König, W. and Popp, M., Siliziuminfiltriertes Siliziumkarbid wirtschaftlich Läppen, Industrie-Anzeriger, 1986, **108**, 24-26.

7. Salmang, H. and Scholze, H., Keramische Werkstoffe, Keramik, 1983, **2**, Springer-Verlag, 1983.

8. Kitagawa, T., Plasma hot machining for engineering ceramics, Vortrag IPT, 11 November 1988, Aachen.

9. Uehara, K. and Takeshita, H., Cutting ceramics with a technique of hot machining, Annals of the CIRP, 1986, **35**, 55-58.

JOINING OF CERAMICS

E. LUGSCHEIDER, M. BORETIUS and W. TILLMANN
Material Science Institute,
Aachen University of Technology,
D-5100 Aachen, Germany.

ABSTRACT

The development of techniques for the reliable joining of ceramics to ceramics or metals is an essential prerequisite to the broader usage of advanced ceramics for engineering applications. Current methods are described including form and force fit joints, glueing, brazing and welding. Both current and emerging industrial applications are used as illustrations. There are needs for improved, more-refractory brazes for high temperature thermo-mechanical uses, and also for better understanding of the stresses near the joint and how these relate to appropriate material properties.

INTRODUCTION

The economical and ecological necessity of reducing the consumption of energy and resources as well as the reduction of pollution, lead to more and more demands on materials and technical processes, and hence the increasing attention and importance of ceramics. Examples are the emission reducing catalyst in exhaust pipes of cars and in power station chimneys, and the use of ceramic components in the engine and turbine industry.

If components of complex geometries are manufactured from ceramics, monolithic shaping is limited. On the one hand the risk of failure rises with the size of the components through a reduction of the mechanical strength. On the other hand machining is becoming more expensive and complicated. Manufacturing in modules offers many advantages concerning the realization of complex shapes and economic production, but the production of ceramic metal components need special joining techniques. Intensive research and development work is in progress in universities and industry to adapt joining techniques to match the properties of modern, highly specialized materials as well as to the desired application.

FORM FIT AND FORCE FIT JOINTS

As for metallic materials, there are many possible joining techniques for ceramics, that can be divided according to their working principle (Figure 1).

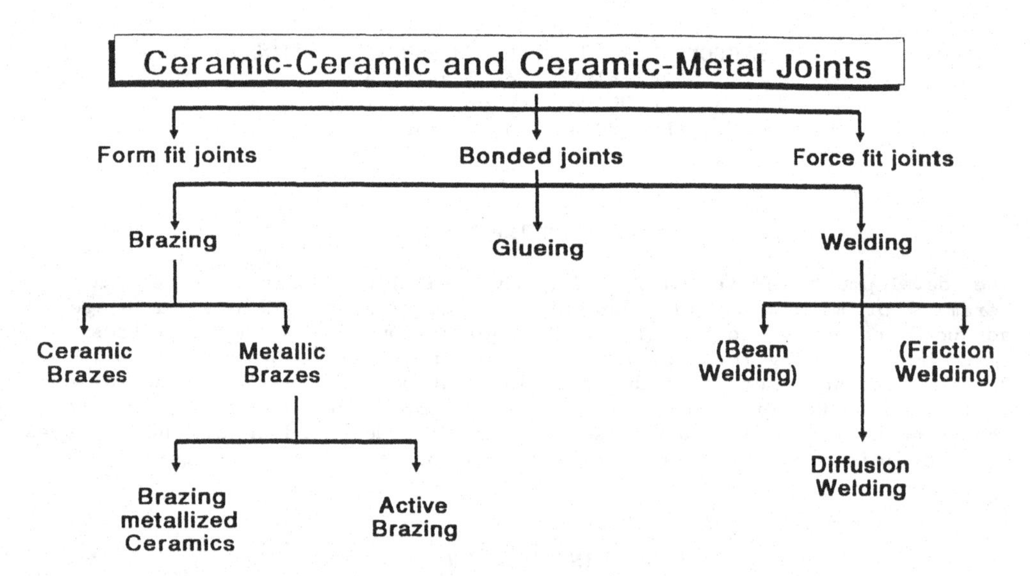

Figure 1. Techniques for joining ceramic to ceramic and ceramic to metal.

Form and force fit joints and their combination can generally be produced quite easily with the added advantage that they can be loosened if necessary. These joining techniques are applied where parts have to be replaced or changed, for example, ceramic inserts in cutting tools. The two outer inserts in Figure 2 are fixed by a combination of form and force fitting, whereas the insert in the middle of Figure 2 is attached to the holder by force fitting only (1).

Apart from these fairly simple parts, components that have to withstand highest demands even under long duration are joined in the same manner. An example is the hip-endoprostheses (Figure 3). The artificial unit consists of a metal shaft made from a CoCrMo or titanium alloy, a polished ceramic ball from alumina and a socket from polyethylene. Main demands on the ceramic-metal joints are the bio-compatibility of the materials, a well fitted position, corrosion resistance and a service life of more than ten years in a physiological environment. This can be achieved using a conical joint. The cone of the shaft is pressed into the ball by a force of 10 kN producing a form and force fit connection. Although this joint does not follow the construction rule not to load ceramics with tensile stresses, the joint performed with a high reliability (2).

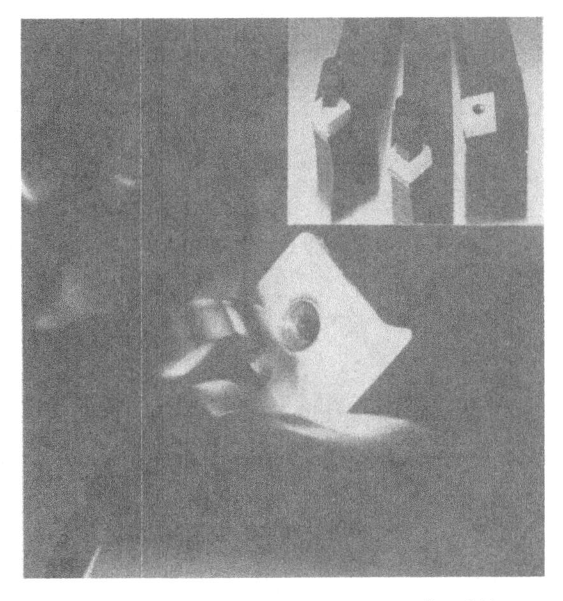

Figure 2. Ceramic indexable inserts for cutting tools (1).

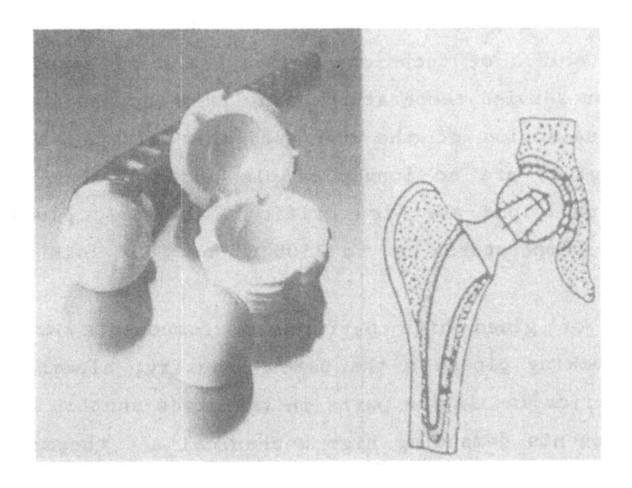

Figure 3. Hip-Endoprostheses (1,2).

Examples for the application of ceramic components in completely different fields are thermal insulating parts of portliners and pistons made of alumina-titania or zirconia used in combustion engines (Figure 4). The ceramic is cast directly into the engine component. The working principle is determined by a form and force fit joint caused by shrinkage of the metallic partner (3).

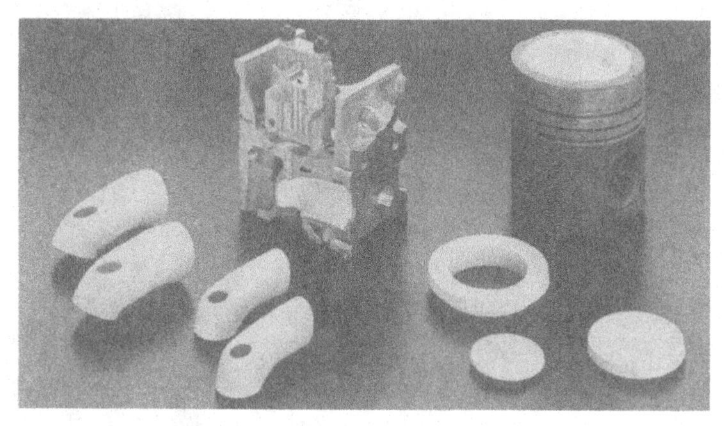

Figure 4. Ceramic components for thermal insulation (3).

BONDING OF CERAMICS

Glueing

Typical bonding techniques for ceramics are glueing, brazing and special welding processes (Figure 1). Glueing is a universal technique for joining nearly all materials of technical interest and is also applied to ceramic joints. Higher service temperatures can limit the application of glueing as well as the selection of the most suitable glue (Table 1). At higher temperatures glues are no longer stable and for some glues applications under humid conditions are also critical. Ceramic glues and mortars can allow service temperatures up to 1500°C, but the joints cannot be loaded mechanically.

Examples of glued high performance components are piezo-electrical exchangers, sucking pipes in the paper industry, alumina covers for radar aerials and silica insulation parts in the space shuttle (4).

For components demanding high mechanical and thermal stresses welding and brazing are the preferred joining techniques.

TABLE 1
Glues for joining ceramics (4)

Glue	Curing	Service Temp. [°C]	Joint Behaviour	Shear Strength
Epoxy resin	Warm	170-220	Brittle	High
Polyurethane	Warm + cold	120-180	Brittle - flexible	Medium
Phenolic resin	Warm + cold	80-150	Brittle - flexible	Medium
Silicone	Cold	180-220	Flexible - elastic	Low
Cyanoacrylate	Cold	150-250	Brittle	Medium
Elastomer	Warm	90-110	Flexible	Medium

Welding

Only beam welding, friction welding and diffusion welding can be used to join ceramics. Electronbeam and laser welding are restricted to small ceramic parts. The processes can be used only for ceramics with defined melting points (e.g. Al_2O_3), and not for ceramics that sublime such as SiC and Si_3N_4. A further disadvantage is the high stresses caused by severe temperature gradients, which can easily damage ceramic joints.

Ceramics can be joined to themselves or to metals by diffusion welding. The process can be adapted for joining small parts and simple geometries. It is rather complicated to apply the necessary pressure for larger components. High equipment costs and long joining times lead to high production costs. Therefore only small numbers, prototypes or high performance components, where production costs only play a minor role, are joined by this method. Because of the intense research on diffusion welding, especially for high temperature applications, an increasing importance of the technique can be presumed for the future.

Examples for diffusion welded components are electro-chemical cells for future high temperature electrolysis. The component shown in Figure 5 produces direct current from thermal energy and has been used for powering satellites for many years. Apart from conventional welded and brazed joints, these components contain diffusion welded joints between alumina and a FeNiCo alloy (Figure 5, at A).

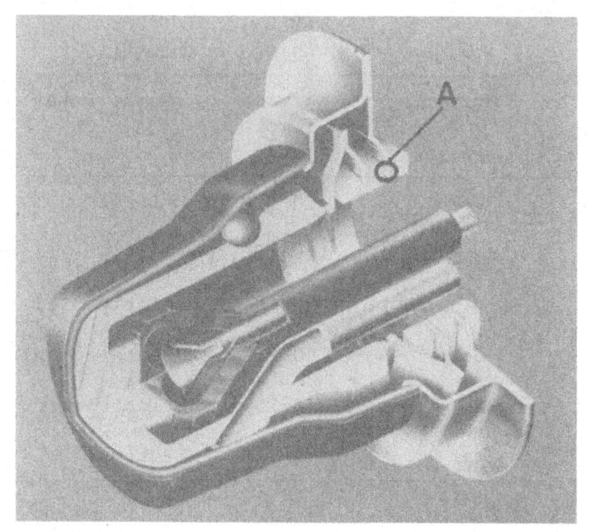

Figure 5. Thermionic energy converter (5).

Brazing

Brazing is the most important joining technique for high performance ceramics especially with regard to economic aspects. Brazing techniques are versatile and both ceramic and metallic brazes are used.

Brazing with ceramic brazes: Ceramic brazes are mixtures of oxides that wet a ceramic surface directly when the filler material is melted. They consist of Al_2O_3, CaO, MnO, Y_2O_3 and SiO_2 and are mainly used for joining oxide ceramics. Because of their poor flow behaviour and the fact that the joints brazed with ceramic brazes are very brittle, the technical importance is fairly restricted.

If brazes contain a high amount of SiO_2 that solidify as a glass they are called glass brazes. According to their properties during the brazing process, either stable or crystallizing filler materials may occur (Figure 6). Unlike stable glass brazes, crystallizing brazes transform to a ceramic-like, polycrystalline state. Those joints can be stressed thermally under service conditions up to the range of brazing temperatures. The selection of a suitable glass braze requires the thermal expansion behaviour of the braze to be adapted to the ceramic, because thermal stresses cannot be accommodated by the brittle glass brazes and thus cracks can arise.

The selection of a suitable glass braze is restricted because the brazing temperature and thermal expansion behaviour of the glass brazes are

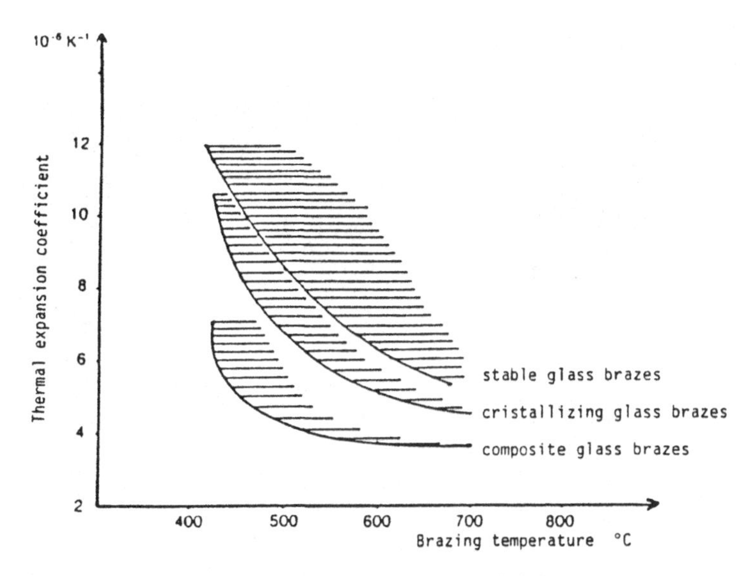

Figure 6. Glass brazes for joining of ceramics (6).

inversely related. That is why additive-containing composite brazes have been developed. These brazes have a constant expansion behaviour within a certain temperature range (Figure 6).

Applications for glass brazes can be found in gas discharge displays, electron flashes and clinical thermometers. A larger device is the pure alumina torus of a fusion reactor for research investigations (Figure 7). All segments are joined by glass brazes. One hundred and twenty-two vacuum-tight joints, with a brazing seam length of total 66 m, have been brazed in 26 joining operations (7).

Brazing with filler metals: Although glass brazes show a good wetting behaviour on oxide ceramics, metallic brazes are of much greater importance for joining ceramics. Due to their ductility they are able to accommodate thermal stresses and this is important especially when ceramic to metal joints are brazed. Figure 8 shows two processes that enable ceramics to be wetted. On the one hand this can be carried out by metallizing the ceramic before brazing and on the other hand the ceramic can be directly brazed by using active filler metals (8).

Figure 7. Al_2O_3-torus of a fusion reactor brazed with glass brazes (7).

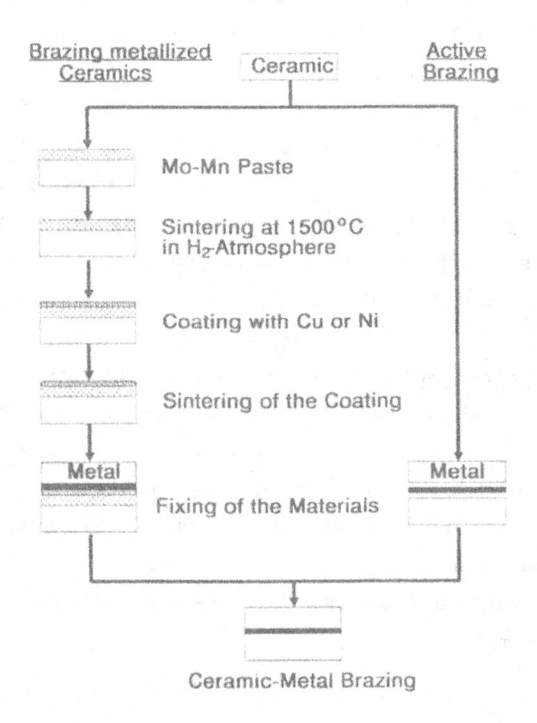

Figure 8. Techniques for brazing ceramic-metal joints (8).

Brazing of metallized ceramics: Amongst numerous metallizing processes, the molybdenum-manganese-process for brazing alumina has achieved a wide field of application. The whole process is carried out in three steps. In the first step a Mo-Mn powder mixture is sintered under wet hydrogen atmosphere onto the ceramic surface. In the second step the metallization is coated by nickel or copper in order to improve the wetting behaviour of the surface. And finally the brazing process is carried out using conventional filler metals.

Most commonly used filler metals for brazing metallized ceramics are ductile Ag-Cu filler metals, or Ag-Cu-Pd filler metals when higher service temperatures have to be realised (Table 2). The bond strength values that can be achieved by the Mo-Mn-process are fairly high. Brazed alumina joints reach up to 70% of the strength of the bulk material. When Al_2O_3 is brazed to FeNiCo alloy values of 50% of the bulk material strength can be reached. The values are measured by a four-point bending test (40 mm and 20 mm load points) with a test specimen geometry of 4.5 x 3.5 x 50 mm.

TABLE 2
Filler metals for brazing metallized ceramics

Brazing Alloy	Chemical Composition wt.%				Solidus °C	Liquidus °C	Brazing Temp. °C
	Cu	In	Pd	Ag			
Ag-Cu-In	25.0	14.5	-	61.5	630	705	755
Ag-Cu-In	27.0	10.0	-	63.0	685	730	780
Ag-Cu	28.0	-	-	72.0	780	780	830
Ag-Cu-Pd	31.5	-	10.0	58.5	824	852	900
Ag-Cu-Pd	21.0	-	25.0	54.0	901	950	1000
Ag-Pd	-	-	5.0	95.0	970	1010	1050

Figure 9 shows a micrograph of an Al_2O_3-Vacon 70 (Fe-Ni-Co alloy) joint brazed after metallization of the ceramic. The Mo-Mn metallization of the Al_2O_3 ceramic can be clearly seen followed by a thin nickel layer (black), the Ag-Cu filler metal zone and finally the metal Vacon 70.

Due to the fact that the metallizing process strongly depends on technical experience, usually only larger companies are able to produce metallized ceramics. Up to now metallizing processes are only available for Al_2O_3: there are no metallizing processes in industrial use for non-oxide ceramics.

Figure 9. Cross-section of a metallized brazed joint $Al_2O_3 \rightarrow$ Vacon 70; filler metal: Ag28Cu, $T_{Braz} = 850°C$; $t_{Braz} = 10$ min.

Figure 10 shows examples of brazed metallized components, such as feedthroughs, diodes and thyristor cases. Parts like these are brazed for the electrical and electronic industry on a large scale (9). A component that is surely not in large scale production is a high-power tube shown in Figure 11. The component is used for heating a plasma for nuclear experiments. Two amplifiers with a power of 1.5 MW each supply the necessary high frequency energy (10).

Active brazing: A direct wetting of oxide as well as non-oxide ceramics can be achieved using so called active filler metals which contain reactive agents like titanium, zirconium or hafnium that are able to reduce the boundary energy to such an extent that a wetting reaction can occur with the ceramic.

When oxide ceramics are joined with active filler metals a small reaction layer is formed at the interface between the ceramic and filler metal (black zone in Figure 12) that mainly consists of titanium oxides (11). When silicon carbide or silicon nitride are brazed, titanium carbide and titanium nitride as well as several silicides are formed within

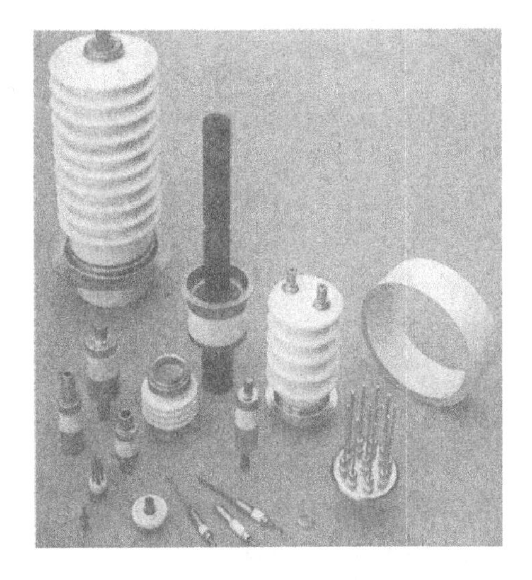

Figure 10. Metallized and brazed electrical and electronic devices (9).

Figure 11. Metallized and brazed high-power tube (10).

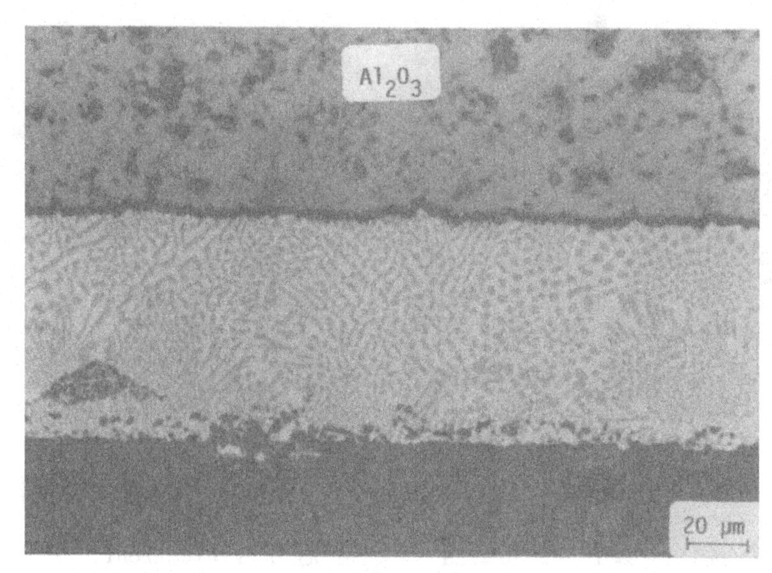

Figure 12. Cross-section of an active brazed joint Al_2O_3 → Vacon 70; filler metal: Ag26, 5Cu3Ti, T_{Braz} = 900°C; t_{Braz} = 10 min.

the reaction zone (12). A characteristic feature of active brazing is the titanium enrichment at the ceramic surface (Figure 13).

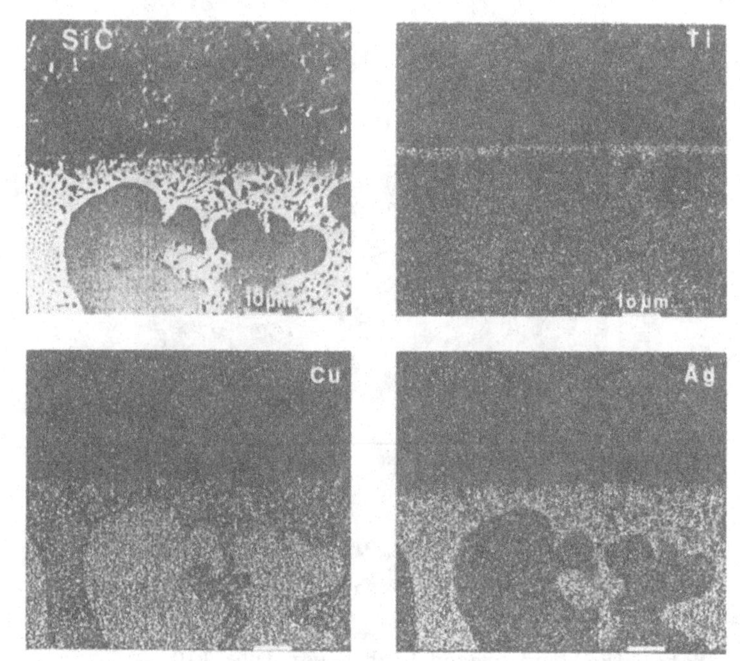

Figure 13. Cross-section and element distribution of an active brazed joint SiC → Cu/(Steel); filler metal: Ag27, 5Cu2Ti, T_{Braz} = 850°C; t_{Braz} = 10 min.

Although the first experiments on active brazing were more than 40 years ago, active brazing has not been commonly used in industry up to now. The reason is the brittleness of the high titanium-containing powder filler metals used in the past, resulting in low mechanical strength of the joints. Active brazing gained more interest after ductile foils were developed that could be manufactured conventionally by casting and rolling. These filler metals contain very low titanium, and the silver-copper eutectic is used as an alloy basis.

Table 3 shows active filler metals currently available in the market (13,14). The "active solders" also included have low liquidus temperatures but the brazing temperatures are as high as for the other active filler metals. Due to thermodynamic reasons the titanium component within the active solders can react with the ceramic only at high temperatures. The advantage of active solders is that the joints are almost free of residual stresses. The mechanical strength values of active soldered joints are fairly low.

TABLE 3
Properties of commercially-used active filler metals (13,14)

Active Brazing Alloy	Chemical Composition wt.%				Solidus °C	Liquidus °C	Brazing* Temp. °C
	Cu	In	Ti	Ag			
Ag-Cu-In-Ti	23.5	14.5	1.25	60.75	605	715	760
Ag-Cu-In-Ti	19.5	5.0	3.0	72.5	730	760	900
Ag-Cu-Ti	27.5	-	2.0	70.5	780	795	840
Ag-Cu-Ti	26.5	-	3.0	70.5	780	805	900
Ag-Cu-Ti	34.5	-	1.5	64.0	770	810	900
Ag-In-Ti	-	1.0	1.0	98.0	950	960	1030
Ag-Ti	-	-	4.0	96.0	970	970	1030
Sn-Ag-Ti	86-10-4				221	300	900
Pb-In-Ti	92-4-4				320	325	900

* According to the manufacturers.

Today many R&D projects are in progress to develop and test new active filler alloys. Examples from the Material Science Institute at Aachen University are the development of hafnium containing active brazes and the development of high temperature active filler metals for service temperatures higher than 800°C.

Active filler metals on the basis of the Ag-Cu eutectic containing hafnium as a reactive agent are very interesting for joining Si_3N_4 (Figure 14). Hafnium forms a thin hafnium-nitride layer that acts as a diffusion barrier for further reactions between hafnium and the ceramic. Very thin and stable reaction zones of HfN with more favourable expansion properties than TiN, when using titanium containing active brazes, result in lower stresses within the joint and consequently very high strength values even for critical material combinations. An average bending strength of 150 MPa with a standard deviation of 40 MPa could be achieved for the combination Si_3N_4/austenitic CrNi-steel. The geometry of the test specimen was 3.5 x 4.5 x 50 mm (15).

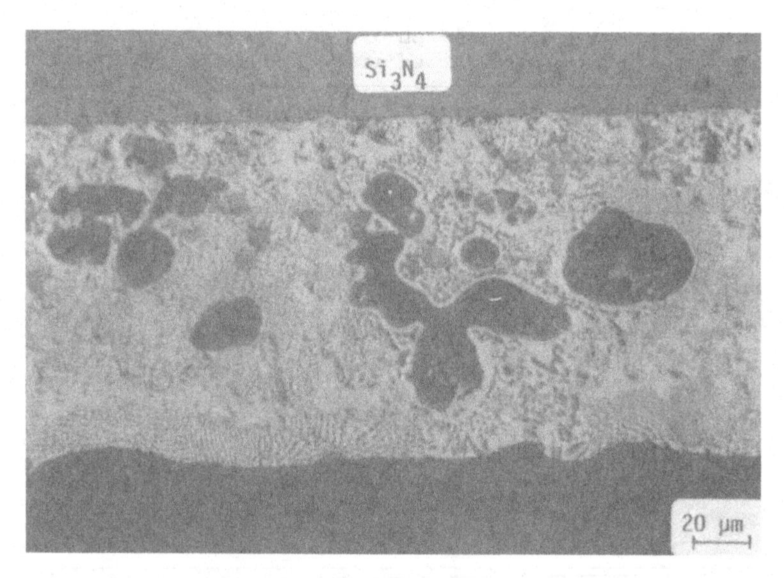

Figure 14. Cross-section of an active brazed joint $Si_3N_4 \rightarrow$ AISI 304; filler metal: AgCu5Hf, $T_{Braz} = 1050°C$; $t_{Braz} = 5$ min.

Because of the fact that modern high performance ceramics like SiC and Si_3N_4 are very appropriate for high temperature applications the development of high temperature active filler metals is a necessity. Recently active brazes with melting temperatures < 1350°C have been developed to allow service temperatures > 800°C. These brazes also have to be fabricated as brazing foils. Figure 15 shows a cross-section of a ceramic-ceramic joint made of Si_3N_4 and brazed with a high temperature active braze development of Aachen University.

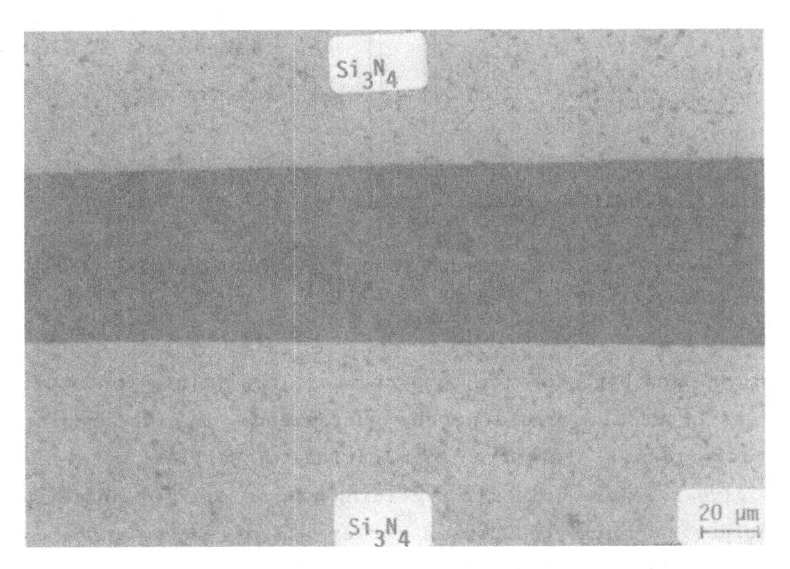

Figure 15. Cross-section of an active brazed joint $Si_3N_4 \rightarrow Si_3N_4$; filler metal: PdNiTi, T_{Braz} = 1225°C; t_{Braz} = 10 min.

Apart from well known problems concerning the brazing techniques at high temperatures, the problematic behaviour of the ceramic under high vacuum conditions and the reaction between filler metal and ceramic at high temperatures are topics of current research (16).

Two examples of industrial developments of active brazed components are found in automotive industry: rocker arms plated with ceramic parts and the joining of ceramic turbocharger rotors onto metal shafts.

The slide faces between the rocker arm and camshaft are exposed to high wear and mechanical stresses. To increase the service life of such a system, slide faces of ZrO_2, dispersion strengthened Al_2O_3 and Si_3N_4 have been attached to the rocker arm and have been tested. Figure 16 shows results of wear tests. The wear rate has been measured at 4 different points in the system after a testing time of 600 h. The results especially with Si_3N_4 show almost no wear (17).

To join ceramic plates to the rocker arms glueing was tested first. But due to extreme conditions like water contents up to 20% in the

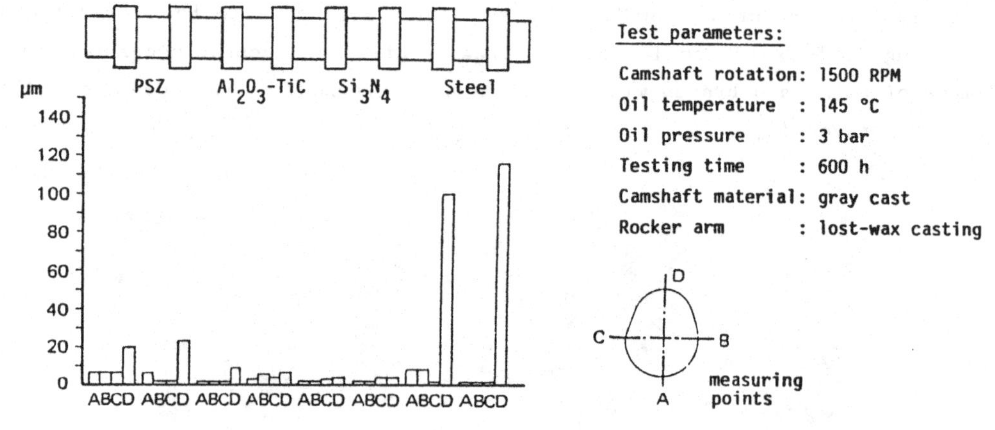

Figure 16. Wear rates of camshafts in combination with ceramic plated rocker arms (17).

surrounding oil and high service temperatures, the joints tend to fail and a reliable mass production could not be guaranteed. Active brazing leads to superior properties. Further optimization of filler metals and joint geometries give reliable joints for gliding plates of ZrO_2 and Si_3N_4 and the metallic holder. Figure 17 shows a rocker arm of a 6-cylinder engine with an active brazed Si_3N_4 gliding plate (18).

Another interesting application for active brazing is the turbocharger with a Si_3N_4 ceramic rotor on the exhaust side (Figure 18). From economic

Figure 17. Rocker arm with an active brazed Si_3N_4 slide face (18).

and safety points of view it is not favourable to manufacture the rotor as one integral piece, so that the ceramic rotor must be joined to a metal shaft. Due to the service temperatures of 400-500°C within the joining zone only brazing can serve as an economic joining technique (Figure 19). Direct brazing of Si_3N_4 to the metal shaft cannot be recommended because of the different thermal expansion behaviour of the materials resulting in thermal stresses within the joint and the danger of fracture within the ceramic.

Figure 18. Model of turbocharger with a ceramic rotor on the exhaust side (19).

Figure 20a shows a finite element (FE) calculation for such a joint. Theoretically the induced stresses can reach values up to 1250 MPa. An example of a possible approach to reduce these stresses is the use of interlayers (Figure 20b). The stress maximum can thus be reduced to 210 MPa which can be withstood by the ceramic.

In general these interlayers can be made of ductile materials or materials whose thermal expansion behaviour is similar to the ceramic. FE calculations and practical experiments show that a combination of both type of interlayers gives the greatest effectiveness.

The turbocharger design presented in Figure 21 shows the use of a metal sleeve made from Incoloy with a similar thermal expansion behaviour to the ceramic. First the metal shaft is joined to the Incoloy sleeve by friction welding and then the sleeve is brazed onto the Si_3N_4 rotor by active

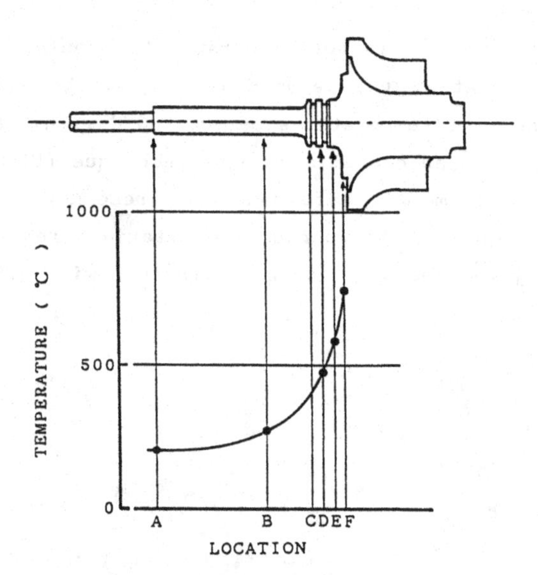

Figure 19. Temperature distribution within a turbocharger rotor in service (19).

brazing. Figure 22 present a turbocharger that has been brazed in such a manner.

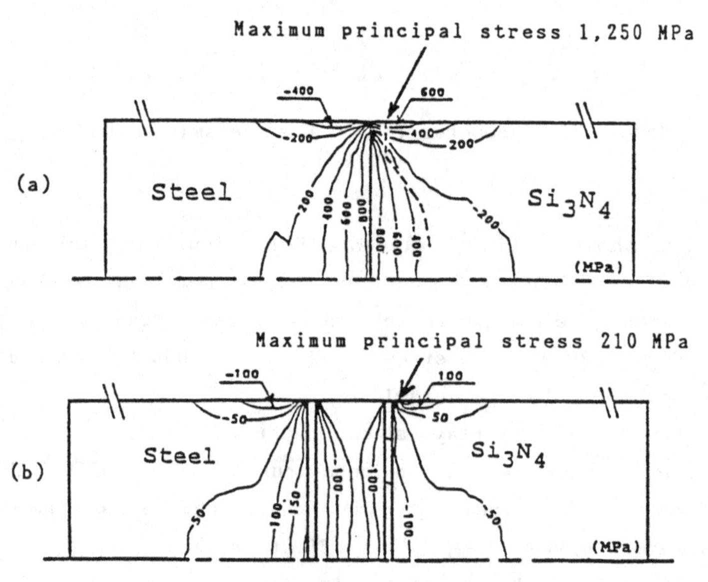

Figure 20. Residual stress distribution of an active brazed Si_3N_4 → Steel joint:
(a) Direct joining.
(b) Joining with a metallic multi interlayer (19).

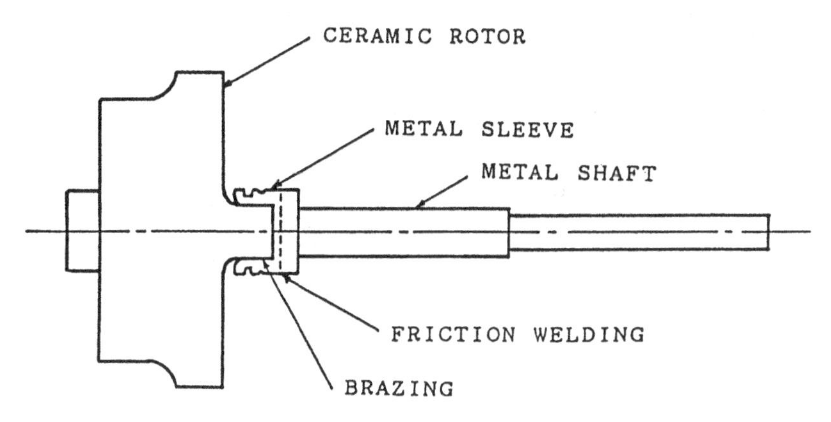

Figure 21. Active brazed joint between a Si_3N_4 rotor and a metal shaft by the use of a low-expansion metal sleeve (19).

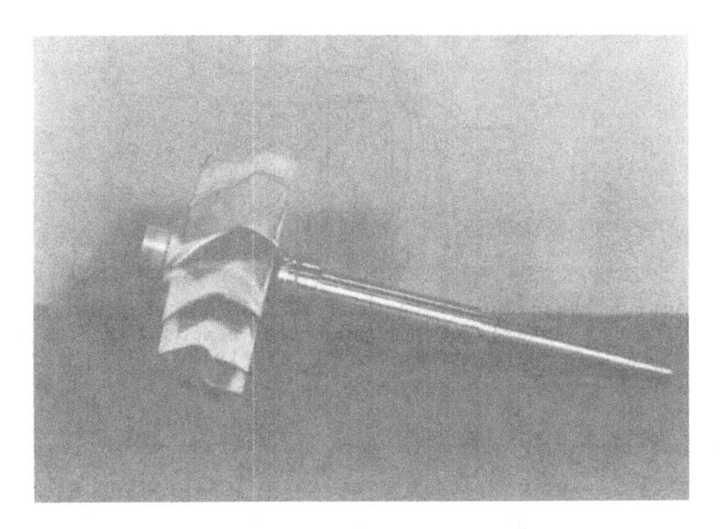

Figure 22. Active brazed turbocharger according to the design in Figure 21, for a 6.5 tonne truck diesel engine (19).

Figure 23a shows a more optimized design for a turbocharger. The metal shaft as well as the ceramic rotor are joined to a multi interlayer which consist of ductile nickel foils and of tungsten alloy with a low expansion coefficient. The connection between ceramic rotor and multi interlayer has been realised by active brazing. The multi interlayer itself and the connection to the metal shaft were realised by conventional brazing. By this fairly complicated construction, maximum stresses are diverted from the ceramic into the metal. High stress peaks can be reduced drastically by plastic flow of the metal (Figure 23b).

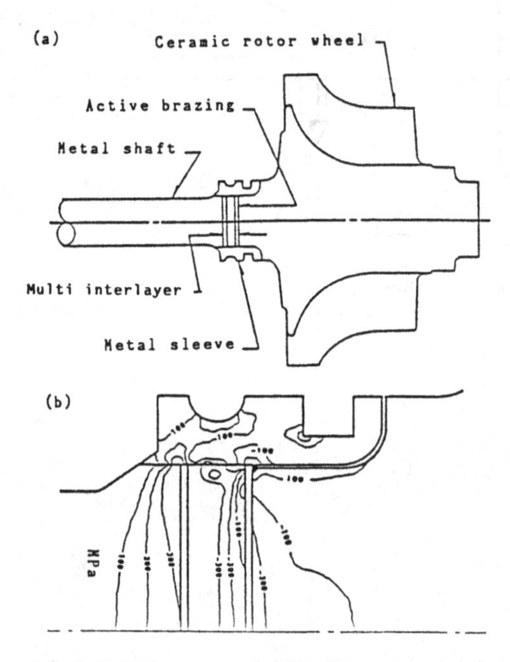

Figure 23. Active brazed joint between a Si_3N_4 rotor and a metal shaft by the use of a multi interlayer (a) and the residual stress distribution within this joint (b) (19).

Figure 24 shows the joint in more detail where it is covered by an additional Incoloy sleeve which serves as a seal, as an oil retainer and protects the interlayer materials from oxidation (19).

Another interesting example for active brazing is the fabrication of an ion-acceleration component (Figure 25). Metal rings have to be attached between alumina ceramic insulating rings. These metallic rings have to be connected to acceleration electrodes. The problem was solved by joining thin niobium rings to the insulator rings by active brazing. Niobium was selected because of its almost identical expansion behaviour compared with Al_2O_3. The size of the ceramic insulator rings are 710 mm in diameter and 100 mm in height (20).

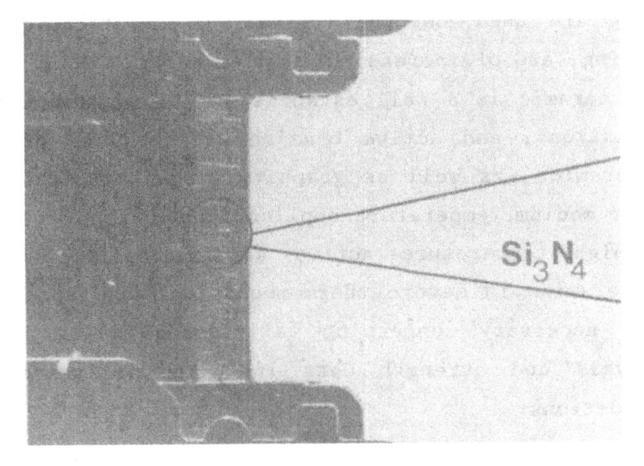

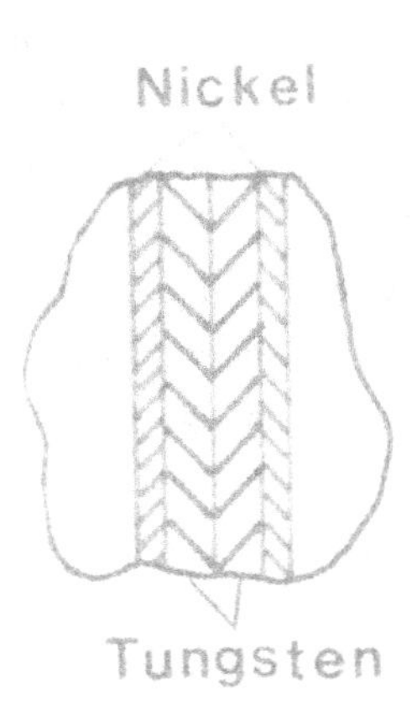

Figure 24. Cross-section of an active brazed Si_3N_4 rotor with a metal shaft using a metallic multi interlayer (19).

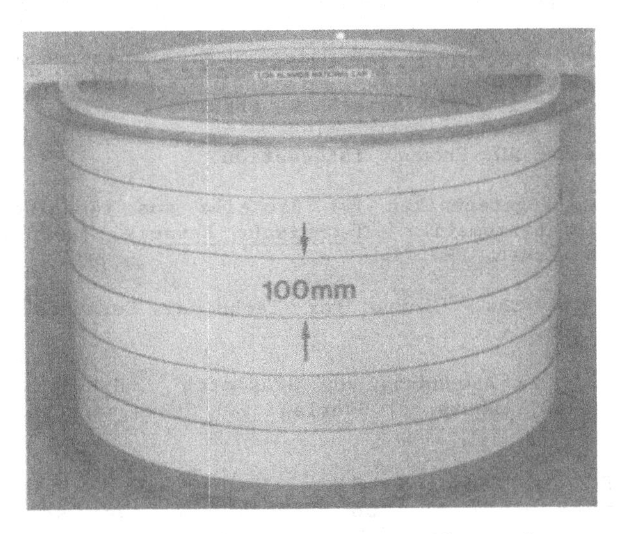

Figure 25. Active brazed Al_2O_3 insulator of an ion-accelerator (20).

CONCLUSIONS

Advanced ceramics stimulate new applications where materials are stressed under high thermal, mechanical, chemical and wear conditions. Design aspects influenced by the specific properties of these materials require adequate joining techniques for ceramic-ceramic and ceramic-metal combinations.

For many applications adapted techniques are available. Form and force fitting as well as glueing are used and the thermal joining techniques, welding and especially brazing, are of increasing importance.

Brazing of metallized ceramic is a well established joining technique for low temperature applications, and active brazing which allows direct wetting of all types of ceramics, as well as graphite, cemented carbides, etc., shows promise even for medium temperature applications.

The development of high temperature active brazing fillers is a necessity for joining of ceramics in severe thermomechanical applications. Further investigations are necessary concerning tailor-made filler metal developments, stress analysis and strength data including influence by different environmental conditions.

REFERENCES

1. Feldmühle AG, Product information.

2. Burghardt, H., Krauth, A. and Maier, H.R., Form- und kraftschlüssige Keramik-Metall Verbindungen. DVS-Berichte, (1980), **66**, pp. 75-81, Deutscher Verlag f. Schweisstechnik, Düsseldorf, FRG.

3. Hoechst CeramTec AG, Product information.

4. Schäfer, W., Fügetechniken für Bauteile aus technischer Keramik-Übersicht und Systematik. Technische Keramik, (1988), pp. 142-149, Vulkanverlag, Essen, FRG.

5. Information of the "Centre for Technical Ceramics", Eindhoven, The Netherlands.

6. Paschke, H., Die Anwendung von Glasloten. DVS-Berichte, (1980), **66**, pp. 45-48, Deutscher Verlag f. Schweisstechnik, Düsseldorf, FRG.

7. Hauth, W.E. and Stoddard, S.D., Joining of technical ceramics. DVS-Berichte, (1980), **66**, p. 49-52, Deutscher Verlag f. Schweisstechnik, Düsseldorf, FRG.

8. Lugscheider, E., Krappitz, H. and Mizuhara, H., Fügen von nichtmetallisierter Keramik mit Metall durch Einsatz duktiler Aktivlote. Fortschrittsberichte der Deutschen Keramischen Gesselschaft, (1985), **1**, No. 2, Bauverlag Wiesbaden-Berlin, FRG.

9. Friedrichsfeld, Product information.

10. Brown Boveri Technology for High-Energy Physics and Fusion Research, Product information.

11. Nicholas, M.G., Valentine, T.M. and Waite, M.J., The wetting of alumina by copper alloyed with titanium and other elements. J. of Mat. Sci., (1980), **15**, pp. 2197-2206.

12. Kofi, J., Yano. T. and Iseki, T., Brazing of pressureless-sintered SiC using Ag-Cu-Ti alloy. J. of Mat. Sci., 1987, **22**, pp. 2431-2434.

13. Degussa AG, Product information.

14. Wesgo, Product information.

15. Lugscheider, E. and Tillmann, W., Development of hafnium containing filler metals for active brazing of non-oxide ceramics. Final report of a National R&D Project (DFG), 1989, FRG.

16. Lugscheider, E. and Boretius, M., Development of high-temperature active brazing alloys for service temperatures above 800°C. Report of a National R&D Project (BMFT), 1990, FRG.

17. Dworak, U., Einsatz keramischer Wekstoffe im Motorenbau. Technische Keramik, 1988, pp. 260-276, Vulkanverlag, Essen, FRG.

18. Krappitz, H., Thiemann, K.H. and Weise, W., Herstellung und Betriebsverhalten gelöteter Kramik-Metall Verbunde für den Ventiltrieb von Verbrennungs-kraftmaschinen. Proceedings of the 2nd Int. Conference "Brazing, High-temperature Brazing and Diffusion Bonding", pp. 80-85, Essen, FRG, 19-20th September 1989.

19. Sasabe, K., Brazing of ceramic turbocharger rotors. Proceedings of the 2nd Int. Conference "Brazing, High-temperature Brazing and Diffusion Bonding", p. 164-167, Essen, FRG, 19-20th September 1989.

20. Ballard, E.O., Meyer, E.A. and Brennan, G.M., Brazing of large-diameter ceramic rings to niobium using active metal TiCuSil process. Welding J., 1985, pp. 37-42.

NONDESTRUCTIVE TESTING OF CERAMIC ENGINEERING COMPONENTS BY X-RAY, ULTRASONIC AND OTHER TECHNIQUES

W. ARNOLD and H. REITER
Fraunhofer-Institut for Nondestructive Testing (IzfP),
Building 37, The University,
D-6600 Saarbrücken 11, Germany.

ABSTRACT

A review is given of nondestructive testing methods which can be applied for quality assurance of ceramics. We discuss micro-focus X-ray testing, computer tomography, high-frequency ultrasound, Scanning Acoustic Microscopy, and testing of surfaces by means of surface waves and dye penetration techniques.

INTRODUCTION

Characteristics of ceramic materials are high hardness, low density, low thermal and electrical conductivity, as well as high resistance to aggressive media. During the last years ceramics have become especially interesting, as compared to metals, through the excellent combination of properties of low specific weight, wear resistance, as well as corrosion resistance, even at high temperatures (1). Fundamental disadvantages of high strength ceramics are their brittleness, and thus the incapacity to reduce by plastic deformation the stress occurring at defects, as well as a large scatter in the distribution of the strength values, from which a corresponding failure probability can be calculated. The reliability that can be achieved plays a dominant role in the use of ceramic materials in engineering applications. This reliability can be increased considerably by quality assurance measures.

Here, nondestructive testing techniques are required, first during the production phase of a new product to contribute to production optimization, and second during the manufacture in order to sort out defective parts before they are put into service.

From fracture mechanics investigations (2,3) it can be discerned that defects, dependent on type and position, with dimensions clearly below 100 μm, can lead to the failure of a component. This size is smaller at least by one order of magnitude than is generally assumed for metallic materials. Thus, for the detection and resolution of such small defects very high demands are made on nondestructive testing techniques which in most cases go beyond the state of engineering. In addition to material defects there are further factors that determine decisively the quality of materials, viz. the structural composition and the existing internal stress. The strength is dependent on the microstructure, the grain size and the grain size distribution, and the porosity homogeneity and anisotropy. The local stress situation in a component part is given by the sum of internal stress and load stress. The applicable loads are thus reduced by the existing internal stress. For example, a non appropriate surface machining may reduce the strength by 40% due to local stresses (4).

Nondestructive testing (NDT) must therefore concentrate on these three factors: the structure, the defects and the stress.

ULTRASOUND VELOCITY MEASUREMENTS

The ultrasound velocity is a function of several factors. To a first approximation density, linear elastic modulus and temperature have to be mentioned. In addition, stresses and textures are superimposed. Therefore, a quantitative assignment of a single effect being responsible for a change of the sound velocity is very complex. By using different types of waves and measuring under differing propagation directions, this problem may be solved. Figure 1 shows an example of the sound velocity being a sensitive indicator of local structure modifications. In this case, sound velocity modifications are caused by an incomplete densification of the HPSN disc in the centre area. The advantages of sound velocity measurements are: (a) low ultrasound frequencies can be used; (b) automation of the process (5); (c) rapid technique; (d) characterization of the material at an early stage before further costly processing. Such an example is in Figure 2, where Young's modulus is displayed as a function of porosity (6). Within a certain range of validity, the Young's modulus is linearly proportional to sound velocity, and hence the porosity can be directly monitored by simple sound velocity measurements.

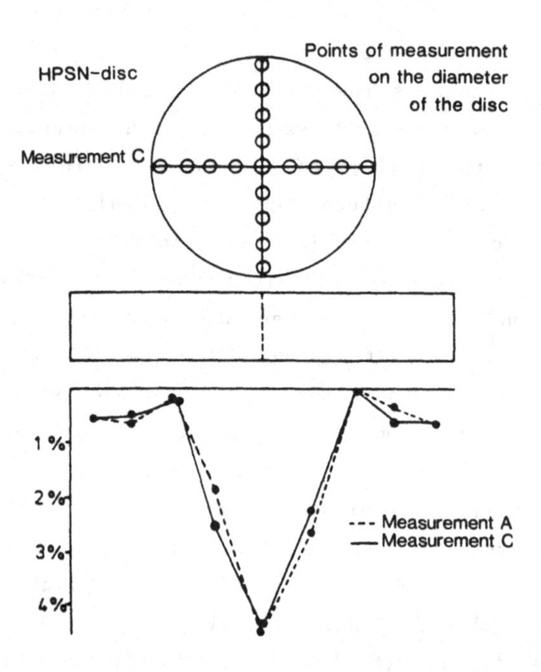

Figure 1. Relative change of the sound velocity over the cross-section of a HPSN-disc (longitudinal waves).

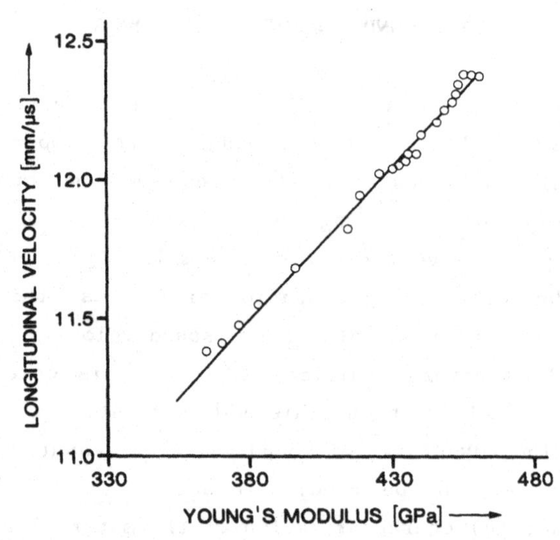

Figure 2. Plot of longitudinal ultrasonic velocity against Young's modulus of α-silicon carbide.

Anisotropy, conditioned by microstructure (texture) or stresses, leads to direction-dependent sound velocities, especially of linearly polarized transverse waves. While the texture can lead to relative sound velocity differences in the order of some percent, the stresses show effects which are smaller by one order of magnitude (7).

X-RAY INSPECTION AND X-RAY COMPUTED TOMOGRAPHY

For the testing of internal defects of complex shaped ceramic structural parts, the microfocal X-ray testing of high resolution should be mentioned first (8). The small focus spot (approximately 10 µm) makes it possible to work in projection geometry. This results in a direct magnification and a reduction of the scattering radiation. The latter leads to a rise of the signal-to-noise ratio. This technique with high resolution, high contrast and depth of focus permits to a high degree the imaging of structural homogeneities and defect structures. The X-ray film has high potential as an image-recording medium because of its excellent image quality (senstivity, contrast, resolution). Other image-recording media like X-ray image intensifiers, X-ray vidikon and diode arrays, have for example the advantage of on-line inspection and assessment, and investigation of moving parts or the rapid inspection from differing angles with differing enlargements (manipulator). Digital processing of the video signals can also be applied to increase the image quality. Next to adding individual images for signal-to-noise improvement, algorithms are used for the correction of zero images, or increases of contrast and edges, relief presentation and false colour presentation. In the case of high-contrast defects in metallic structure parts, automated evaluation can nowadays be extended such that only flaws are presented as images on the screen. Automated decision on the useability of the structure part is possible according to given evaluation criteria. Thus, in contrast to the subjective result of visual evaluation by an observer, an automated and objective evaluation and assessment are possible with the aid of the quantitative measurement technique and computer-aided image analyses. However, in the case of small flaws in ceramic materials, which are relatively poor in contrast, such an objective evaluation is very difficult because of the small signal-to-noise ratio.

Figure 3 outlines the principle of application of the microfocal X-ray technique. One example for the investigation of ceramic radial turbine rotors by means of this technique is presented in Figure 4, with reference to porosity, and crack-like indications in the blade area and in the hub region of the rotor.

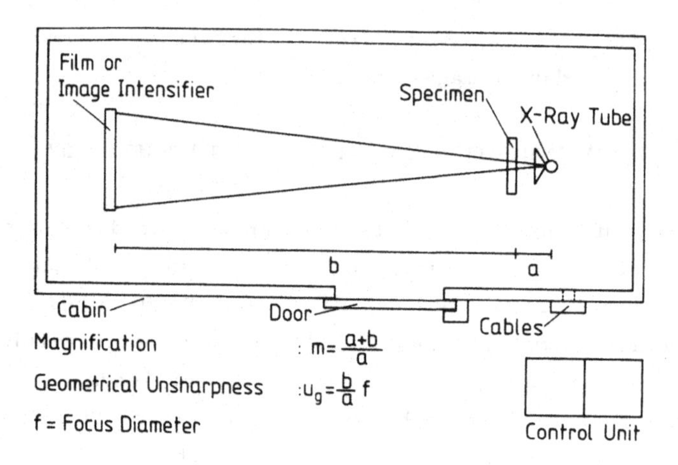

Figure 3. Principle of X-ray microfocus inspection technique.

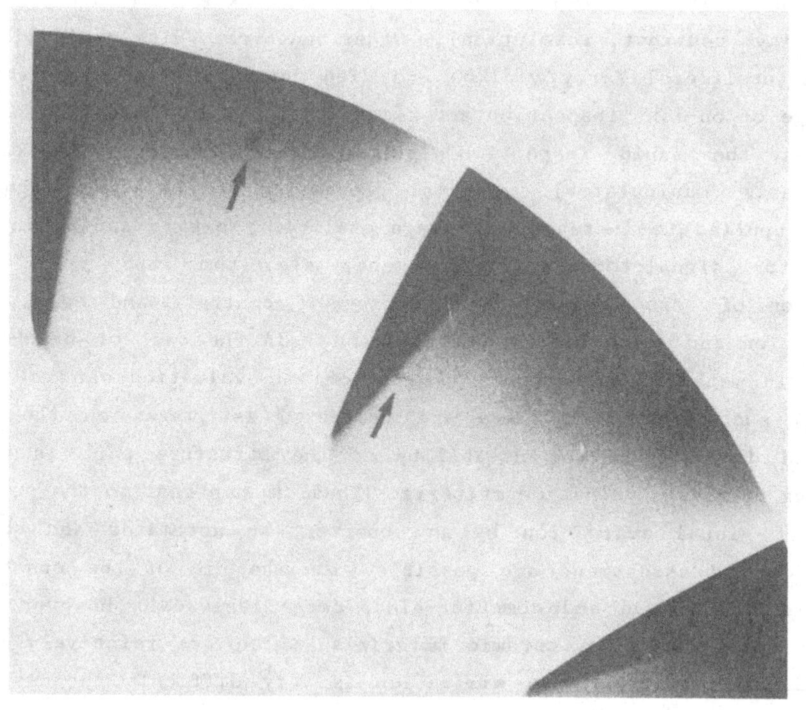

Figure 4. Microfocal radiography of blades of a turbocharger. Defect indications are marked by arrows.

Microfocal X-ray testing is to be considered as an essential element of a concept for the quality control of ceramic components. In addition to the detection of flaws there is the possibility of presenting structure differences as a consequence of differences in local density or absorption. It is true, though, that the information on flaw orientation or depth extension is lost as a consequence of the projection of a three-dimensional object onto a two-dimensional imaging medium. From a single image the position of a flaw with regard to the structural geometry cannot be discerned correctly. This can in particular lead to non-detection of flaws in certain directions, in particular crack-like and extended defects. A realistic detection probability for such flaws can only be obtained when investigations in several radiation directions are carried out. This leads finally to the X-ray computed tomography (CT), realized in the seventies for medical diagnosis. The principle is shown in Figure 5. The attenuation profile for each angle (i.e. for each projection) is registered over all angles and a sectional image is calculated from the individual projections of the object by means of reconstruction algorithms.

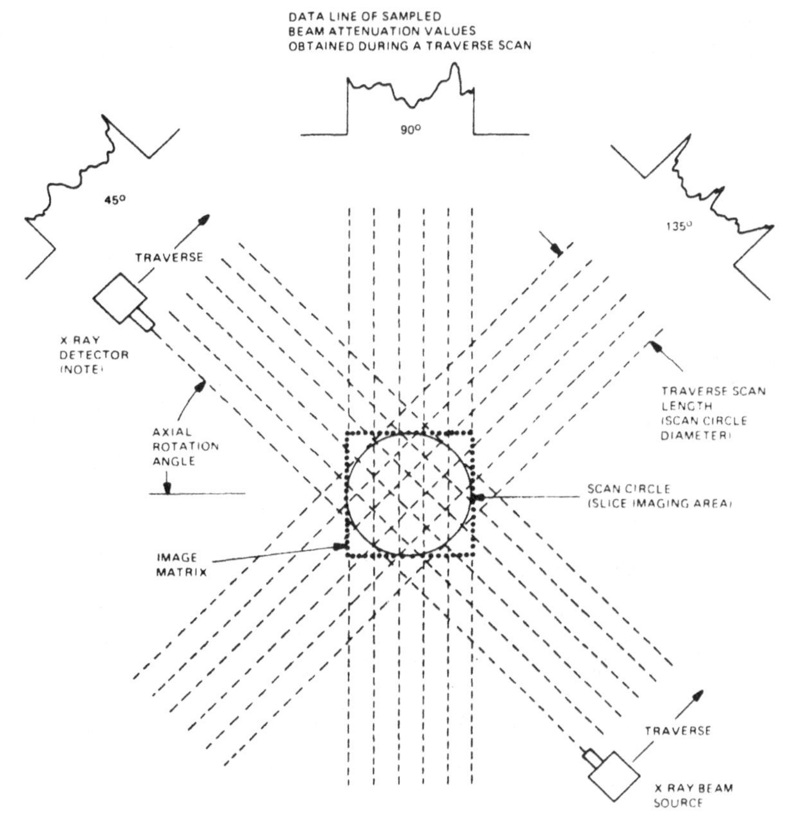

Figure 5. Principle of X-ray computed tomography.

The high demands on detection probability and resolution for the investigations of high performance ceramics can only partly be satisfied by medical CT scanners. With a modified CT scanner the smallest Voxel attainable is $(200 \times 200 \times 5000\ \mu m)^3$ (9). For high resolution X-ray CT with a Voxel size down to $(10\ \mu m)^3$, we have to use the microfocal X-ray technique with an image intensifier or an X-ray sensitive diode line as detection system (10,11). Figure 6 shows an inhomogeneous density area in a sample of green ceramic (diameter approximately 8 mm), and a crack. Two iron inclusions of approximately 50 µm diameter are situated at a distance of 1 mmm to the left of the crack. During the first sintering process the iron reacts with the surrounding matrix. A pore has developed surrounded by more absorbing material. The Pixel width and the sectional height are 24 µm. This highly resolving tomography can be regarded as sectional-image X-ray microscopy.

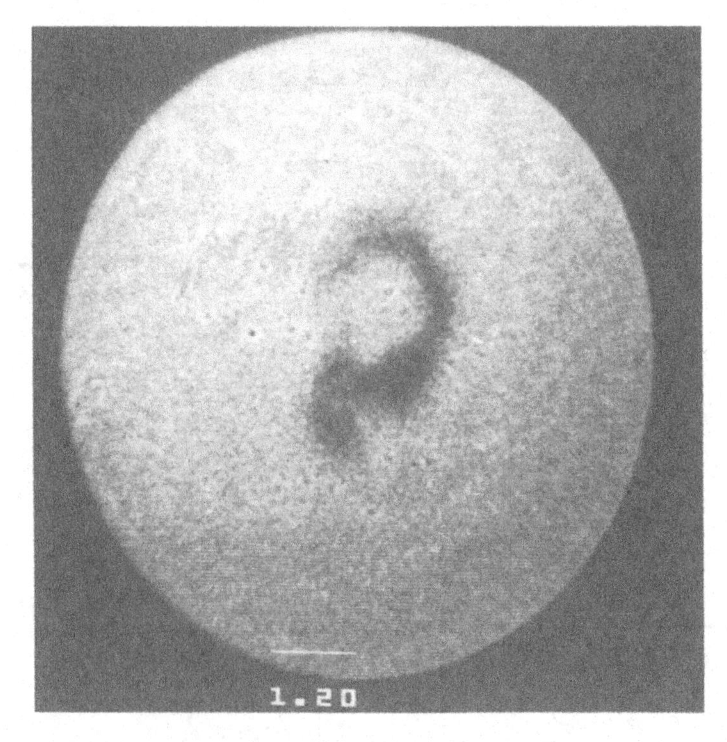

Figure 6. Micro X-ray computed tomogram of a green ceramic part (diameter 8 mm) with crack-like indication.

HIGH-FREQUENCY ULTRASOUND TECHNIQUE

C-Scan Imaging

In the IzfP a "C-Scan" imaging system has been set up specifically to investigate ceramic components. For the generation of ultrasound we have at our disposal a broad-band transmitter (USH 100 of Krautkrämer), generating an exponentially decaying electrical pulse of about 100 MHz band width, and a narrow-band high-frequency ultrasound apparatus (12) which has been developed in-house. The HF-US apparatus generates a carrier signal which can be adjusted between 10 and 200 MHz. The receiver signal, after rectification or peak detection is transformed into a C-Scan image that can be printed on a plotter (Figure 7). In this arrangement integration is obtained through the pulse form in a time interval given by a gate, and the obtained value is assigned to a colour by means of a software programme.

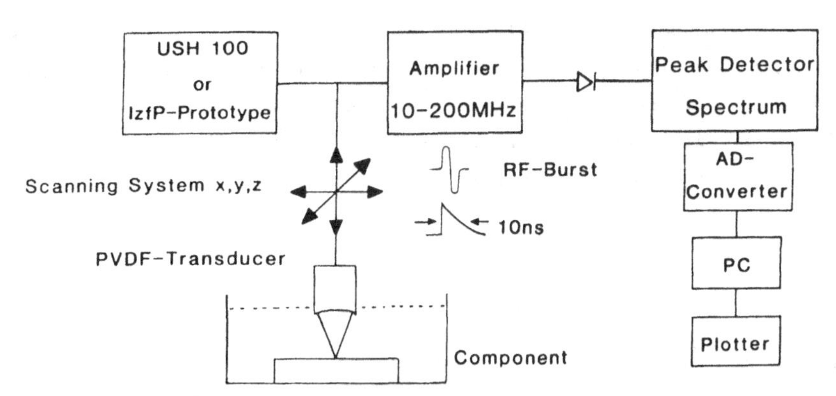

Figure 7. Block diagram of the system used for high-frequency ultrasonic C-Scan imaging.

An important element of the apparatus is the probes. In this case focusing PVDF and P(VDF-TrFE) probes of Krautkrämer have been used and also focusing probes of Panametrics with respectively differing radiuses of curvature. PVDF, a piezoelectric polymer is suited for use in immersion techniques because of its low acoustic impedance (12) which is similar to that of water. When using P(VDF-TrFE) probes, the transducer losses employing broadband electric impedance matching are even lower by 12 dB because of their higher electromechanical coupling factors (13,14,15). Focusing probes are also used in order to increase the signal-to-noise ratio (SNR). The 3 dB width d_{sc} of the probe is given by (16):

$$d_{sc} = 1.03\ \lambda\ Z/D \qquad (1)$$

where λ is the wavelength in the coupling medium, Z the focusing length, and D_c the transducer diameter. Inserting values for a frequency of about 30 MHz, yields for d_{sc} a value of 400 µm for a PVDF-probe having a radius of curvature of 25 mm and D_c = 3 mm. When, on the other hand, a focusing probe with a shorter focusing length is used (for example Panametrics V 390, 50 MHz mean frequency, Z = 0.5", D_c = 0.25"), the result at 50 MHz is d_{sc} = 60 µm. In this way finer structures also can be resolved, i.e. we have a higher lateral resolution capacity at out disposal. In contrast, the penetration capacity is reduced as verified through examples.

With broadband excitation the focusing spot is enlarged through the low-frequency parts of the frequency spectrum that is transmitted by a probe. Thus, the lateral resolution capacity is reduced. On the other hand, with narrow-band excitation the focusing spot is defined by the given frequency according to Eq. (1). As an example for the different lateral resolution obtained with the two excitation techniques, the C-Scan of a soldered layer of Si_3N_4 ceramic/steel sample was examined, where it was possible to distinguish between well soldered and badly soldered areas (12) similar to the one shown in Figure 8. The narrow-band excitation technique

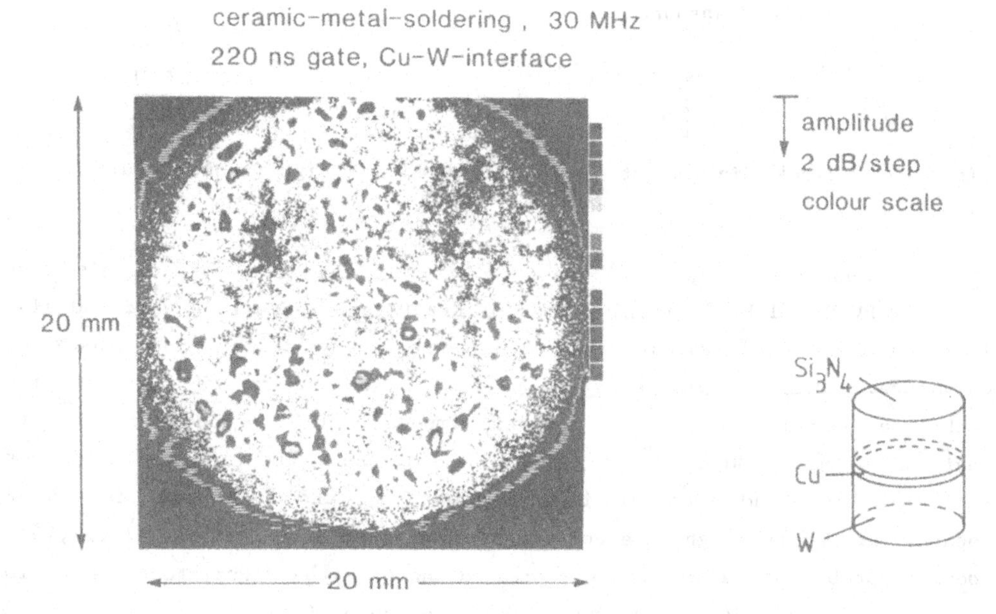

Figure 8. High-frequency ultrasonic inspection of a ceramic-metal joint.

reduces, however, the axial resolution capacity, because the pulse is longer. The dynamic range for narrow-band excitation for frequencies above 20 MHz is higher by at least 10 dB compared to broadband excitation because the voltage of the exponentially decaying broadband excitation pulse is decreasing with higher frequencies. For producing a C-Scan image, which at present is composed of 480 x 480 points, the sample is scanned in the form of a meander. The scanner used of COGENT has a position accuracy of about 10 μm. When taking the data bidirectionally the scanning duration is a few minutes for an area of 20 x 20 mm^2. This can be reduced considerably by using faster scanning systems, and we plan to use robotic systems in future.

Ceramic as well as powder-metallurgically produced components were investigated. For ceramic components with an ultrasound attenuation of about 1 dB/cm the smallest detectable flaw size is 30 μm diameter in 3 mm depth when using P(VDF-TrFE) probes with a centre frequency of 35 MHz (12). Flaws of this size could also be detected in powder-metallurgically produced samples with a SNR of 12 dB.

On SiC and Si_3N_4 samples with defined flaw sizes the SNR was determined for different types of flaws (12). On the basis of the backscattered amplitude no statement can be made on the type of flaw, as the values for the different types of flaws hardly differed from one another. Flaws that are smaller than 100 μm are within the Rayleigh scattering regime for an excitation frequency of 50 MHz and thus a wavelength of about 250 μm in ceramics. Here, the amplitudes that are backscattered by the flaws are $\alpha\ a^3$ and are similar for different flaws (17,18), and it is therefore difficult to determine the type of flaw. It has been proposed to use the Born approximation to determine the flaw size of different types of flaws (19) even if the requirements for this procedure (i.e. weak scatter) are not met. We found that this can lead to quite unreliable results (20).

Volume Representation of Ultrasonic Data

Analysing large volume data sets by loading separate C- or B-Scan extracts from a file to the graphic screen would be time-consuming and would make an interpretation of the relative position of features in different planes difficult. Advanced volume visualization software packages have become available which allow an efficient, interactive and real-time representation of volume data (20). We are employing a SUN 370 SPARC workstation with a TAAC-1 graphic processors (application accelerator), Figure 9. The graphic processor is equipped with 8 Mbytes of internal memory and allows 2-D and

3-D addressing modes for fast volume data access. The video output of the processor is inserted by hardware into a window on the graphic screen of the workstation. The measurement data are transferred by an Ethernet link from the PC to the workstation. Different visualization techniques are available, for which examples are given in the following.

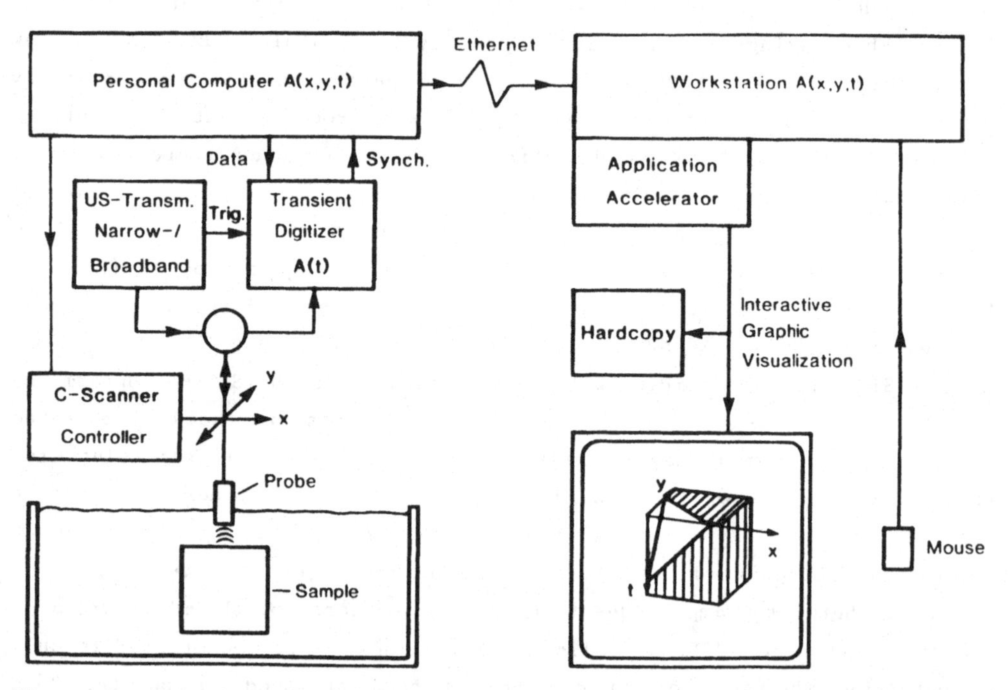

Figure 9. Schematic diagram of the high-frequency ultrasound equipment used for acquisition and visualization of ultrasound volume data.

As an example we present a volume measurement of a slab of a ZrO_2-ceramic (50 mm x 50 mm, thickness 4 mm), which has been measured by using a 50 MHz polymer probe head and broadband excitation. The sample has three grooves of 0.5 mm depth at its back side. In the left halves of Figure 10, the measured volume is represented by a cube in perspective view. The left surface of the cube corresponds to the 50 mm x 50 mm xy-scanning plane, the axis going from left to right is the time (or depth) axis, with a length of 2 μs. Ultrasound was incident from the left side.

In the slicing technique, the absolute amplitude values are represented by a colour scale. At the outer surfaces of the cube, the entrance echo and the backwall echo are visible at the left or right side of the front

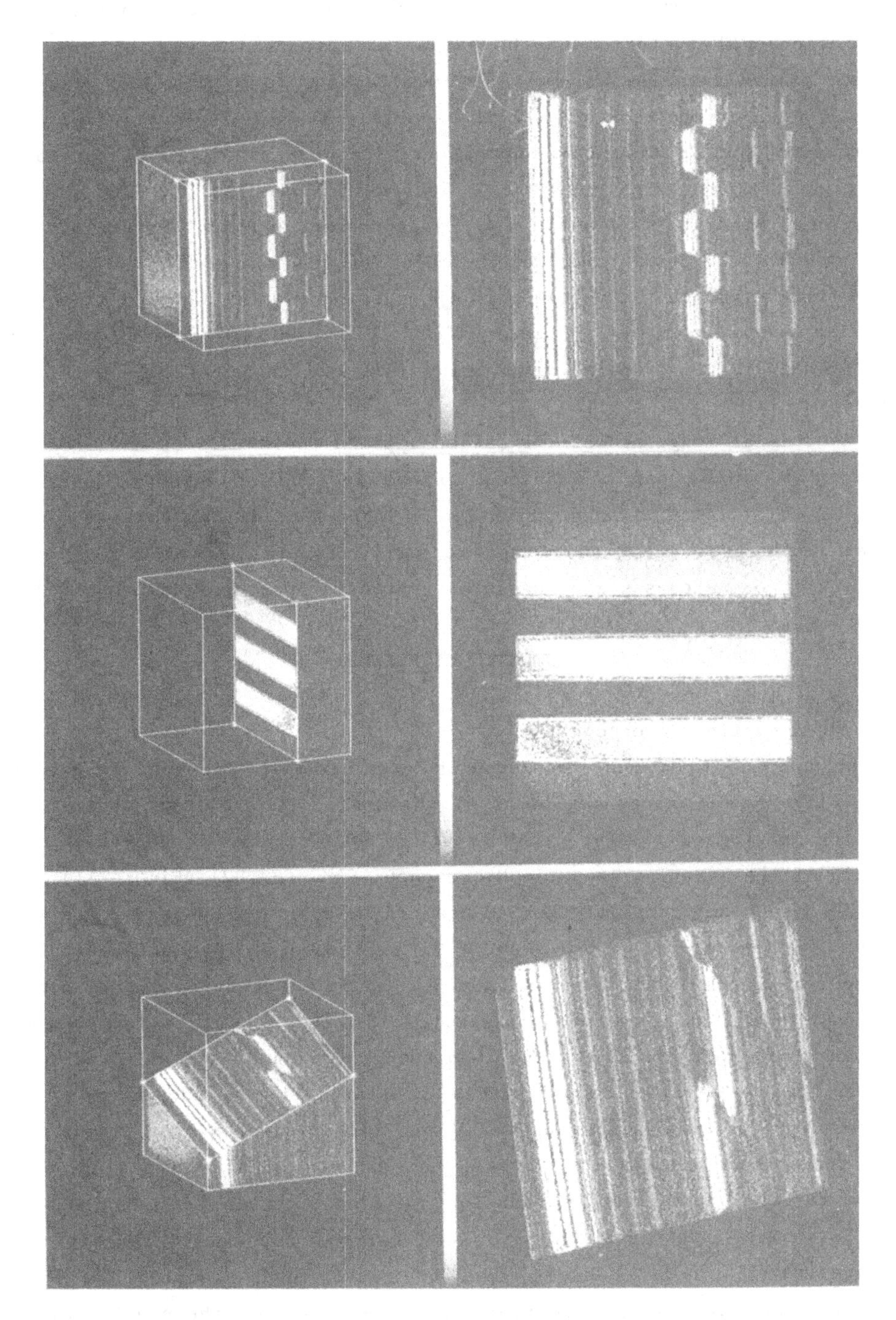

Figure 10. Ultrasound data representation of a ZrO_2-ceramic by using the slicing technique: (a) B-image (upper part), (b) C-image (middle part), (c) image under an arbitrary cutting angle (lower part).

surface, respectively. The cube can be turned by mouse control in any orientation. The interior of the sample is inspected by moving one of the surface planes into the volume. Figure 10a shows a B-image cut making the three grooves on the back side of the sample visible. The cutting plane can be extracted and is displayed separately without perspective distortions in the right halves of Figure 10. Turning around the cutting plane by 90 degrees immediately results in a C-image representation (Figure 10b). In a short time all possible B- and C-images of the sample can be studied by "travelling" through the volume with the cutting plane. Cutting is possible under arbitrary angles (Figure 10c), allowing correction for small misalignment of the sample or to bring flat and inclined defects into the image plane.

Additional image processing can be applied to the slices obtained. The measurement volume can be reduced to the regions of interest by several cuts. A cursor permits quantification of defect positions as well as local signal amplitudes. Furthermore, we will use in future reconstruction algorithms for defects sizing based on Synthetic Aperture Techniques (21).

ACOUSTIC MICROSCOPY

Scanning Acoustic Microscopy

In the surface treatment of ceramics (for example lapping or grinding) damage like crack traces can occur at the surface which under strain can be starting points for further crack growth. Such cracks can be detected with the aid of the Scanning Acoustic Microscope (SAM) (22,32) (Figure 11).

The frequency range of the SAM used is between 100 MHz and 2 GHz with a lateral resolution of ∿ 15-0.5 µm. The influence of the surface roughness of the ceramic is troublesome concerning the interpretation of the acoustic images. As a result we found that with SiSiC and Al_2O_3 the lapped samples show shorter cracks on an average than the ground samples, while the density of the crack remains comparatively constant, i.e. independent of the treatment technique; whereas with ZrO_2 the lapped samples show longer cracks than the ground samples. Likewise, the density of the cracks is clearly larger for the lapped samples than for the ground ones.

Extensive investigations have been carried out concerning the stable propagation of cracks, which aimed at better understanding the R-curve effect in ceramics (24) which describes an increase of the crack resistance with increasing crack length. This is more a question of fracture mechanics

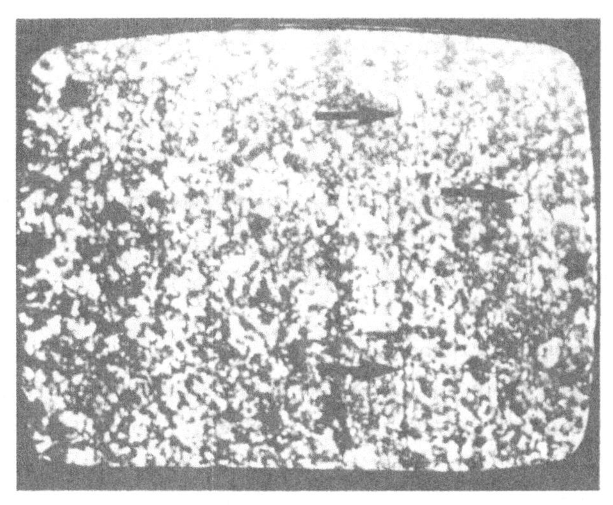

Figure 11. SAM-image of a ground SiSiC plate with surface cracks (arrows); field of view 750 x 500 μm^2.

than of quality assurance of technical ceramics, but is briefly mentioned here in order to demonstrate the possibilities of SAM. For understanding the R-curve different models have been developed which explain the additional energy consumption that occurs with stable crack growth (25-27). In order to record the process occurring during loading, a three-point bending device for investigations in the acoustic microscope has been constructed with which in situ measurements at the crack flanks and observations of the crack growth were carried out. In addition, we recorded V(z) curves (28,29) in front of the crack tip and as a function of the load, which made possible a determination of the elastic properties in a range limited by the focusing spot (-10 µm). The load delivery was adjusted in such a way that stable crack growth occurred. When comparing the crack process during loading and after unloading, it can be discerned that in the range behind the crack tip processes occur that contribute to an additional energy consumption with crack growth. For example after unloading an additional crack path can be formed. Thus the two crack flanks remain connected by this "bridge" and additional energy is consumed due to interlacing and friction of the crack flanks ("bridging") (30). Concerning the V(z) measurements a modification of the periodicity has been established from which a modification of the surface modulus G_R can be concluded (31,32). After controlled crack growth the acoustic lens was moved to the crack tip and then the different curves were recorded as a function of the load. Under loads up to 10% G_R modulus modifications were observed at the crack tip. The G_R modulus in front of the unloaded crack tip was, however, equal to that of the other material.

Summarizing these results: according to the microcrack model (24,25) a reduction of the elastic modulus can indeed by observed around the crack tip. Additional effects can be observed whose interpretation is dependent on the crack flank model (27).

"Scanning Laser Acoustic Microscopy"

In the case of the "Scanning Laser Acoustic Microscope" (SLAM) the object investigated is being subjected to plane waves, Figure 12a (33). The sound waves generate displacements on the sample surface which are modulated by scattered waves produced by flaws inside the object or at the surface, respectively. A laser system scanning with video frequency is being used for the imaging of the signals onto a TV screen. The microscope works in the frequency range of 10 to 500 MHz.

Figure 12b shows an example of the detection of a surface 70 μm pore, in silicon nitride, at an investigation frequency of 100 MHz. It should also be mentioned that SLAM images are acoustic holograms. Thus it is possible to apply reconstruction algorithms in order to reconstruct the flaw geometry under certain conditions (34).

SURFACE WAVES

For the investigation of near surface zones surface waves can be employed with good results. They have a penetration depth of about one wave length. Because of the energy concentration a very high detection capacity can be achieved with them for flaws that are smaller than the ultrasound wave itself. Therefore, relatively low frequencies can be employed; thus the ultrasound attenuation due to scattering is smaller and the wave propagation is less affected by surface roughness.

DYE PENETRATION TECHNIQUES

For NDT of the surface, careful thorough visual investigation is the technique used most, possibly backed up by means of dye penetration. Figure 13 shows a corresponding example. By means of fluorescent colour penetration agents pore-like indications and cracks can be detected in the

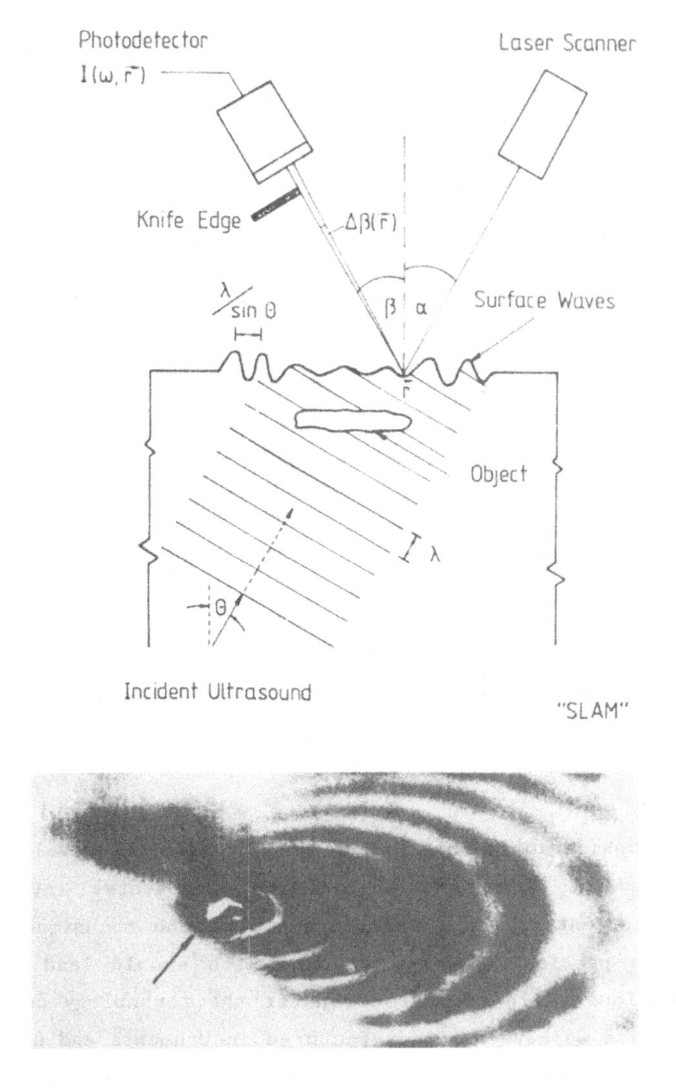

Figure 12. Above: Principle of Scanning Laser Acoustic Microscopy. Below: Acoustic micrograph of a pore in HPSiC; 100 MHz.

blade area, as well as a crack in the transition area between the axis and the hub of a radial turbine rotor. It is true, though, that the colour penetration agents available on the market are not optimal for ceramic materials. Dye penetration testing can make visible the defects in relation with the surface (for example cracks) and quantify their dimensions at the surface (for example crack length) but it does not permit any statement about their depth.

Figure 13. Dye penetration testing. Suitable chemical agents penetrate into open cracks and pores at the surface and make them visible by fluorescenting in ultraviolet light (arrows).

CONCLUSION

In conclusion, considerable progress has been made in the development of NDT for quality assurance of ceramic engineering components. At present the efforts in R&D concentrate on the improvement of the techniques discussed. This concerns in particular software work which should lead to automated inspection techniques. In order to transfer the technology developed into industrial practise we have also manufactured instruments and software based on the research described here. These instruments are either manufactured by us or by industrial partners and are commercially available.

ACKNOWLEDGEMENT

This work was supported by various research grants from the German Ministry of Science and Technology (Materials Research Programme) and by the European Community. We are also indebted to the companies Krautkrämer-Brenson, Cologne, Leica, Wetzlar, and Philips, Hamburg for support and collaboration.

REFERENCES

1. Ziegler, G., Technische Mitteilungen, **80** (1987) 203.

2. Evans, A.G., in: "Progress in Nitrogen Ceramics", Ed. R.F. Riley, Martinus Nijhoff Publ., Boston (1983) A595.

3. Munz, D., Rosenfelder, O., Goebbels, K. and Reiter, H., in: "Fracture Mechanics of Ceramics", **7**, Eds. R.C. Bradt, A.G. Evans, D.P.H. Hasselman and F.F. Lange, Plenum Press, New York (1986) 265.

4. Marshall, D.B., in: "Progress in Nitrogen Ceramics", Ed. R.F. Riley, Martinus Nijhoff Publ., Boston (1983) 635.

5. Herzer, R. and Schneider,E., Proc. 3rd Int. Symp. Mat. Charact., Eds. P. Höller, et al., Springer, Berlin (1989) 673.

6. Panakkal, J.P., Willems, H. and Arnold, W., J. Mat. Sci., **25** (1990) 1397.

7. Klein, P., Staatsexamensarbeit, University and IzfP Saarbrücken, (1989), Unpublished.

8. Sharpe, R.S. and Parish, R.W., in: "Microfocal Radiography", Ed. R.V. Ely, Academic Press, London (1980) 43.

9. Reiter, H. and Vontz, T., DGZfP-Proceedings, **10**, Lindau (1987) 630.

10. Maisl, M., Reiter, H. and Höller, P., J. Eng. Mat. Tech., **112** (1990) 223.

11. Maisl, M. and Reiter, H., 12th World Conference on Nondestructive Testing, Eds. J. Bogaard and G.M. van Dijk, Elsevier Science Publishers, B.V., Amsterdam (1989) 1667.

12. Pangraz, S., Simon, H., Herzer, R. and Arnold, W., Proc. 18th Int. Symposium Acoustical Imaging, Ed. H. Lee, Plenum Press (1991), To be published.

13. Ohigashi, H., J. Appl. Phys., **47** (1976) 949.

14. Ohigashi, H. and Koga, K., Jap. J. Appl. Phys., **21** (1982) L445.

15. Pangraz, S. and Arnold, W., Ferroelectr., **93** (1989) 251.

16. Gilmore, R.S., Tam, K.C., Young, J.D. and Howard, D.R., Phil Trans. R. Soc. Lond., **A320** (1986) 215.

17. Ermolov, I.N., Non-Destr. Test., **5** (1972) 87.

18. Hirsekorn, S., IzfP-Report No. 790218-TW (1979), Unpublished.

19. Thompson, D.O., et al., Rev. Sci. Instr., **57** (1986) 3089.

20. Netzelmann, U., Herzer, R., Stolz, H. and Arnold, W., Proc. 19th Int. Symp. on Acoustical Imaging, Bochum 1991, To be published.

21. Müller, W., Schmitz, V. and Schäfer, G., in: "Proc. 3rd German-Japanese Seminar Research Struct. Strength and NDE in Nucl. Eng.", (1985) No. 2.5.

22. See for example Atalar, A. and Hoppe, M., Rev. Sci. Instrum., **57** (1986) 2568.

23. Matthei, E., Vetters, H. and Mayr, P., DGZfP-Proceedings, **13** (1989) 231.

24. Buresch, F.E., Mat. Sci. and Eng., **71** (1985) 187.

25. Evans, A.G. and Faber, K.T., J. Am. Cer. Soc., **67** (1984) 255.

26. Buresch, F.E., Materialprüfung, **29** (1987) 261.

27. Knehans, R. and Steinbrech, R., Fort. deutsch. ker. Ges., **1** (1985) 59.

28. Briggs, A., "An Introduction to Scanning Acoustic Microscopy", Oxford University Press (1985), and references contained therein.

29. Weglein, R.D., Appl. Phys. Lett., **34** (1979) 179.

30. Swanson, P.L., Fairbanks, C.J., Lawn, B.R., Mai Yiu-Wing and Hockey, B.J., J. Am. Ceram. Soc., **70** (1987) 279, and ibid. (1987) 290.

31. Quinten, A. and Arnold, W., Mat. Sci. Eng., **A122** (1989) 15.

32. Pangraz, S., Babilon, A. and Arnold, W., Proc. 19th Int. Symp. on Acoustical Imaging, Bochum 1991, To be published.

33. Reiter, H. and Arnold, W., Beitr. Elektronenmikroskop. Direktabb. Oberfl., **17** (1984) 129.

34. Morsch, A. and Arnold, W., Proc. 12th World Conf. on NDT, Eds. J. Bogaard, G.M. van Dijk, Elsevier Science Publishers, (1989) 1617.

THE PROCESSING OF STRUCTURAL CERAMICS

R.J. BROOK*
Max Planck Institute for Metals Research,
D7000 Stuttgart 80,
Germany.

ABSTRACT

The prospects for the cost effective and reliable processing of ceramics have improved dramatically as a consequence of the progress that has been made in refining experimental methods (powder quality, clean rooms, sophisticated characterisation) and in increasing theoretical capability (computer modelling of complex processes); the fact that the two approaches can now work in effective synergy has greatly improved the opportunities for systematic innovation. Examples can be found in powder design, in chemical control and in proposals for near net shape processes and powder-free processing.

INTRODUCTION

The processing of structural ceramics has now been recognised as one of the key technologies; the promise of ceramics as represented by their properties and their suitability for applications can only be realised if successful paths to their fabrication can be identified. Success in this context requires the reliable production of homogeneous and flaw-free components at cost levels which are often dictated by those of rival materials.

The relatively modest pace at which ceramics are managing to find a place in structural applications can to a large extent be attributed to the processing problem. The brittle nature of ceramics and their resulting flaw sensitivity means that high standards of structural perfection must be attained; at the levels of toughness typical for monolithic (unreinforced) microstructures (< 3 MPa $m^{\frac{1}{2}}$), flaws of 30 μm or more in size are already

* Now with Department of Materials, University of Oxford, OX1 3PH, UK.

detrimental at the stress levels required for structural components (300 MPa). Attempts can be made to avoid this by using, for example, fibre or whisker reinforcement to enhance the toughness; the processing of such composite microstructures can, however, bring additional difficulties as a consequence of their heterogeneous nature and flaw sizes are commonly increased. There is therefore no avoiding the processing problem as ways are sought to bring ceramics to the markets which their properties would justify. Fortunately, the developments of recent years have indicated the levels of progress that can be made and the promise for further advance remains good. The purpose of this short review is to summarise some of these developments.

EXPERIMENT: SIMPLIFICATION AND PRECISION

The complexity of the basic ceramic process, namely the conversion of a powder into a solid component, has made it difficult to achieve total control and hence reliability in the end product. The original studies of densification mechanisms in powder compacts have been fruitful in identifying driving forces and the contributing mechanisms and therefore in indicating the importance of such variables as powder particle size, temperature and, at a later stage, the character of additives. A difficulty has been, however, that this understanding relates to the mean or average behaviour of the system and not to the exceptions or accidents, for example, at the extreme of the particle size distribution, which determine the occurrence of the larger structural faults or flaws.

In a similar way, the earlier work gave attention to characterisation methods which tended to average out behaviour, either because methods were used such as dilatometry which by their very nature consider the total sample or because the resolution in the methods of direct observation was often, as in optical microscopy, relatively limited.

Progress in both these aspects has resulted in dramatic property improvements (strength, toughness) and in greater understanding of the physical basis on which such improvements have been won. Two notable causes for the introduction of flaws into components lay in the existence of exceptions or accidents in the powders that were used or in the accidental contamination of the system during processing. The great improvements in powder quality that have been achieved by the introduction of chemical preparation methods (1) have brought about corresponding changes in the final products. Control of powder shape, size distribution, chemical homogeneity and degree of agglomeration has limited the ability of accidents in the powders to produce corresponding flaws in components. Similarly, the use of clean processing or, at the least, the curtailment of the more common forms of contamination by making use of deionised water, polymer vessels or by using gloves during handling, has limited the occurrence of accidents introduced from the environment (2). These two procedures, simple in concept but more complex in practice, have the effect that the average behaviour of the material during processing becomes a reliable measure of its significant behaviour since the accidents responsible for flaw origins have been avoided. Changes in processing parameters can then be more confidently interpreted and process optimisation more systematically conducted.

The improvements in the resolution of characterisation methods both passive as in transmission electron microscopy or in surface analysis and active as in the use of lithography to prepare desired model microstructures (3) or to probe the performance of specific microstructural features (the electrical resistance of a single interface (4)), have been critical in ensuring that mechanistic or kinetic arguments can be more accurately focused onto the phenomena known to exist. The early widespread use of solid state sintering arguments to systems which have not always been free from amorphous or, at the sintering temperature, liquid grain boundary phases is an example. The current tendency to process materials with the deliberate intent of producing a given microstructural configuration (followed by close characterisation of the result to ensure that correspondence with the target has in fact been achieved) is a firm and persuasive footing for subsequent kinetic analysis and comparison with models.

THEORY: PERTINENCE AND COMPLEXITY

The early stages of the modelling of processing events such as drying or sintering consisted of the analysis of simple and idealised geometries considered to represent the ceramic system. As noted earlier, these were highly successful in providing qualitative accounts for the phenomena involved and in identifying the process variables and their influence. More recently, progress has been made by studying systems deliberately fabricated to provide a close correspondence between the theoretical analysis and the real system. The use of monodisperse, spherical powders, even if, in retrospect, technologically difficult to justify, has the advantage that it corresponds to the conventional powder models. Similarly, the use of polymer inclusions of known size distribution to yield ceramic pieces with known pore size distribution where the influence of given distributions can then be modelled (5) has succeeded in producing unambiguous results relating to process mechanisms. Modelling what is known by measurement to exist has proved in this instance more effective than modelling what has been assumed to exist.

A second ground for progress has been the availability of more sophisticated computer codes allowing the modelling of more complex systems. The Monte Carlo studies of grain growth under different sets of conditions (6) and more recently of sintering (7) have been helpful in showing the consequences that arise in terms of microstructural development from basic process assumptions. The indication of stability in grain size distributions in clean systems free from boundary phase where relatively isotropic grain boundary energies can then be expected has thrown into question one of the conventional assumptions in processing, namely that narrow size distribution powders must be used.

The current position is one where theoretical and experimental contributions can now work together with very much more effect than has been the case in the past. As represented in Figure 1, initial studies were plagued by the fact that experimental systems were complex (risk of contamination, diversity of powder size and shape) and indistinctly characterised (presence or absence of grain boundary phases, degree of segregation of dopant or impurity ions); the theories, in contrast, mostly dealt in terms of simplified and idealised models. The consequences were that there was limited opportunity for close interaction between the two contributions and conclusions remained for the most part both qualitative and ambiguous. With growing control of the experimental conditions, however, and with better characterisation coupled with the growing capability of the theoretical models, the two approaches have been able to come together and, as indicated in the examples noted earlier, there is now an excellent opportunity for the close matching of experimental and theoretical contributions so that deliberate and systematic innovation can take place.

The future trend as indicated in the figure is for the design of microstructures which are more precisely fitted to the intended applications. New degrees of planned complexity (composites, controlled grain boundary phase structure, multifunction monolithic components) can be introduced on the basis of the sound and quantitative understanding that has been gained. This is a most significant change; it has occurred through advances made in disparate fields (chemical processing, physical methods of characterisation, computer modelling capability) and the conditions for rapid progress in processing are as a consequence unusually favourable.

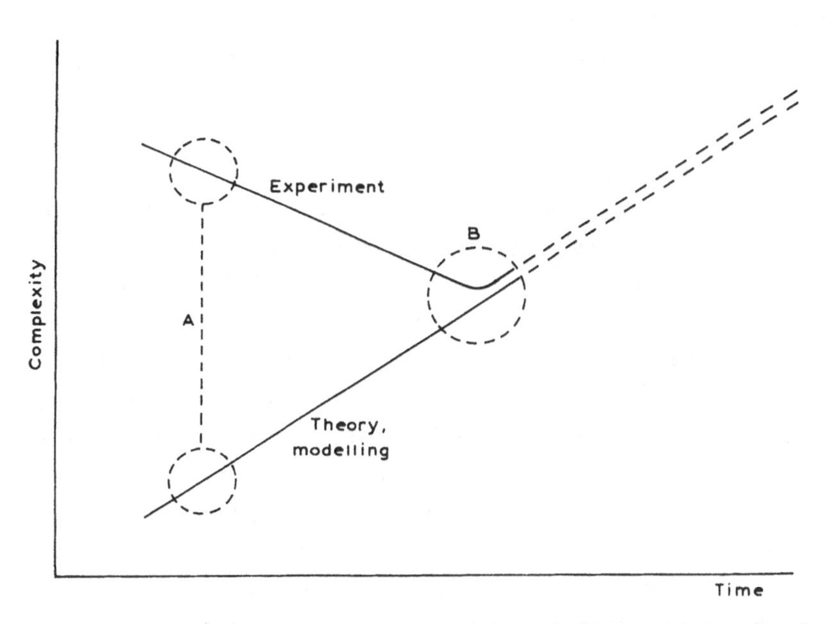

Figure 1. Initial studies of processing (A) were hampered by the fact that the experiments were insufficiently refined to match the idealised and simplified pictures that could be modelled. The current position (B) allows for synergy between the two approaches so that a basis for the design of precisely defined microstructures is now available.

DEVELOPMENT DIRECTIONS

Aside from the general pattern for work as outlined above, there are a number of more specific themes that can be identified. Brief notes on a selection of them are given in the following paragraphs.

Complexity in Powders

Great progress has been made by removing structural accidents from powders and by making available a range of compositions (Al_2O_3, ZrO_2, Si_3N_4, SiC,

AlN, $BaTiO_3$) in the form of equiaxed, narrow size distribution, chemically homogeneous particulates. With this step accomplished, there is now the opportunity to construct powder feeds more specifically designed to match processing requirements. The use of coated powders to provide an optimised distribution of sintering aids or an enhanced microstructural stability (8) is one instance. The use of wider size distributions to achieve better packing densities in pure systems, where on the basis of the computer modelling the abnormal grain growth problem is not expected, is another. Coating processes are available without excessive cost penalties and the opportunities for this aspect of powder development provide a clear research direction.

Grain Boundary Chemistry

The precision of grain boundary characterisation as for example in TEM observations of intergranular phases or in surface analysis observations of segregation in single phase systems has allowed a closer identification of the roles of additives during ceramic processing. This at first led to a recognition of the enormous potential for complexity in additive function, but with more widespread use and with systematic comparative studies, the detailed picture can be established. This then allows the exploration of concepts such as the equilibrium boundary phase thickness (9) or the influence of charge and size differences on segregation (10); it has led to renewed consideration of grain boundary engineering methods such as volatilisation of second phases (11). The development of structural ceramics for use in high temperature, high stress and corrosive conditions will require taking full advantage of the opportunities now available for grain boundary design.

Near Net Shaping

The substantial fraction of total product costs that can be associated with machining and the great advantages that arise during the fabrication of composites when matrix shrinkage can be avoided have led to wide recognition of the importance of near net shaping. This explains the interest in such methods as injection moulding (12); it also emphasizes the need for densification with minimal shrinkage. This has always been an attraction of reaction bonded silicon nitride. The possibility is also exploited in the proposals for reaction bonded aluminium oxide (13). Here mixtures of alumina and aluminium powders are shaped and then fired in air; the resulting oxidation of the aluminium fraction provides an expansion which can be tuned to compensate for the shrinkage from pore elimination. The combination of densification processes and chemical change (self propagating reactions are a further example (14)) offers a promising line of development provided that sufficient homogeneity can be retained.

Powder-free Processing

One response to the difficulties posed by powders has been to seek fabrication methods for ceramics which avoid totally the use of powders. Examples are the directed metal oxidation process (15) or the methods involving sol/gel reactions for oxide systems or organic precursors for non-oxide systems. Progress with such methods has been hindered by the very large shrinkages encountered in passing from the low density starting material (1000 kg m^{-3}) to the ceramic product with a density lying typically in the range 3000-6000 kg m^{-3}. A consequence has been that the methods have been most fruitful when applied to the preparation of low-dimensional forms such as films, fibres or powders. Opportunities exist, however, either for hybrid processes where classical powder methods and the newer processes are combined or where compensating reactions can be proposed to limit the extent of shrinkage. This is the concept behind the so-called AFCOP process (active filler controlled polymer pyrolysis) where the incorporation, for example, of titanium particulates in an organic precursor yields titanium carbide on reaction with the precursor breakdown species. Again, with suitable choice of reactants, the total shrinkage can be brought to low levels (16). The possibilities of combinations of methods, as indicated by this example, are many; an important step will be the study of the intermediate stages, both chemical and structural, that arise on passing from the initial condition to the product. This is a large task but the promise that lies in processing sequences precisely matched to performance and cost requirements suggest that it will be rewarding.

REFERENCES

1. Segal, D., 'Chemical Synthesis of Advanced Ceramic Materials', Cambridge University Press, Cambridge (1989).

2. Alcock, J.R. and Riley, F.L., 'An Investigation of Dust Particles found in a Ceramic Processing Environment', J. Eur. Ceram. Soc., **6** (1990) 339-350.

3. Rödel, J. and Glaeser, A.M., 'Post Drag and Pore-boundary Separation in Alumina', J. Am. Ceram. Soc., **73** (1990) 3302-12.

4. Olsson, E., 'Interfacial Microstructure in ZnO Varistor Materials', Ph.D. Thesis, Chalmers University of Technology, Göteborg (1988).

5. Zhao, J.H. and Harmer, M.P., 'Effect of Pore Distribution on Microstructure Development: First and Second Generation Pores', J. Am. Ceram. Soc., **71** (1988) 530-9.

6. Srolovitz, D.J., Grest, G.S. and Anderson, M.P., 'Computer Simulation of Grain Growth: Abnormal Grain Growth', Acta Met., **33** (1985) 2233-47.

7. Hassold, G.N., Wei Chen, I. and Srolovitz, D.R., 'Computer Simulation of Final Stage Sintering: Model, Kinetics and Microstructure', J. Am. Ceram. Soc., **73** (1990) 2857-64.

8. Dransfield, G.P., Fothergill, K.A. and Egerton, T.A., 'The Use of Plasma Synthesis and Pigment Coating Technology to Produce a Yttria Stabilized Zirconia having Superior Properties', in: "Euro. Ceramics, Vol. 1", Eds. G. de With, R.A. Terpstra and R. Metselaar, Elsevier, Barking, (1989).

9. Clarke, D.R., 'On the Equilibrium Thickness of Intergranular Glass Phases in Ceramic Materials', J. Am. Ceram. Soc., **70** (1987) 15-22.

10. Hwang, S.-L. and Wei Chen, I., 'Grain Size Control of Tetragonal Zirconia Polycrystals using the Space Charge Concept', J. Am. Ceram. Soc., **73** (1990) 3269-77.

11. Ruckmich, S., Kranzmann, A., Bischoff, E. and Brook, R.J., 'The Description of Microstructure Applied on Thermal Conductivity of AlN Substrate Material, Submitted to J. Europ. Ceram. Soc., (To be published).

12. Edirisinghe, M.J. and Evans, J.R.G., 'Review: Fabrication of Engineering Ceramics by Injection Moulding', Int. J. High Tech. Ceram., **2** (1986) 1-31, 249-78.

13. Claussen, N., Le, T.Y. and Wu, S.X., 'Low Shrinkage Reaction-Bonded Alumina', J. Eur. Ceram. Soc., **5** (1989) 29-35.

14. Munir, Z.A., 'Synthesis of High Temperature Materials by Self-Propagating Combustion Methods', Amer. Ceram. Bull., **67** (1988) 342.

15. Newkirk, M.S., Urquhart, A.W., Zwicker, H.R. and Brevel, E., 'Formation of Lanoxide Ceramic Composite Materials', J. Mater. Res., **1** (1986) 81.

16. Seibold, M. and Greil, P., 'Composite Ceramics from Polymer Metal Mixtures', in: Advanced Materials & Processes, Eds. H.E. Exner and V. Schumacher, DEM Informationsgesellschaft mbH, Oberursel (1990).

CERAMIC COMPOSITE DEVELOPMENT

P. LAMICQ, C. BONNET and S. CHATEIGNER
Société Européenne de Propulsion,
24 rue Salomon de Rothschild,
92150 Suresnes, France.

ABSTRACT

The background to ceramic matrix composite materials based on carbon or silicon carbide fibres in a silicon carbide matrix is presented. Some important applications are summarised such as components for rocket motors, turbojet engines and the space plane.

INTRODUCTION

A family of ceramic matrix composites (CMC) has been generated by a Chemical Vapour Infiltration process to deposit silicon carbide inside preforms of carbon or silicon carbide fibres. These CMC are able to sustain high temperatures with good strength retention.

The preforms are made with small diameter fibres (typically 10 to 35 µm) to keep the fibre stiffness low enough and be able to generate complex shapes by conventional composite fabrication routes.

Ceramic fibres are of microcrystalline structure, and exhibit grain growth at high temperature. This phenomenon generates defects and reduces strength which is a limiting factor to high temperature performance, especially for long term exposure.

The SiC matrix is well crystallized and behaves like a ceramic: thermally and chemically strong, stiff and with low strain to failure. So, the composite is quite different from polymer matrix composites, where fibres may deform to their maximum elongation, the polymer matrix having higher strain to failure than the fibres.

In a CMC, the ceramic matrix breaks first, and a large number of microcracks are generated. If the propagation of these cracks were as easy as in sintered ceramics, the composites would have poor strength and be of little interest. In fact, through a good choice of fibre-matrix interface it is possible to deflect and multiply the cracks so that they are dispersed in the bulk of the composite. Damage is then very progressive and needs very high energy absorption.

CMCs are strong, stiff and tough. They do not break catastrophically like sintered ceramics, as they are damage tolerant. They are much less sensitive to internal or surface defects, design errors or local over-stresses.

DEVELOPMENT OF CERAMIC MATRIX COMPOSITES

These new composites come from a common research programme implemented jointly by SEP, CNRS and University of Bordeaux 1, with a sponsorship by the French administration.

They were created towards the end of the seventies on a laboratory scale. They developed quite rapidly and were tested for a series of uses to become a major development for the Bordeaux plant of SEP. At first, the purpose of the programme was to improve the oxidation sensitivity of carbon-carbon composites, by means of silicon carbide, deposited by CVI. Unexpected mechanical results were observed - specifically a very high crack propagation resistance. Local damage can be accommodated with no catastrophic failure of the composite. A CMC bar can be bent without rupture, whereas a sintered ceramic breaks with no internal distortion.

After these preliminary results, various research programmes were implemented on a number of subjects: sounder understanding of basic phenomena, improvement of processes, development of other types of ceramics, fibres and matrix, industrial facilities for manufacturing of parts.

Only ten years after the first laboratory samples, CMC composites have been tested successfully as many parts subjected to very severe mechanical, chemical and thermal environments. They will be used in the SNECMA M88 tubro for the RAFALE fighter. They are manufactured with industrial facilities and large furnaces are now being commissioned for the aft parts of the Hermes European Space Plane.

APPLICATIONS

CMCs are well adapted to uses in high temperature oxidizing environments, such as all types of engine but also thermal structures for space planes. Some examples are given below:

Liquid Propellant Rocket Motors

Uncooled large scale exit cones can sustain the temperature of liquid oxygen/hydrogen combustion. Such a part has been manufactured and tested at the scale of HM7 motor, third stage of Ariane launcher, Figure 1. Two successive ground firings were performed on the same part, with full success: one for the normal burntime of 750 s, then a second one of 900 s with partially increased temperatures.

On smaller motors, complete assemblies of chamber and nozzle have been designed at various sizes and thrusts, from 5 N to 6000 N with N_2O_4/MMH biliquid propellant, Figure 2. Numerous tests have been performed, including long-term duration up to more than 50 h; pulses resulting in stress cycling, and cold starts after cooling resulting in thermal shock and thermal cycling. Results have demonstrated equal or better lifetime than the refractory alloys used before, while being achieved at higher temperatures, resulting in better performance.

Turbojet Engines

Several parts have been developed in co-operation with SNECMA, mainly for the afterburner area of the engine. The main advantages of CMCs are:

- Lower weight compared to metals, with a 40% reduction.
- Higher lifetime resulting in less maintenance, mainly for parts submitted to high thermal shock.

Actual flights have been performed on the MIRAGE 2000 using the M53 engine in 1989, Figure 3. Since then, the M88 engine for the RAFALE fighter has been flown with CMC parts, Figure 4.

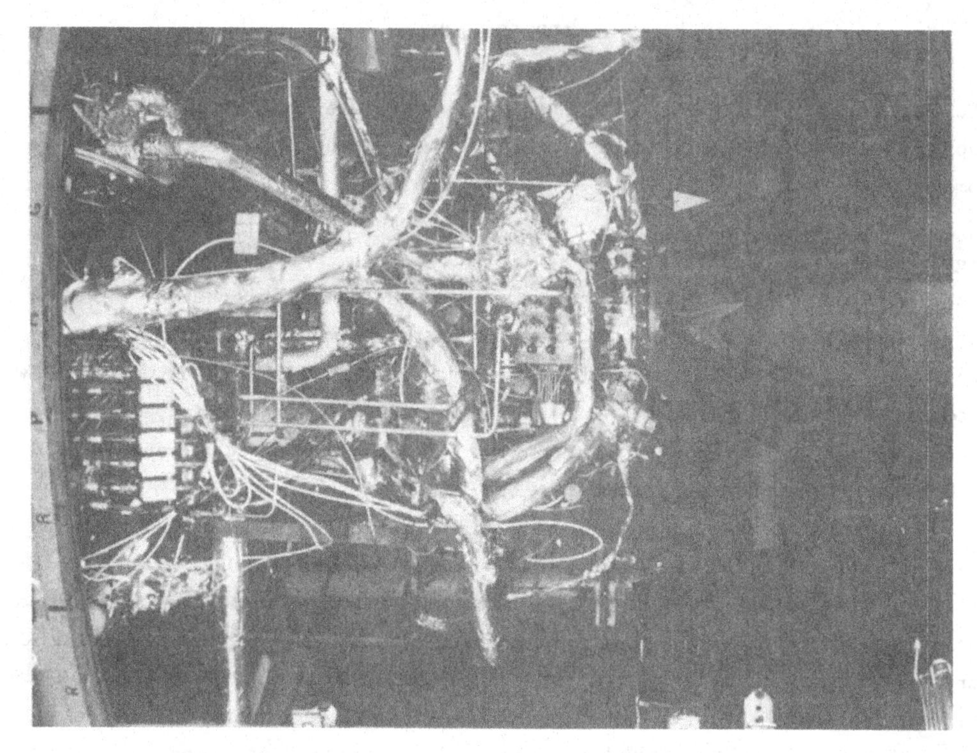

Figure 1. Silicon carbide exit cone on Ariane third stage engine.

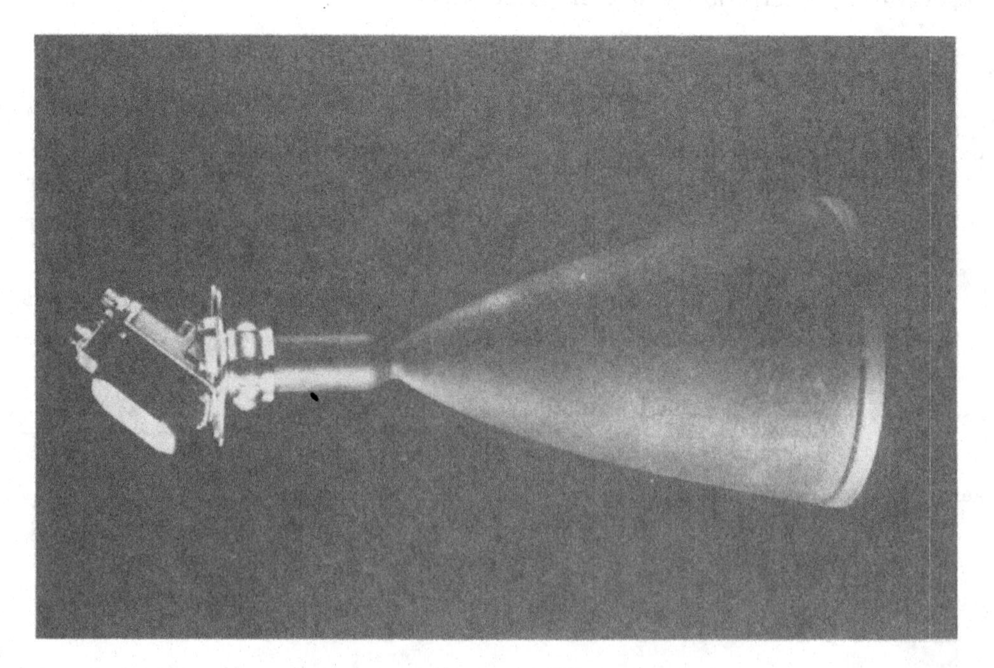

Figure 2. Liquid propellant 20 N motor for satellite. Combustion chamber and nozzle in SiC/SiC.

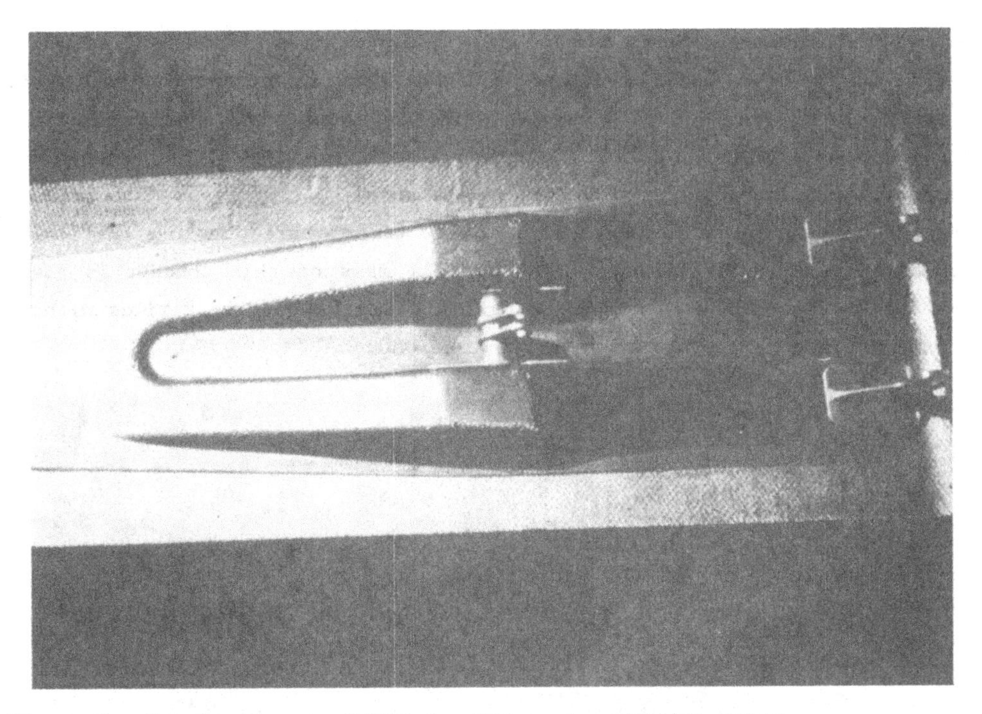

Figure 3. Nozzle flap in C/SiC for M53 engine (MIRAGE 2000).

Figure 4. Hot nozzle flap for M88 engine (RAFALE).

Thermal Structures for Space Plane

Several parts have been developed for the thermal protection system of Hermes Space Plane. Such a system involves panels of CMC at external surfaces of the plane, where the thermal fluxes are high. The aerodynamic shape of the plane is formed by the panel assembly, Figure 5. The primary structure of the space plane remains at low temperature due to a multilayer insulation of high efficiency. The overall mass of this concept is lower than the US Space Shuttle system using light weight but thick tiles without an external strong protective material like CMC.

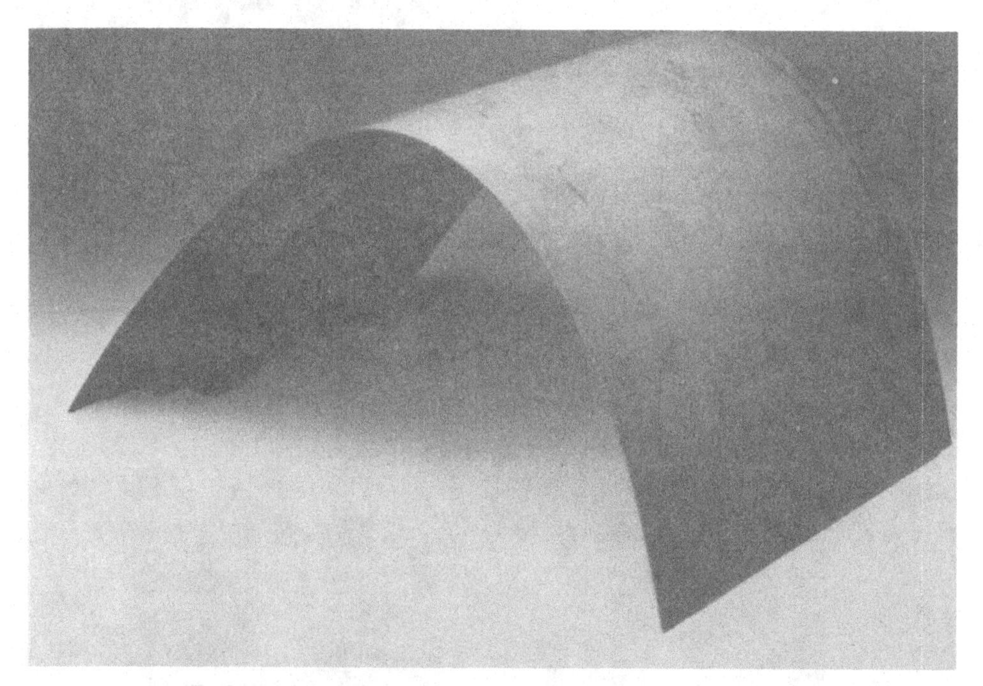

Figure 5. C/SiC leading edge for Hermes.

Several types of part have been designed for different locations in Hermes, Figure 6. The more spectacular ones are large rear parts like winglets, rudders, elevons and body flap. Dimensions may reach 2.50 x 1.80 m. A demonstration part is being made within the limits of the present manufacturing facility (1.80 x 0.80 x 0.30 m). Larger furnaces will soon be available for full scale parts.

Tests have been performed to demonstrate the performance of CMCs on ground simulation of re-entry heating. Specifically, a leading edge has been tested with oxidizing atmosphere in a solar furnace simulating the heat

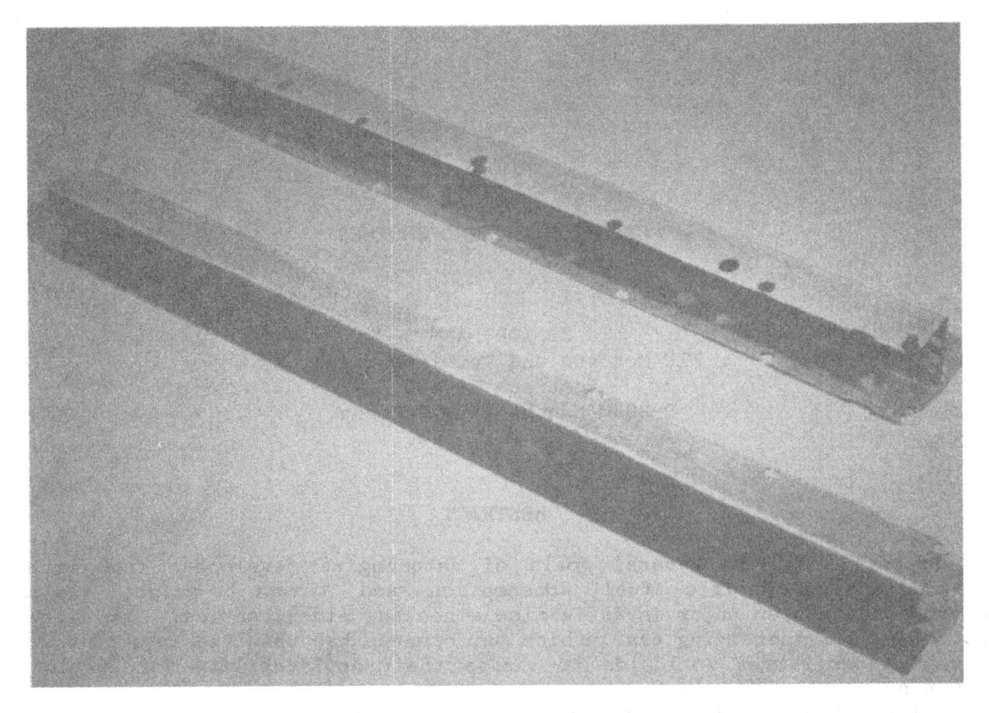

Figure 6. SiC/SiC stiffeners for large parts of Hermes.

fluxes and maximum temperature of 1550°C. Tension and compression loads were applied alternately during the tests. After 16 re-entry simulations, and then 2 tests at a higher temperature of 1700°C, the part was still able to sustain specified loads.

CONCLUSION

A family of Ceramic Matrix Composites has been developed. These carbon and ceramic composites are tough, and not brittle like sintered ceramics. Applications have been started in many areas, specifically in engines and the Hermes Space Plane. Industrial facilities are now available to prepare for the first series production. A large extension of CMC activities is anticipated.

CERAMICS IN AERO-ENGINES

J. WORTMANN
MTU Motoren und Turbinen Union,
Postfach 500640,
D-8000 München 50, Germany.

ABSTRACT

In keeping with the general goals of aero-engine development to improve significantly specific fuel consumption and thrust-to-weight ratios, ceramics have been under investigation since the mid-seventies. Their great potential for increasing gas-turbine performance has resulted in a number of research programmes worldwide to assess their applicability for static or rotating components.

There has been remarkable progress in improving the material strength, in particular by process control and the clean room approach. This development has been accompanied by revised design and lifeing methods. Thus static components such as hybrid vanes and combustor rings have withstood temperatures of < 1900K for several hundred hours and some ten cycles. On the other hand the essential aspect concerning all aero-engine materials of the need to guarantee defect tolerance, i.e. the ability to reduce stresses in a controlled manner, is lacking in the form normally associated with metals.

Attention is therefore focused on methods of controlling the inherent lack of defect tolerance of ceramic materials by appropriate design concepts (small volumes, simple shapes, constant wall thickness, compression stress state) and strengthening and toughening of the material itself by fibre reinforcement.

Apart from the ambitious application of monolithic or fibre-reinforced ceramics as structural components in the hot gas path in the highest temperature environment, there are applications such as thermal barrier coatings, bearings and flaps which are already in series production.

INTRODUCTION

Projected thermodynamic cycle temperatures for advanced aero-engines run 100K to 200K above present levels. Contemplated temperatures (> 2000°C) are in the stoichiometric combustion range. The objective - intended to benefit civil engines - is to boost specific power (thrust-to-weight ratio) and improve overall efficiency to reduce specific fuel consumption. Key importance here attaches to cooling air savings, turbine cycle calculations anticipating a 30% economy in specific fuel consumption if the need for turbine cooling is obviated.

Material development efforts, accordingly, are largely targeted at specific hot strength to give high cycle temperatures and efficiencies at moderate cooling requirements and low weights. This is where ceramic materials (Figure 1) hold promise; they possess high mechanical strength to elevated temperatures and are chemically stable. Having a mere one-third of the specific weight of superalloys, they are dimensionally stable, hard and wear-resistant, and they do not fall in the strategic materials category (1). Eligible component specimens for ceramic use in engine applications have been studied in depth in the U.S. DARPA/NAVSEA project in the seventies (Figure 2) (7).

ALTERNATIVE CERAMIC MATERIAL OPTIONS

The engine applications study deals largely with silicon compounds. Si_3N_4 and SiC derived from a variety of processes for diverse service temperatures have therefore been widely evaluated for suitability ever since the mid-seventies. Complementing this range of materials are fibre-reinforced ceramics (C/SiC and SiC/SiC) and C/C. Oxide ceramics, however, which uniquely have potential for use at stoichiometric combustion temperatures, have so far attracted very restricted attention in the investigations (2).

ASSESSMENT CRITERIA

High hot strength at low specific weight, and resistance in an aggressive environment are not the only characteristics guiding the selection of an aero-engine material. The various properties need judicious weighting to suit the respective stationary or rotating component for use either in a hot gas stream or in a more moderate environment. A combustor element primarily needs thermal shock and oxidation resistance; a turbine rotor blade also needs creep and fatigue strength; a bearing demands special tribological properties (1).

Demonstrated suitability for reproducible manufacture and non-destructive testing is of cardinal importance, as is a failure pattern that can be controlled and modelled - all properties that can be grouped under

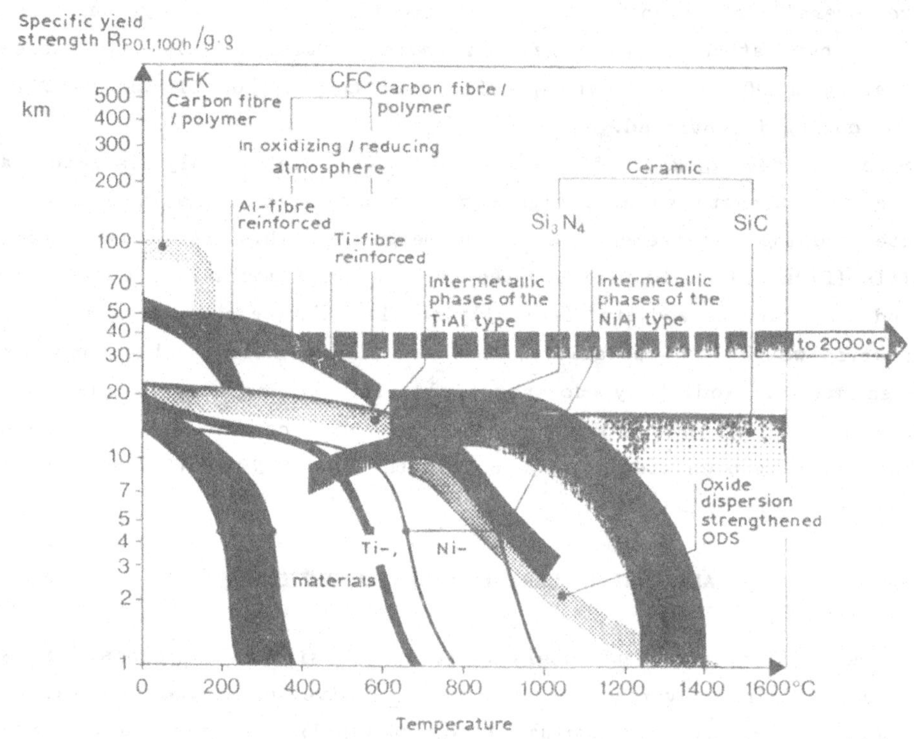

Figure 1. Specific yield strength of different materials for engine application.

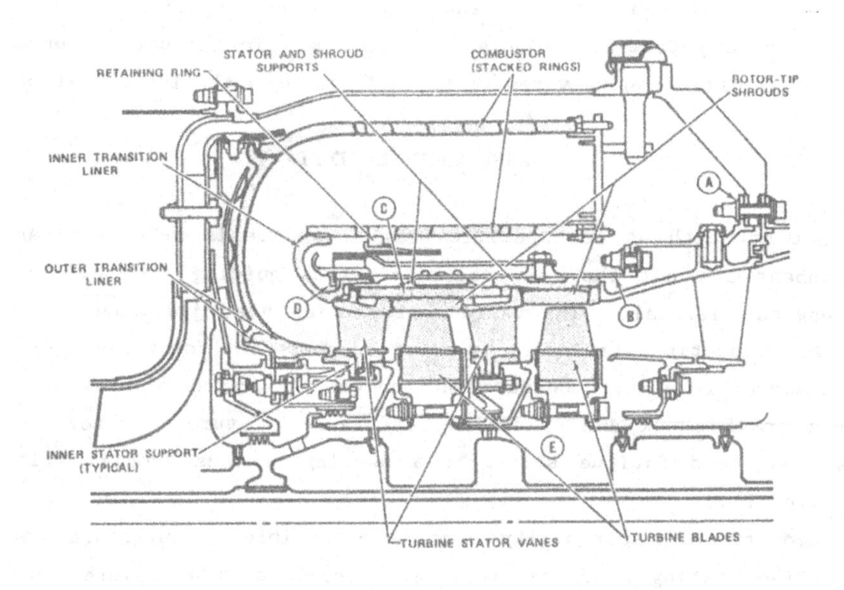

Figure 2. Candidate components for turbine application (9).

the general reliability heading. For a ceramic, being a highly rigid material very sensitive to contact stresses in attachments, the capability to integrate into a dynamically highly-stressed all-metal machine is a prime consideration.

OBJECTIVE OF PAST DEVELOPMENT EFFORTS

The mechanical behaviour of ceramic materials is essentially governed by inhomogeneities in volume and surface. Their one consequential peculiarity, however, is an inability to relieve stress peaks by plasticity. In a two-pronged effort, development therefore focuses on potential reductions in defect size and incidence by improving manufacturing (e.g. clean-room technology) and machining processes, and on building a design methodology that accounts for the typical brittle fracture behaviour of ceramics. Relevant factors here would be small volume, simple shape, constant wall thickness, unrestricted thermal expansion and compression stress state (3).

APPLICATIONS IN ENGINES

The introduction of novel, untried aero-engine materials and manufacturing methods follows established procedures to gain in-flight experience at minimum risk, and to corroborate or identify specific failure characteristics arising on test specimens. The introduction of ceramic materials will follow the same routes, although on account of the inherent brittleness of ceramics and the persisting lack of experience in their handling, they will take more time to become established. At present the introduction of new technologies takes 15 to 20 years even for metallic materials. But whereas the temperature gain achieved by, for example, single-crystal technology is $< 50K$, the potential for ceramics is $> 500K$. This is the horizon against which engineers must exercise the long term patience needed to perfect ceramics for these applications.

In this approach the route to new material technologies is marked by the selection of increasingly significant components for increasingly complex loads, and by incrementally complex testing on test facilities and ultimately in the engine.

In this manner, ceramic coatings are already finding use in engine applications, although for the time being mostly on stationary components. In contrast structural components are still undergoing rig testing with very few exceptions.

Ceramic Coatings

Sprayed abradable coatings on rotating compressor spacers have long proved their value; the material used is Al_2O_3. The selection criteria are dictated by the high speeds (adhesion) on the one hand and by the rubbing action of compressor blades (tribological properties) on the other.

Thermal barrier coating technology has not, as yet, evolved sufficiently to the full potential for use on rotating components. The intended function of the thermal barrier coating derives from the low thermal conductivity of the ZrO_2 used, which is stabilized with 6% to 8% Y_2O_3. An additional advantage is its coefficient of thermal expansion (7 to 10 x 10^{-6} K^{-1}), which is relatively high for ceramics and approaches those of the metallic substrates (nickel-base alloys typically 15 x 10^{-6} K^{-1}).

These coatings serve to extend the lives of thermally stressed components by lowering the base material temperatures and the thermal stresses. The potential reduction in base material surface temperature is $\sim$ 150K, depending on the thickness of coating. This translates into commensurate gas temperature increases, life extensions, or reductions in cooling air requirements. On casing segments these coatings lower the material temperature and control thermally induced casing expansion and, with it, compressor and turbine clearances. Successful use of these coatings on stationary components is developing.

The ZrO_2 coating on the walls of the turbine casing (Figure 3) is 3.5 mm thick which reduces the cooling air requirements to 30% of that of the uncoated casing.

Next to thermal conductivity, thermal shock resistance ranks high among selection criteria for insulating applications. It is quantified by the factor R, which is used to compare diverse materials and material structures. This is an extremely important property for ceramics generally and thermal barrier coatings specifically. Since the latter should have a high R the material needs optimization to the specific application. R is defined by (12):

$$R = \frac{k\sigma\ (1-\mu)}{E\alpha}$$

where E = modulus of elasticity

μ = Poisson ratio

k = thermal conductivity

σ = tensile strength

α = coefficient of thermal expansion.

The choice of a thermal barrier coating for a component exposed to hot gas requires familiarity with its thermal shock behaviour (Figure 4). Is it the compressive stress set up in the coating plane during heating, and the

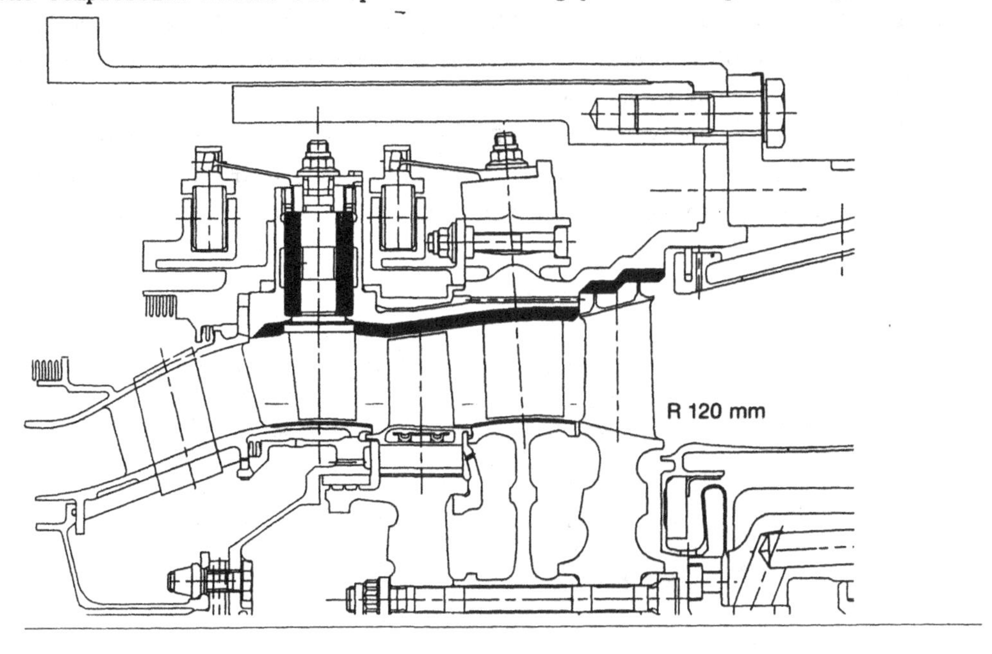

Figure 3. Application of ZrO_2 in a power turbine.

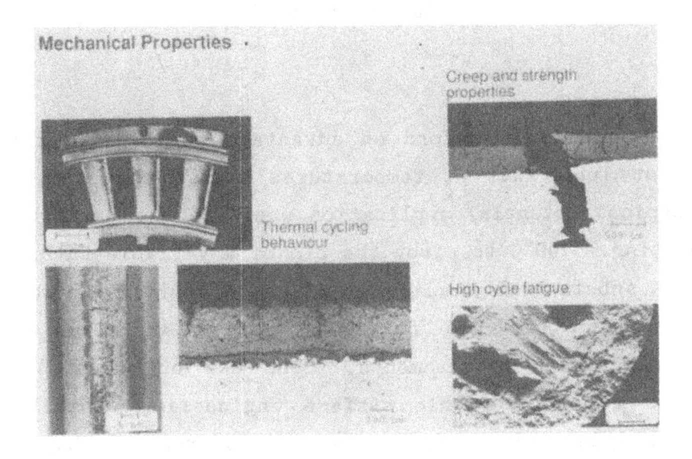

Figure 4. Typical failure mechanisms of ceramic coatings (Peichl and Schneiderbanger).

resultant tensile stress normal to the boundary with the base material that (especially on convex surfaces) constitutes the critical load controlling the failure pattern (4)? Or is it perhaps the tensile stresses induced in the coating in the steady-state operating condition or during the cooling phase due to differential coefficients of thermal expansion between the coating and the base material (5)? The lack of a valid life reduction model based on a range of material failure mechanisms including also effective loads other than thermal and contact stresses, such as corrosion, erosion, bond layer oxidation and vibrations (Figure 5), presently limits broader use of these coatings on components where failure would jeopardize the safety of the engine, such as the leading edges of highly stressed turbine blades.

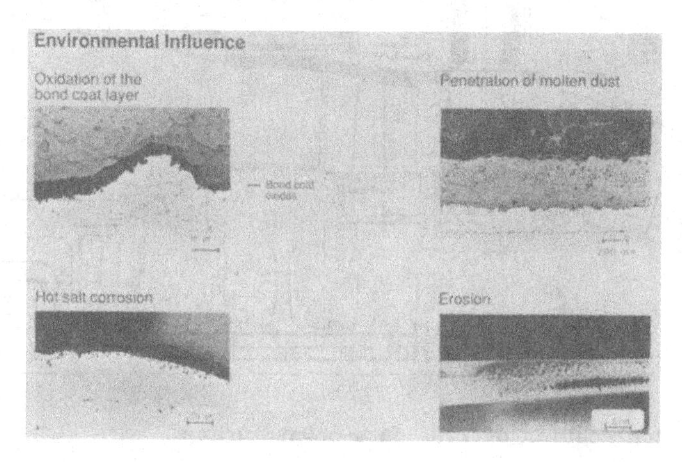

Figure 5. Typical failure mechanisms of ceramic coatings (Peichl and Schneiderbanger).

Bearings

Ceramic bearing materials afford an advantage through their capability to operate without lubricants at temperatures higher than those sustained by metals. A typical potential application would be a ceramic bearing bushing of ZrO_2 operating < 700°C to pivot the pin of a variable turbine nozzle vane (Figure 3). A substantially more sophisticated application would be the gas bearing (1), which provides notable benefits in terms of lubricant, scavenging and cooling requirements. The use of ceramics as a bearing material calls for considerable surface engineering capability (wear and thermal shock behaviour, residual stress states, mechanical strength, fracture toughness and roughness) to cover the wide envelope of operating conditions (Figure 6).

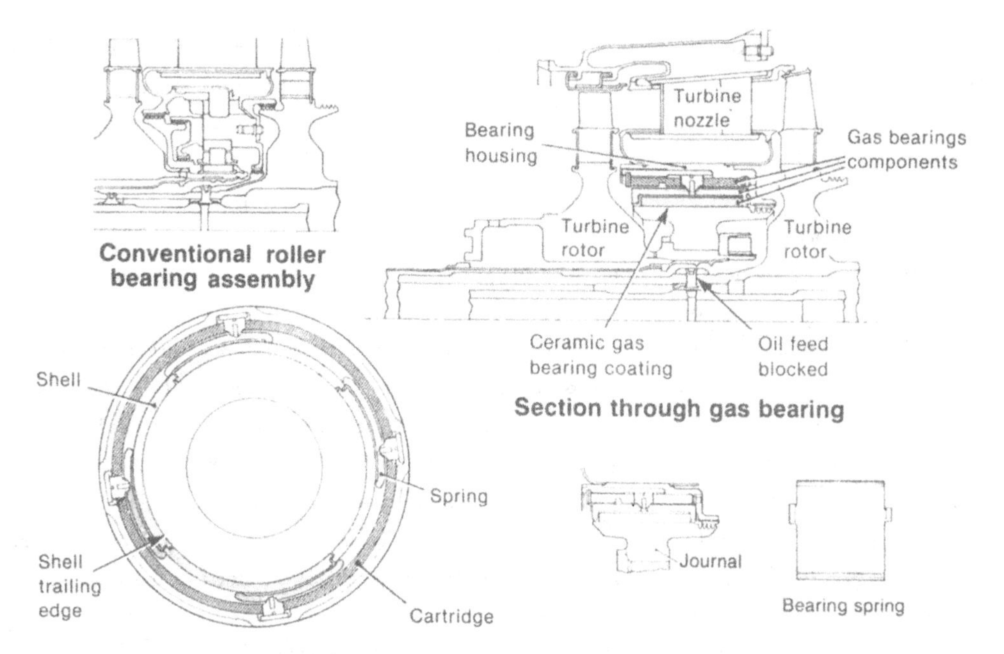

Figure 6. Gas bearing (1).

The increased demands on performance of advanced engines enhances the load on bearings on account of the higher rotational speeds involved. This is where ceramic ball bearings potentially offer a solution. Presently, conventional bearings tolerate a maximum D x N = 2 x 10^6 rpm.mm, largely because of the centrifugal load on the balls, where D is the diameter in mm and N the number of rpm. Balls of Si_3N_4 should sustain D x N = 3 x 10^6 rpm.mm. The three-point contact bearing depicted in Figure 7, survived 2.5 x 10^6 rpm.mm in demonstrations under additional 2000 N thrust loads plus unbalance conditions to simulate blade shedding (6). These experiments,

Figure 7. Roller bearing.

however, also demonstrated that the grade of material was still too inconsistent and the NDT methodology too imperfect to prevent premature ball failures. The use of such bearings is thus currently prevented (material baseline 1976).

Stationary Components

Stationary components of special interest are combustor elements, vanes, shrouds and assembly positioning pieces.

The attractiveness of ceramic materials for combustor applications is readily apparent: high temperature loads in a corrosive-oxidative environment demand very significant cooling of metallic combustor walls.

By increasing further the compression ratio, the air temperature is also increased and its ability to cool the combustion linear walls reduced. Increased pressure will also produce a more luminous flame that transfers more heat to the liner walls resulting in an increased demand for cooling. The use of ceramics for combustors is distinctly advantageous since higher discharge temperatures may be allowed and cooling air requirements may be virtually eliminated. However, the combustor must be designed to allow for high thermal stresses for both steady-state and transient operating conditions and to take into account the existence of hot spots.

Various design concepts have attempted to anticipate critical loads from build up of high ceramic stresses both in the plane of and normal to the wall. The tendency is away from the monolithic design principle and towards rings and ultimately to elements of smallest possible size (Figure 8).

The attachment to the metallic support structure is of major interest and has to allow for the differential thermal expansion behaviour and hold the elements in position to control the gaps. Contact stresses can cause

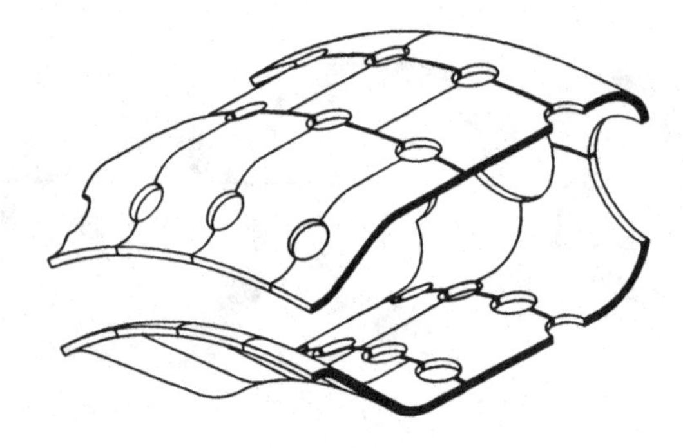

Figure 8. Combustor elements.

failures of the elements (9) but in favourable conditions ceramic elements can survive operating temperatures of 1930°C for some hours.

In the selection of materials for combustor elements, the thermo-physical data controlling thermal shock resistance are, after mechanical strength, crucial criteria. In this context, RBSN and sintered SiC are of interest: Si/SiC has demonstrated limited potential (7,8), because of the separation of free Si and other shortcomings.

In aero-engine applications, however, preference goes to materials exhibiting the temperature capability of monolithic ceramics, but giving better ductility and less sensitivity to contact stresses. These are the properties one hopes to find in ceramic composites. The investigations described in (9) claimed that the greater ductility of composites made for a simpler design (rings in lieu of elements giving an appreciable weight saving in the metallic support structure), and that the geometric integrity was maintained despite cracking. No comment was made however for the reasons that enabled the SiC/SiC material to maintain performance at > 1200°C which is the recrystallisation temperature of the fibre, nor whether the advantages would be maintained for more than a few hours.

Ceramic shrouds, designed as rings or elements, offer superior dimensional stability compared to superalloys. Rings tested at 1000°C in the Advanced Mechanical Engineering Demonstrator of RR (1) proved the effectiveness of maintaining a smaller tip clearance, resulting in a 3%

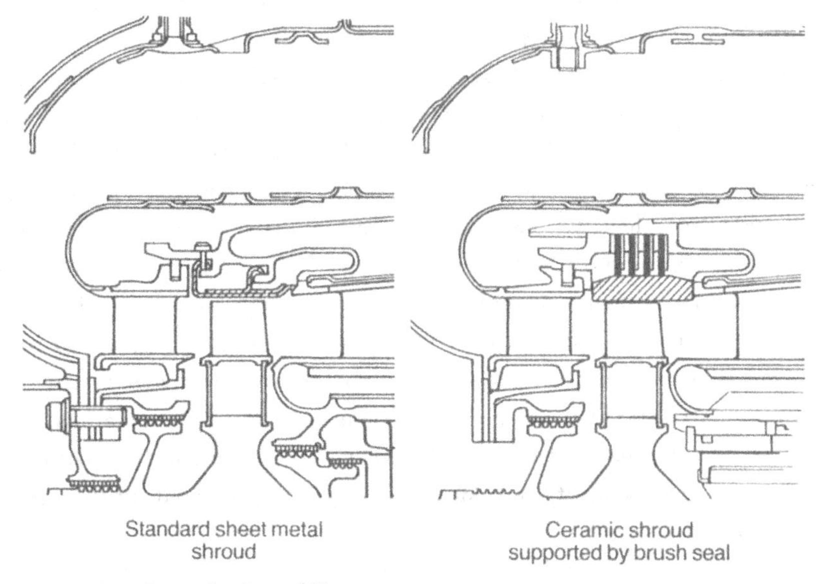

Figure 9. Ceramic shroud ring (1).

specific fuel consumption and 5.6% power improvement. The ring was supported by wire brush seals (Figure 9) thus providing a compliant mounting. An abradable coating will be required to control blade tip wear.

The remaining stationary components investigated as candidates for ceramic composite construction (sintered SiC as well as SiC/SiC) are the exhaust cone, flame holder ring, and hot and cold flaps of the SNECMA M53-P2. Operating temperatures for these components run appreciably below 1000°C. The benefit provided by the use of the composite is seen to lie in its weight saving, which may be around 50% (10). Successful trials warranted the production use of flaps in engine M88.

Nozzle vanes are subjected to stresses similar to those on combustor elements, although creep induced by gas bending loads is an additional concern. MTU conducted rig tests on metal-ceramic vanes that were developed on the principle of minimizing all external forces, except thermal stresses (11). Such a vane consists of an outer ceramic shell and a metal structure within (Figure 10). Only the ceramic shell is in contact with the gas and needs no cooling, whereas the metal core is cooled. Direct heat flow from core to shell is avoided by heat-insulating elements. The material was sintered α-SiC. A cascade of four vanes survived several hundred hours in a hot gas test rig at temperatures $< 1630°C$, but failed when the gas temperatures reached 1700°C. It is felt that with the subsequent improvement in strength of SiC through MTU lab work (Figure 11), even test temperatures of 1700°C should not cause fracture of the material.

Data acquisition and material characterisation is usually carried out using relatively small specimens. How data can be translated into properties of components with differing sizes, geometries and subjected to non standard loading profiles is not well understood. It is in principle agreed that by a probabilistic approach the presence of defects, volume and surface related, has to be taken into consideration. There are finite element programs available, like Ceram 3 D, which permit prediction of the level of survival under static complex loadings. On the other hand it is felt that growth of subcritical flaws with time, as creep or fatigue or stress corrosion cracks, is not sufficiently understood to model life reduction of realistic components under multiaxial stresses. In this context it has to be noted that the state of component production does not permit the assumption of reproducible defect distributions, type, size or orientation.

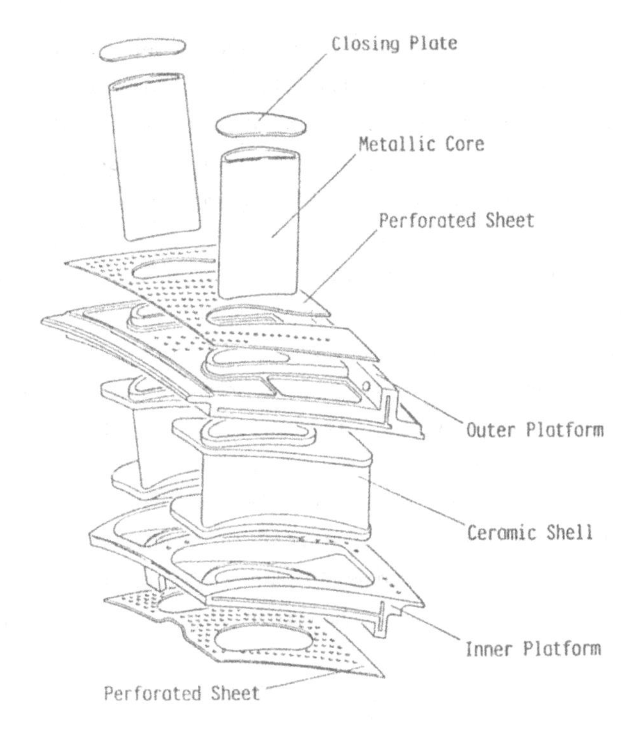

Figure 10. Hybrid concept: Ceramic shell and metallic core.

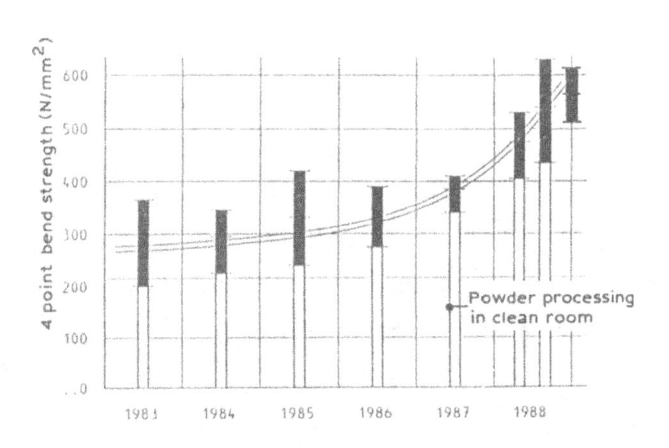

Figure 11. Improvement of SiC strength by strictly following clean room concepts.

Rotating Components

Rotating components - discs and blades - demand the severest requirements for static and dynamic properties. These are aggravated by requirements for corrosion and erosion resistance, and sufficient ductility to absorb impact loads. In addition, there is the necessity of maximum component integrity. It is not surprising that the use of ceramic materials in such applications is some way off. Investigations along these lines are nevertheless underway, as in ceramic blades interlocking in metal discs (Figure 12). The problem limiting this approach is one of force transfer in the blade root, where excessive stress concentrations must be eliminated. To date a considerable number of attachment concepts have been pursued which essentially involve the use of compliant layers (7). Other concepts involve integrally bladed discs or separately bladed discs which could be manufactured in two halves, as discussed in (1). Another alternative approach applies the paired metal-ceramic concept to this component, as generally shown in Figure 10, where the centrifugal force places the ceramic blade shell under compressive stresses, rather than the tensile stresses normally associated with rotor blades. This concept, however, demands a certain minimum size and will therefore not find use for small engines.

CONCLUSION

Owing to their tremendous temperature potential at low specific weight,

Figure 12. Ceramic blades in metallic disc.

ceramic materials will gradually become established in aero-engines. Initially used in non-load-bearing and static applications their reliability and usefulness should be demonstrated. But first a number of essential conditions must be met.

Apart from the requirements for reproducibility of the manufacturing processes and detectability of critical defects, a strong need exists for developing a design methodology that on the basis of detailed knowledge of material behaviour under operational conditions will ensure the production of safe and reliable components. Elements of this design methodology are mechanical models of fracture processes and growth rates as well as statistical models of microstructures and flaw populations. It is especially for use in dynamically highly stressed aero-engines that a host of questions need answering that pertain to integrated designs of brittle materials able to absorb stress peaks by cracking alone. Meanwhile, fibre-reinforced ceramics with their superior deformation behaviour and which are more in line with conventional design criteria will continue to attract attention.

REFERENCES

1. Syers, G., The application of engineering ceramics in gas turbines. Symposium on "Materials and Manufacture in Aerospace", Hatfield, March 1990, Institute of Metals.

2. Simus, T.S., Beyond superalloys. Superalloys 1988, TMS 1988, 173.

3. Kochendörfer, R., Monolithic and fibre ceramic components for turbine-engines and rockets, AGARD-CP-449, 1988, 2.

4. Schneiderbanger, S. and Peichl, L., Wärmedämmschichten für polykristalline und einkristalline Nickelbasislegierungen. Abschlussericht BMVg, Rüfo 4, T/RF44/10019/C1511.

5. Johner, G. and Wilms, V., How to achieve strain tolerant thermal barrier coatings by means of varying spray parameters, ASME 88-GT-313, 1988.

6. Hüller, J., Hochtourige Wälzlager unter Verwendung von Keramik, ZTL-Bericht MTU-M 79/041, 1979.

7. Fairbanks, J.W. Editor, Ceramic Gas Turbine Demonstration Engine, MCIC Report, 78-36.

8. MTU Internal Report.

9. Davis, G.F. and Hudson, D.A., The Demonstration of Monolithic and Composite Ceramics in Aircraft Gas Turbine Combustors, AGARD-CP-449, 1988, 10-1.

10. Mestre, R., Utilisation des Composites Hautes Temperatures dans les Turbo-reacteurs, AGARD-CP-449, 1988, **12**.

11. Hüther, W. and Krüger, Initial Results of Tests on Metal-Ceramic Guide Vanes, AGARD-CP-449, **13**.

12. Lackey, W.J. et al., Ceramic Coatings for Heat Engine Materials, ORNL/TM 8959, 1984.

CERAMICS IN THE AUTOMOBILE INDUSTRY

C. RAZIM and C. KANIUT
Mercedes-Benz AG,
Postfach 600202,
7000 Stuttgart, Germany.

ABSTRACT

Discussion is included of the following topics: application potential of ceramics; past and future development of ceramics; criteria for use and selection of materials; estimation of future developments in material applications; and examples from research and development. It is concluded that difficulties in assuring high levels of quality and reliability, and high costs, will act as severe obstacles to the large scale use of ceramics in automotive engineering. Market penetration will thus be gradual and long term.

INTRODUCTION

Following restructuring of the Daimler-Benz Group, the Mercedes-Benz Division is responsible for volume production of motor vehicles - both cars as well as trucks - and, naturally, has to select the materials to be used both on the basis of functional as well as economic criteria. There thus stems a certain scepticism - but not we stress a fundamental rejection - regarding the application of new materials. This predominates, particularly in the case of ceramics, when we are dealing with materials with which it is not possible to assure operating and manufacturing reliability by applying traditional yardsticks.

In addition, in such a large undertaking as the Daimler-Benz Group, there must also be different opinions and solutions in view of the aims of the differing tasks of the individual group divisions and units. For example, a representative of a material research group would necessarily view things differently than is the case for someone from the user side.

For the Mercedes-Benz Vehicle Division, however, we can claim that the following statements - as far as ceramics are concerned - will be valid for the foreseeable future.

Our remarks relate, in particular, to structural ceramics for applications in automotive engineering. It does not appear to be sensible to describe the range of all the properties and possibilities of ceramics and to deal with the advantages and disadvantages of ceramics beyond essential detail. This is well known and presented in other papers in this volume. It is far more beneficial to illustrate here the philosophy of the "optimal" material choice from the specific aspect of the automobile engineer and to support this with a number of examples. This approach may be helpful in providing scientists, public authorities, businessmen, journalists, etc., with a more realistic and sometimes less euphoric vision of ceramics, and thus protect them from disappointing false estimates, ineffective misdirected expenditure or unjustified investments jeopardizing the company's survival. Only if the ceramics industry, and all those with ceramics interests, restrain their illusions and do not think solely in production terms to give premature or false hopes for ceramics with all the negative arising consequences, can we expect that the undoubted potential of these "high-tech" materials will be exploited on a wider basis in a manner aimed at enhancing products and results.

APPLICATION POTENTIAL OF CERAMICS

With reference to Figure 1, although this puts our Company in the limelight, note that this 50 year old advertisement emphasises the "careful material selection". There are probably only very few similar types of advertisement, and it therefore seems that this advertisement - particularly also in respect of ceramics - is, despite its age, very appropriate.

However, in a time of intensified competition worldwide, careful material selection alone is not adequate. "Optimal material selection" embraces further criteria, which we shall deal with. But first we should like to briefly "illuminate" in advance the present situation of ceramic development and application.

Table 1 shows the proportion of various materials, the quantity in present-day vehicles and those of the immediate future, and we see that ceramics play practically a negligible role. In weight terms, iron and steel dominate, followed by plastics and elastomers. Thus where are the fields of application for ceramics.

Figure 1. Mercedes-Benz advertisement 1939. Emphasis on careful choice of materials.

TABLE 1
Material proportions (wt.%) of medium-upper range cars

Material	Proportion 1986	Estimated Proportion 1996
Iron, steel	67	62
Aluminium	4	6
Mg, Pb, non-ferrous	2.5	3
Glass, ceramic	2.5	3
Plastic, elastomer	12	18
Other	12	8

Figure 2 provides a global answer, without claiming to be complete, showing the principal potential or even actual applications of ceramics in automotive engineering. The list presented here is, of course, not a careful material selection in the manner just emphasized as actual ceramic components would first of all have to be carefully matched to each other and would thus usually not be employed simultaneously. Practically all the examples relate directly or indirectly to the engine sector, which is not surprising in view of the specific properties of ceramics. Volume production applications of ceramics which have already been achieved, are the catalytic converter, the oxygen sensor, spark plugs - which, because they are nowadays a standard feature, are practically never emphasized as applications of ceramics - and the water pump seal. The latter, a silicon carbide seal, is the only component manufactured from a structural ceramic.

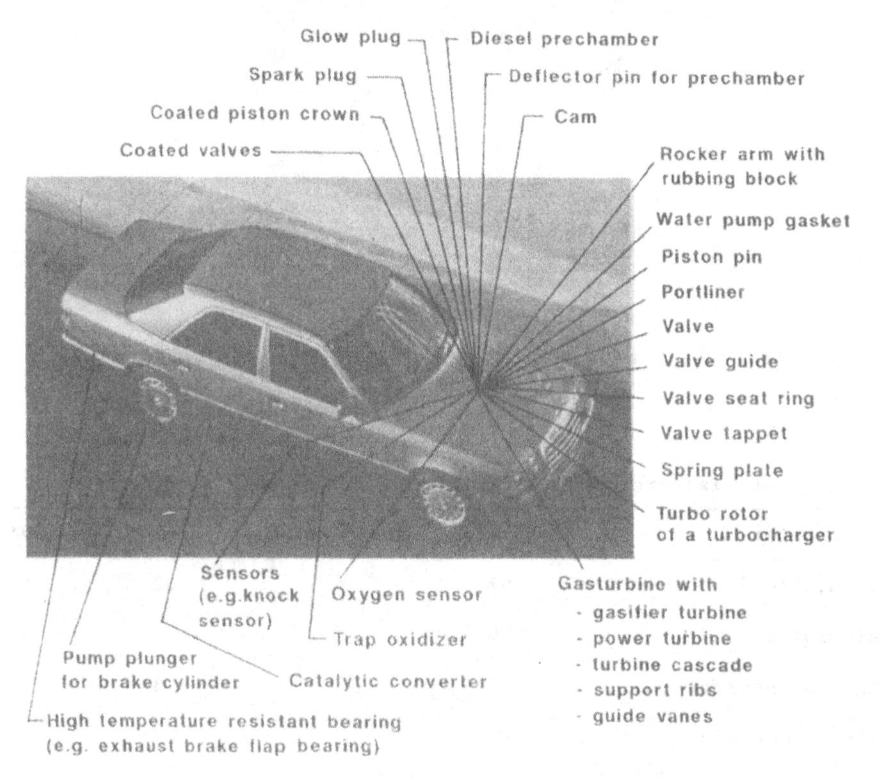

Figure 2. Examples of possible applications of ceramics in automobiles.

Figure 3 takes the example of the Mercedes-Benz 4-cylinder 16-valve car engine to present in concrete terms a number of the ceramic components from the engine sector, mentioned in the previous illustration, and also indicating some of the ceramics investigated.

For the engine sector the principal beneficial properties of ceramics are: low density; high temperature resistance; corrosion resistance; wear resistance in combination with high hardness, low adhesion tendency or non-occurrence of fretting wear, and additionally: high compressive strength; long term raw material availability; low thermal expansion; and low thermal conductivity.

Figure 3. Mercedes-Benz car engine M102E23-16 (cut-away model) as an example of engine parts which could be made of ceramics: portliner (Al_2TiO_5); spring plate (SSN); valves (SSN); valve guides (SSN); valve seat rings (ZrO_2); cylinder liner (ZrO_2); piston crown (coated with ZrO_2); and piston pin (SSN).

This overlooks the fact that ceramics display not only many different, but also extremely contrary properties. Silicon carbide, for example, possesses a very high thermal conductivity whereas zirconium oxide and, even more so, aluminium titanate, possess extremely low thermal conductivities. In addition, the low thermal expansion in some compounds can be a considerable drawback.

Many promoters of ceramics avoid mentioning the disadvantages or play them down with a reference to likely rapid improvements "soon". These include: extreme brittleness; subcritical crack propagation; high scatter of properties; relatively pronounced sensitivity to thermoshocks, and additionally: sensitivity to tensile stresses; close link between production process and component properties; restricted shaping possibilities; difficulty in achieving geometrically complex components; dependence of properties on component volume; problem of translating property data determined on specimen bodies to components; involved and expensive final machining with diamond-tipped tools; complicated joining technology with other materials.

Apart from the production process, the problems of ceramic components are also decisively influenced by the ceramic powder. The aim is to obtain a powder of the highest purity with a defined grain size and shape in combination with the most cost-effective methods of powder synthesis. There are a number of positive developments in this field. Traditional sintering technologies which have been introduced on an industrial scale, such as pressureless sintering (under atmospheric conditions, in an inert gas atmosphere, reaction sintering) or sintering under pressure (hot-pressing, hiping, sinter hiping) are supplemented by more recent methods, which have not yet been introduced, these being "microwave sintering" suitable for oxide ceramics, "exothermic synthesis" ("thermite method") suitable for non-oxide ceramics, or by means of "arc plasma" (ignition of an exothermic reaction) which is likewise suitable for non-oxide ceramics.

PAST AND FUTURE DEVELOPMENT OF CERAMICS

A simplified summary of the statements, opinions, estimations and forecasts propagated in the media regarding the future prospects of ceramics would indicate that ceramics are clearly heading for "rosy times". Figure 4 reveals that, following the practically oppressive importance of metals - most significantly iron and steel - in the last 100 years in particular, a trend appears to be emerging that non-metals might well regain to a considerable extent the importance which was attached to them in previous centuries. This tendency also applies to ceramics, which, of course, at their present stage of development as high performance ceramics bear little relationship to the "old" ceramics.

The reasons for such expectations are based on the enormous progress achieved in recent decades in the development of ceramics. Proceeding from initially disappointing work conducted in the ceramic sector in the United Kingdom from the middle of the 1950's, intensive development activity

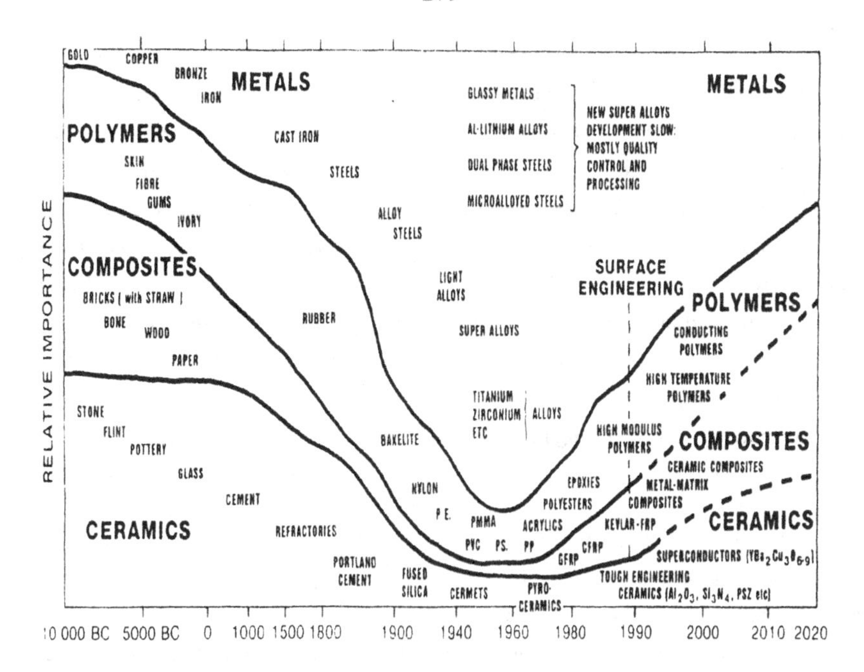

Figure 4. Evolution of engineering materials (after Kelly).

aimed at improving the properties of ceramics began in the 1970's, notably in the USA, a few years later in West Germany and, from around 1977, in Japan. Examples are improvements in fracture toughness from ∿ 4 MPa $m^{\frac{1}{2}}$ for Al_2O_3 in 1950 to ∿ 6 for Al_2O_3-ZrO_2 and ∿ 8 for Al_2O_3 containing SiC whiskers in the 1980's.

Similarly in Figure 5 is shown the flexural strength increases of a number of ceramics. These are based on the one hand on improvements in powder synthesis, new sintering techniques for finer-grained and denser ceramics, quantity-related optimization of sintering aids, etc., on the other hand, however, they are also attributable to the development of dispersion alloys with enhanced toughness, e.g. ZrO_2 particles in an Al_2O_3 matrix.

Fibre reinforcement of ceramics has a positive effect on the toughness of ceramics, whisker reinforcement being of particular interest. The stress-strain curve follows a pseudo-plastic course, Figure 6, despite the presence of microfractures of a brittle nature within the matrix, with different deformation mechanisms being active depending on matrix type, volume and arrangement of the fibre.

Such advances in the ceramic sector, which are presented here as is followed by a general flattening off of this increase and an upper limit is reached asymptotically - comparable with an S-shaped curve.

Unfortunately, this otherwise logical law is often either ignored or insufficiently respected. The result is that the media, often prematurely and glibly, publish unsubstantiated or even untenable statements regarding the future prospects of ceramics on the basis of the theoretically high potential use of ceramics while ignoring, simplifying or inadequately considering the complexly interlinked properties and dependencies of this material. Examples from the range of numerous euphoric headlines in this connection are given below.

- Ceramics increasingly supplant metal.
- The ceramic engine may be available commercially before 1990 (news of 1984).
- Ceramics - a group of materials with a future?
- The ceramic engine - utopia or reality?
- A damper for the ceramic engine.

Sammler companies particularly which are not able to draw on a team of experienced ceramic experts, are naturally exposed, in such a confusion of so-called "information", to the risk of cost-intensive high risk investment.

CRITERIA FOR USE AND SELECTION OF MATERIALS

Quantities of various metals used are indicated in Table 2. Steel is still the dominant material in the mechanical engineering sector. This is based on various factors:

- High stiffness.
- Good plastic deformation characteristics.

TABLE 2
World production of important metals in (1000 t), (after Hauck)

	1900	1950	1986
Raw steel	37,300	191,800	714,489
Aluminium	6	1,507	15,550
Copper	499	2,519	8,261
Lead, zinc tin, nickel	1,448	4,245	13,074
TOTAL	39,253	200,071	751,374

examples, cannot simply be extrapolated into the future. Technical developments are, practically as a matter of principle, subject to logistic laws of growth, with the result that a phase of improvement in performance

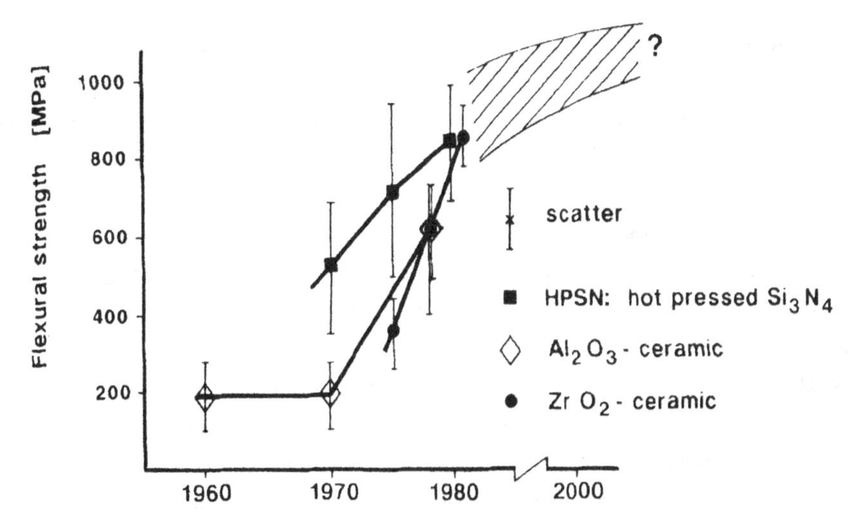

Figure 5. Flexural strength of various ceramic materials (means values and scatter).

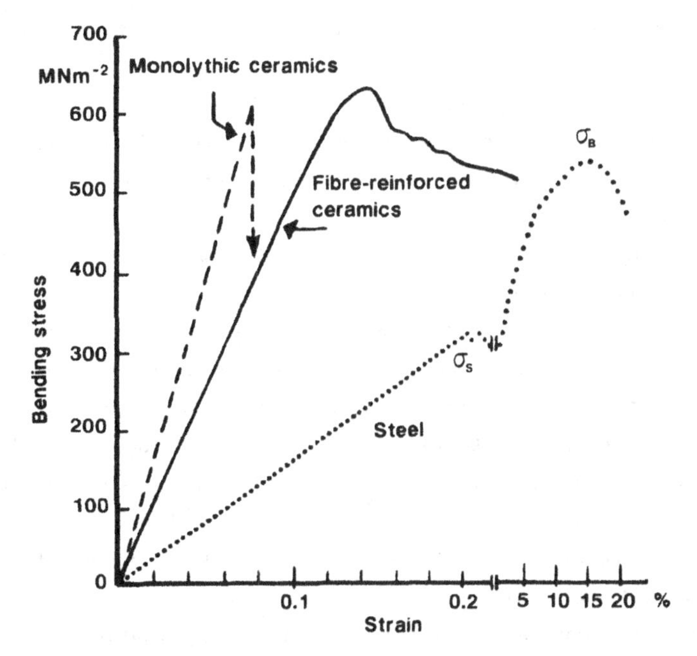

Figure 6. Stress/strain curves of fibre-reinforced ceramics compared with monolithic ceramics and steel (after Ziegler).

- Mastered/masterable production technology.
- Above-average variability of the properties as a result of manufacturing and treatment technologies.
- Favourable material price.
- Raw material availability.
- Recyclability.
- Familiarity in the use of this metal.

The specific advantages which favour ceramics relate particularly to wear, corrosion and temperature resistance and already control the relatively narrow application potential.

In commercial terms, the aspect of experience with a group of materials and the criteria of production reliability, dependability and quality which relate particularly to volume production components, are of major significance.

Possible losses to a company image and concomitant losses in sales resulting from immature material concepts, cannot, as a rule, be compensated even by the attribute of "use of a high-tech material".

The relevant problem area concerning the alternative use of "new" materials, such as ceramics, is illustrated in Figure 7. The potential benefit in respect of the price/performance ratio of the alternative material is countered by a large deficit in designer knowledge in respect of the use of this new material, a deficit which can be reduced vis-a-vis the traditional material concept only to a certain extent by forming a project group. The implementation chances of the new material are practically zero if the resultant risks, particularly for volume production, are estimated to be uncalculable.

A further aspect relating to the use of new materials is the general production pattern. Assuming that the new material is at all accepted by the market it then finds itself in a balance in quantity terms with a large number of other materials. There are still a number of time stages to be overcome, the total duration of which, on the basis of existing experience, may last as long as 20 to 40 years. During this period, initial experience with the material and its basic applicability are translated into rules, fields of application are identified and exploited, production facilities created with the result that the material price drops thanks to the greater available quantities until this material gradually establishes itself. Ceramics thus need not be restricted to niche applications, provided that the prerequisites and outline conditions essential for the increased application of ceramics can be created in line with the necessary time sequence.

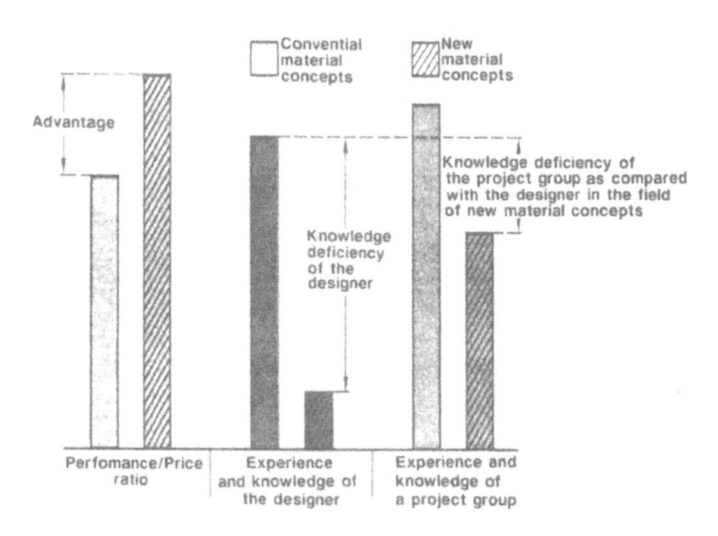

Figure 7. Chances and problems for the use of alternative material variants.

However, this is not likely to be an easy matter, for fundamental reasons due to the nature of this material, irrespective of the possibly lengthy period involved. The first of these reasons relates to the problem of safety factors in the use of ceramics in comparison to steel, as shown in Figure 8. The strength frequency distribution of steel is significantly in excess of the corresponding distribution of the component stress required by the customer, and there is thus a low failure probability. The condition relating to ceramics is far less favourable. This is attributable, on the one hand, to the increased component stress demanded by the customer, and on the other hand, to the brittleness and flaw sensitivity of ceramics, with an asymmetric scatter band (Weibull distribution). Consequently, appropriate and cost-intensive measures must be taken before using ceramics if operating risks, particularly health and safety hazards, are to be confidently ruled out.

A further very important factor is the material price, although this alone is not decisive. The automobile industry, which is cited here as representative for the majority of industrial sectors, has a very tight restriction on material cost (< 10 DM/kg). In contrast, the limits which are set, for example, within the aviation and space industry are

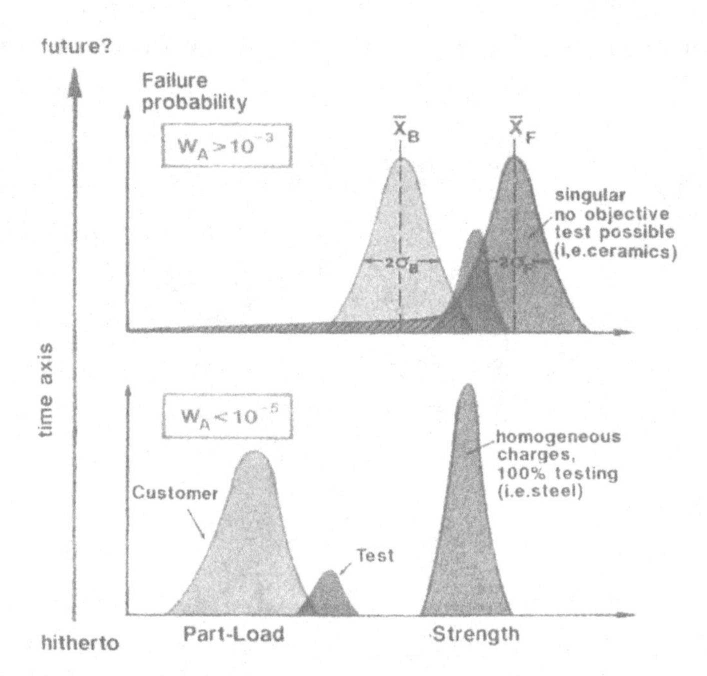

Figure 8. The issue of safety factors: comparison between steel and ceramics.

∿ 100-1000 DM/kg. This industrial branch is thus less prejudiced against new materials.

Figure 9 shows the magnitude of weight-specific prices, taking the example of the dependence on service temperature. Even though prices are dependent on a number of factors and change over time within certain limits, it is nevertheless clear that ceramics, with the exception of the extremely low-stress aluminium titanate, are among the expensive materials. A volume-related approach does not significantly alter this.

To summarise so far - a competitive product can only be achieved by harmonizing the essentially material-inherent parameters of "utility behaviour", "production behaviour" and "basic material price". In view of the complexity of such tasks, this should involve at the earliest opportunity, the component development departments of the company.

Optimal material selection can be symbolized on the basis of an equilateral triangle with the above parameters at the vertices. These are in equilibrium only at the centre. Any other solution overemphasises one parameter and compromise is necessary. In the event that environmental reasons are also of relevance for the use of ceramics, the above approach is

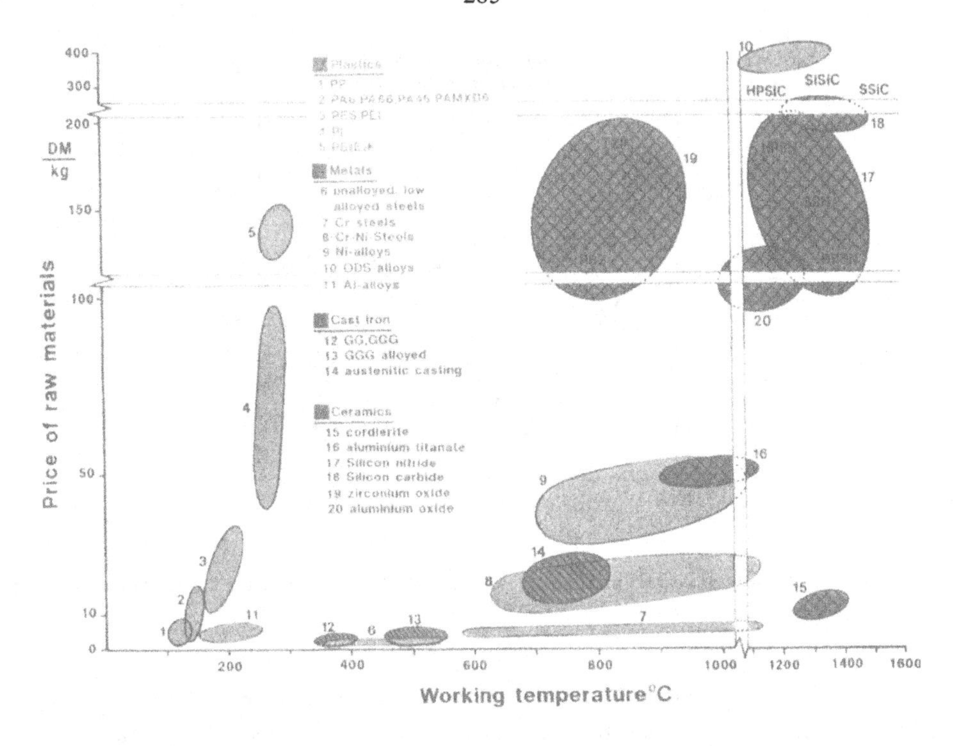

Figure 9. Raw materials: Working temperature and weight related price.

not the triangle just presented but a "strategic tetrahedron of industrial materials engineering" with the fourth equal vertex "environmental".

ESTIMATE OF FUTURE DEVELOPMENTS IN MATERIAL APPLICATION

Translation of the laws of logistic growth curves enables us to make reliably exact trend forecasts for a foreseeable period. The validity and reliability of this translation procedure has been verified by IIASA, Laxenburg in Austria, on the bases of numerous amazingly good agreements between forecast and actual event.

If we now apply this method for predicting the weight-related quantity shares of materials in automobiles, we obtain long term forecast, Figure 10. The historical development of wood through steel indicates the strong growth in synthetics, which could grow to a share in excess of 50%, which may appear unrealistic from the present viewpoint. This, of course, depends on the one hand on what, in the final analysis, we classify in future under the term "synthetics", on the other hand, though, on the breakthrough of further

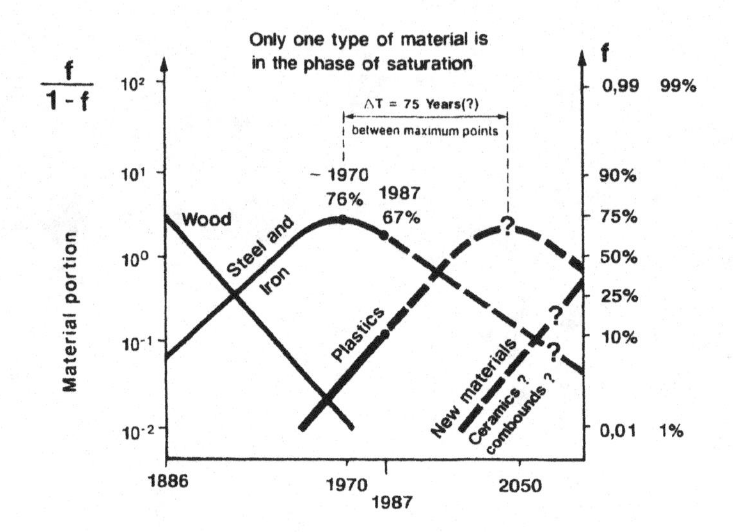

Figure 10. Materials in automotive engineering. Deduction from logistic growth curves.

materials. Whether ceramics will also be among these materials is not a question that can be seriously answered on the basis of the too low threshold which still exists today - i.e. the low share of ceramics in automobiles.

We summarize below the development directions for future materials research:

- Research and development effort and expenditure increase singificantly (e.g. ladle technology, vacuum technology).
- No completely new developments but combination of already known things (e.g. co-polymers, polyblends, composite materials).
- Increasing importance of additive treatments of conventional materials (e.g. surface-treatment technology).
- Adequate testing technology as prerequisite for application of new materials increasingly indispensable (e.g. use of ceramics)
- Increasing importance of environmental protection and economic use of raw materials (e.g. recyclability).

EXAMPLES FROM RESEARCH AND DEVELOPMENT

We shall now illustrate a number of examples relating to R&D work in the ceramic sector at Mercedes-Benz and Daimler-Benz. We begin with the catalytic converter which, despite continuous further development, no longer strictly speaking counts as a development project. Also the low-strength ceramic monolith of cordierite does not actually represent a structural component. However, the catalytic converter in quantity terms is the largest present application of ceramics in automobiles. Figure 11 presents the structure of a three-way catalytic converter with the monolith embedded in temperature-resistant steel wool in a stainless steel housing and the highly porous washcoat applied to the monolith with a catalytic platinum-rhodium coating. The ceramic catalytic converter is presently experiencing competition from metallic converters which reach their operating temperature faster and thus have a lower environmental impact, and which produce a lower exhaust backpressure thanks to thinner cross-sections, which in turn promotes the charge cycle in the combustion chamber.

The driving force for development of ceramics in the last 15 years was also the intended development of the so-called adiabatic engine, which was to offer a number of benefits as a result of greatly reducing the induced heat flow to the outside through the use of ceramic components:

- Improved utilization of fuel, i.e. higher efficiency.
- Suitability for use of alternative fuels.

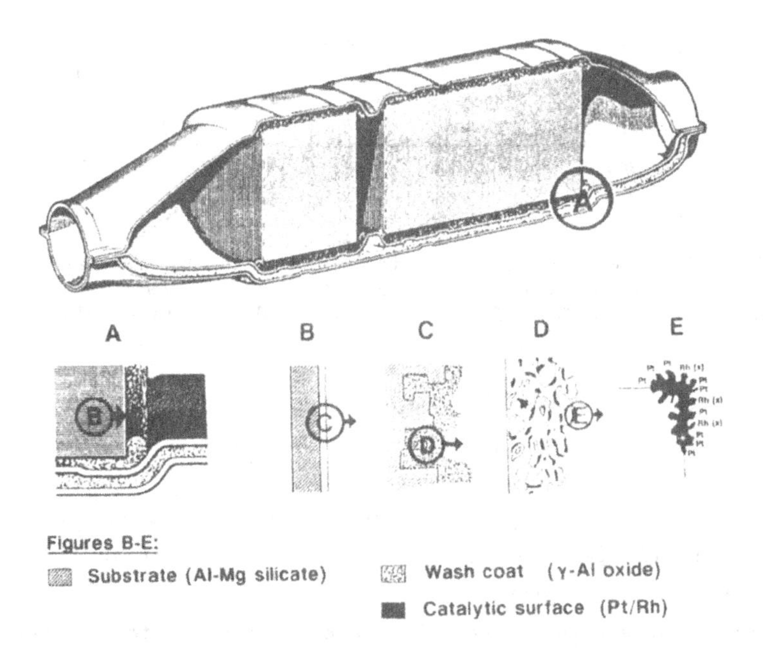

Figure 11. Schematic structure of a monolithic under floor catalyst.

- Reduced exhaust emissions.
- Extended life of components subjected to severe thermal stresses.

By the mid-80's, it was unfortunately necessary to admit that the adiabatic engine actually displayed a greater fuel consumption instead of the hoped-for improvement in efficiency. The cause for this is that - as Figure 12 clearly reveals - the positive increase of the wall temperature in the combustion chamber, which was the desired aim and which actually occurred, leads to a sharp increase in heat transfer coefficients between working gas and combustion chamber, the overall effect of this being extremely negative. For this reason, the adiabatic engine is not a feasible proposition.

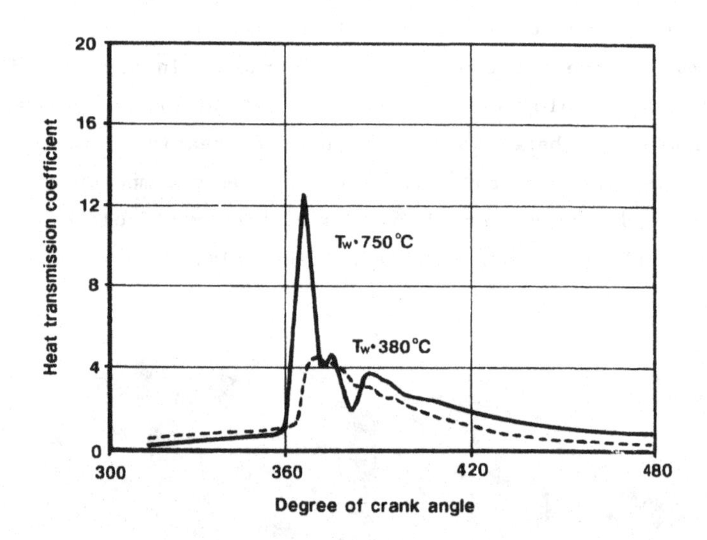

Figure 12. Reciprocating piston engine: Dependence of the heat transmission coefficient on the temperature of the cylinder wall (after Woschni). Mean wall temperature - 380°C. Thermally insulated combustion chamber - 750°C.

Theoretically, benefits accrue if the exhaust gas line is thermally insulated by portliners, Figure 13. The benefits would include a thermal protection of the heat-sensitive components (for example the electronics) in the increasingly packed engine compartment of aero-dynamically styled cars and/or more rapid operation of the catalytic converter as a result of the increased exhaust temperature.

Ceramic liners were fitted to a diesel engine and tested over 200000 km without any problems, Figure 14.

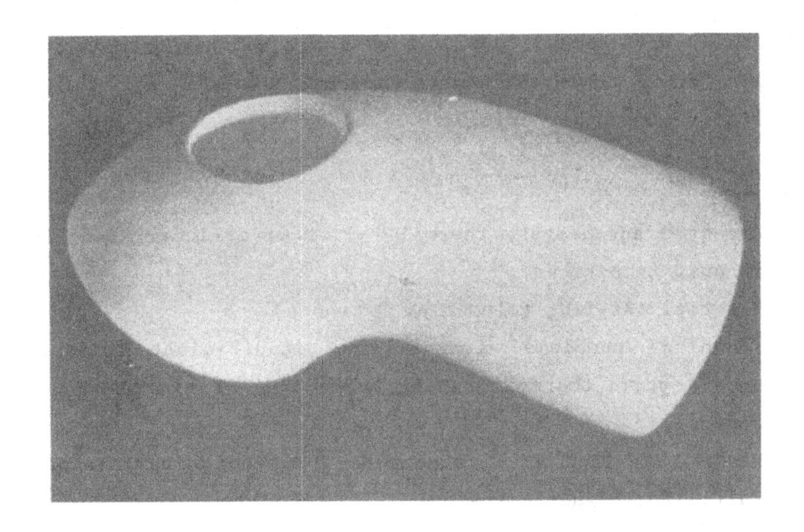

Figure 13. Portliner - Material: Al_2TiO_5.

Figure 14. Ceramic cylinder liner. Material: ZrO_2.

Rocker arms with ceramic rubber blocks are another example, Figure 15.

We list below comments under various headings for the portliner, ceramic cylinder and rocker arms:

Portliner

- Expected advantages: thermal relief of cylinder head; increase of exhaust temperature.
- Material: Al_2TiO_5 (aluminium titanate).
- Technical problems, feasibility: insufficient fatigue strength under cyclic thermal loading; embedding in aluminium alloy can be controlled.
- State of the art: expected functional advantages are not fulfilled; higher costs.

Ceramic Cylinder

- Expected advantages: reduction in wear.
- Material: partially stabilized zirconia.
- Technical problems, feasibility: technical success through skilled engineering.

Figure 15. Rocker arms with ceramic rubbing block.

- State of the art: higher costs compared with the standard version (material: factor 4; production: factor 2); lower costs for improved metallic materials.

Rocker Arms

- Expected advantages: improved friction and wear behaviour.
- Materials: Al_2O_3 (with dispersed TiC phase); SSN (sintered Si_3N_4).
- Technical problems, feasibility: machining of the ceramics based on economic considerations and function; development of a reliable joining technique; assuring component quality.
- State of the art: positive wear test results; status of development close to standard production.

We turn now to joining: for more than ten years, Mercedes-Benz have studied ceramic materials used for testing purposes in the controlling elements of various engines. The main problem was to find a reliable technology to join ceramics and carrier materials for application temperatures up to 400°C. These problems were to a large extent overcome in a three year research project of the Federal Ministry for Research and Technology (BMFT) in which several companies and institutes co-operated under the guidance of Mercedes-Benz AG. The priorities were as follows:

- Development of ductile solders with good adhesion properties and a suitable soldering method.
- Exact knowledge of residual stresses in the ceramic material and the solder layer.
- Controlling undesirable changes in the basic material during soldering.
- Long term stability of the soldered joint.
- Non-destructive testing of the soldered joint.

The results of the work are as follows:

- Availability of a number of active solders, some of which are very ductile, on the basis of AgCu or Ag which yield composite strengths of more than 100 MPa in vacuum through furnace soldering.
- Meticulous individual analysis of the residual stress must be carried out on every component in order to reduce residual stress to a minimum.

- The long term stability of soldered components depends on the propagation of cracks which can grow in the solder layer. Oxide ceramics hardly pose any problems in this respect, whereas for silicon nitride the soldering method must still be further developed.
- Ultrasonic testing is suitable for showing solder defects but does not yet allow a differentiation between stable cracks and cracks which can propagate.

Figure 16 shows another possible control component for which ceramics can be used, the camshaft. Variants and the resulting weight advantages are illustrated without mention being made of the costs. Figure 17 shows test samples. Comments are listed below.

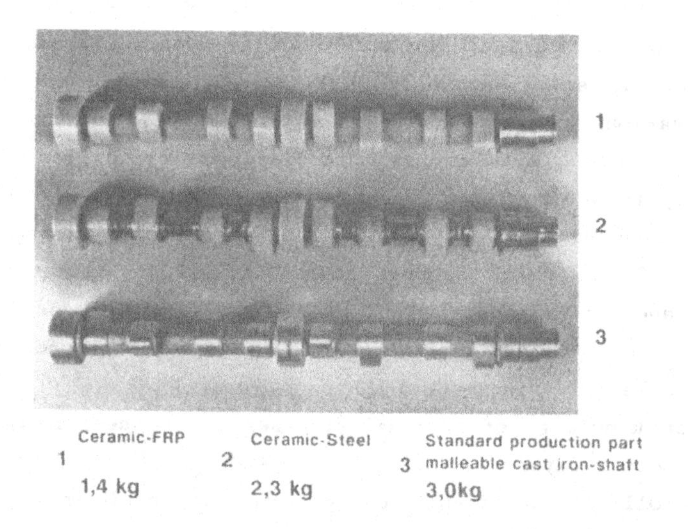

Figure 16. Weight reduction and good wear characteristics by composite design.

Camshaft with Ceramic Cams

- Expected advantages: higher stressability of valve timing gear.
- Materials: PSZ (ZrO_2; PSZ = Partially Stabilized Zirconia); Al_2O_3 (with dispersed TiC phase); SSN (sintered Si_3N_4).
- Technical problems, feasibility: only suited for basic investigations; problematic joining technique; grinding of ceramics.
- State of the art: production of functional specimens for tribological investigations.

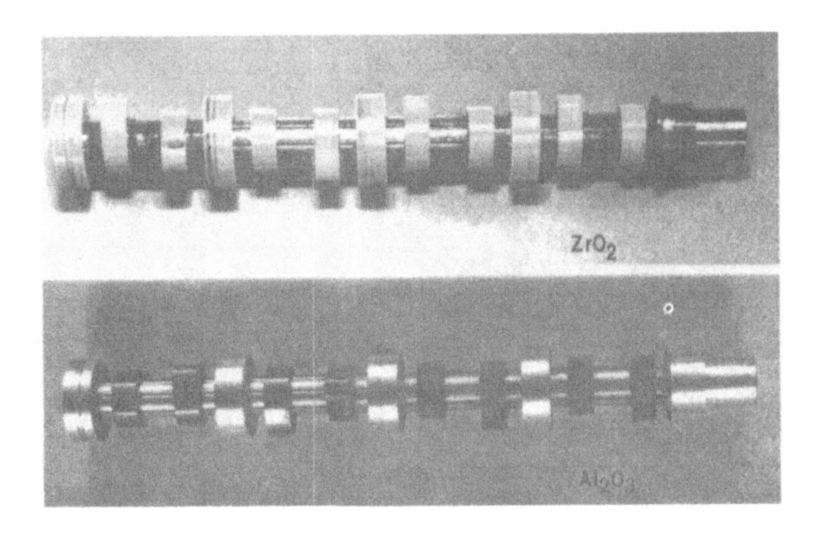

Figure 17. Camshaft with ceramic cams.

A final, highly interesting field of application is the ceramic gas turbine, Figure 18. Environmental damage caused by conventional passenger-car engines and problems emerging for the future supply of fossil fuels are reasons for investigating the potential of using gas turbines for automobiles. In order to achieve acceptable fuel consumption values similar to those of current automotive power units, the temperature must be increased by approximately 300K compared with the present-day gas turbine process, which would necessitate the use of ceramics.

The research gas turbine comprises one gasifier turbine wheel and one power turbine wheel, made from hot pressed silicon nitride. The various problems arising during design, production and service of highly brittle ceramic materials will not be dealt with in detail.

Figure 19 shows that a relatively large gasifier turbine wheel and an even larger diameter power turbine can actually be produced. However, when conducting power unit tests, where these components are subjected to thermal stress and very high centrifugal forces, individual blades may break off due to their brittleness when being hit by large particles; this causes the remaining blades to break off.

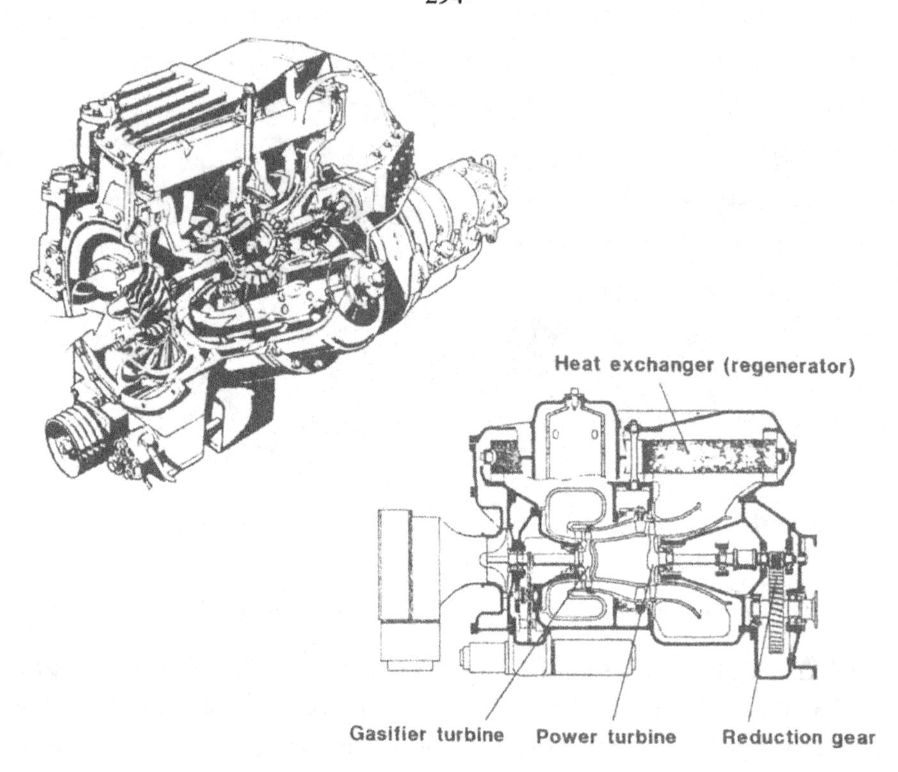

Figure 18. Mercedes-Benz passenger-car research gas turbine PWT 100. Top: Sectioned view. Bottom: Cross-section (after Mörgenthaler).

Figure 19. Mercedes-Benz passenger-car gas turbine PWT 110. Left: Injection-moulded HPSN gasifier turbine wheel ($\phi \sim 120$ mm). Right: Completely damaged turbine during test by impacting particles and overspeed.

CONCLUSION

We believe firmly that problems involved in the use of ceramics, such as ensuring high levels of quality and reliability, or the high price of ceramic components, will continue to be major obstacles towards a tangible increase in the use of ceramics in large scale automotive engineering. There is another problem: the development of ceramics often stimulates the further development of traditional materials. Therefore, in our opinion, success in the use of structural ceramics in automotive engineering can only be on a step by step long term basis, and only to a restricted extent.

CERAMICS IN ENERGY GENERATION SYSTEMS

F. FORLANI
Eniricerche SpA,
S. Donato Milanese,
20097 Italy.

ABSTRACT

The relevant background to the development of engineering ceramics for advanced applications is presented. A number of areas for ceramics in energy production units are considered: heat exchangers, hot gas cleaning, MHD generators, solid electrolyte fuel cells, boilers and gas turbines. Prospects for future markets are good.

INTRODUCTION

Advanced ceramic materials are characterized by properties of major importance to many fields of application, from engine components to cutting tools, from electronic devices to energy production units. In the field of functional applications, mainly electronics, the market for advanced ceramics has been established for a few years: in 1989 the US market of functional ceramics was in the order of 2,500 M$ and the Japanese one roughly double. In the field of structural advanced ceramics, although development effort started some 30 years ago, today their mass production and use are still to be realised. In the frequently quoted graph of Figure 1 (1) it is seen that the major development efforts moved from one country to another and that in some countries, after a period of 10 years characterized by a progressively increasing level of engagements, a slowing down began. Only in Japan and in Germany was no change in trend observed, only fluctuations in the level of effort. Notwithstanding that, very interesting results and practical applications of structural ceramics were obtained, even if several problems remained unresolved.

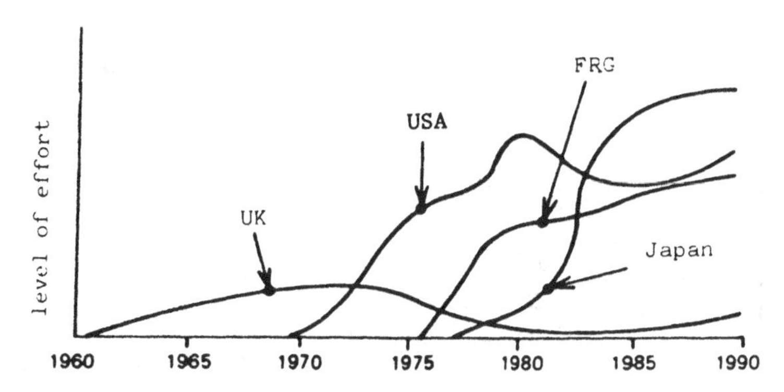

Figure 1. Variation of the effort on ceramics science and technology in different countries.

In fact, optimal balance between costs and reliability was not easily obtained, at least for those ceramic parts to be used under conditions of high stress, in particular mechanical stress and severe temperature excursions. In other words, the still unsolved problem of ceramic brittleness and of the low predictability of their fracture behaviour or, the high processing and inspection costs necessary to ensure a reliable performance, act as deterrents to the widespread use of advanced ceramic parts, especially in those applications where fracture resistance at reasonable cost is mandatory.

PROSPECTS OF CERAMICS IN ENERGY PRODUCTION UNITS

On the other hand, it is not by chance that significant effort has been, and is being, spent to use ceramics in applications where materials must survive mechanical, thermal and chemical stresses. Generally speaking, to break the molecular bonds of a ceramic material in a mechanical (by indentation) or in a thermal (by melting or vaporisation) manner requires more energy than to break the equivalent bonds of metals and alloys. Support to this statement is given by Figure 2 (2) and in Table 1 (3). In a general sense, the inertness of ceramic materials to erosive environments is better than that of metals and alloys.

Finally, ceramics provide advantages in applications where there are rotating parts, especially if they are subjected to bursts of acceleration. This is the case in internal combustion engine turbochargers used to compress the combustion air in the engine. Such turbocharge rotors have been developed in Japan (4,5). The commonly used nickel-based alloys were

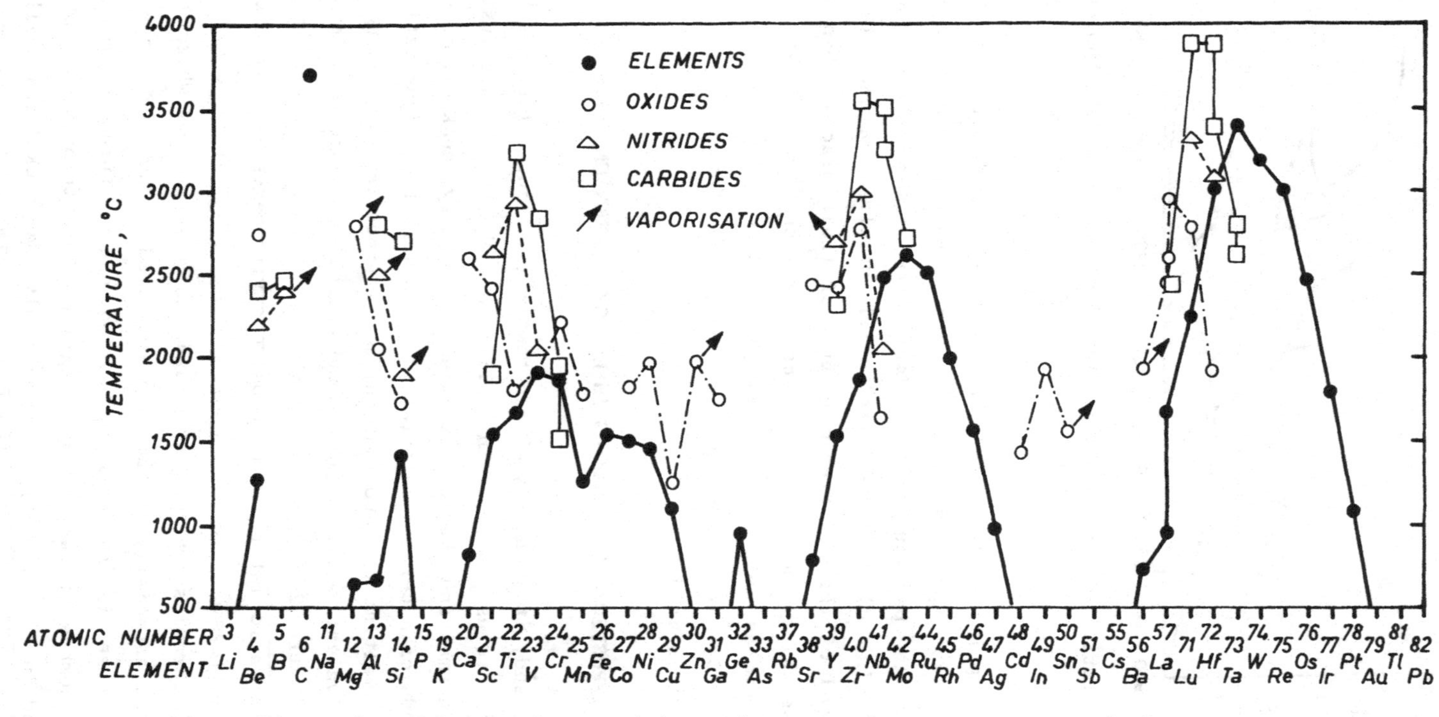

Figure 2. Melting, decomposition and evaporation temperatures of elements and their compounds.

TABLE 1
Comparative properties of engineering ceramics and metals

	Alumina	Partially Stabilized Zirconia	Silicon Carbide	Silicon Nitride	Stainless Steel (18/8)	Annealed Copper
Density (gm/cm^3)	3.8	5.7	3.1	3.1	7.9	8.9
Mean coefficient of Thermal Expansion ($x10^{-6}/K$)	7.7	8.6	4.4	2.6	19.1	16.1
Thermal Conductivity (W/mK)	34	2.2	110	33	16.3	388
Modulus of Elasticity (GPa)	407	205	400	200	206	120
Hardness (GPa)	180	1100	2100	1500	2.5	0.5
Melting Point (°C)	2050	2715	2700	1900	1410	1083

replaced by ceramics, mainly silicon nitride, and an improvement in the moment of inertia of the wheel was obtained. In fact, the mass of the ceramic part is about 40% of the equivalent nickel alloy part. It has been estimated (6) that the introduction of ceramic turbocharger rotors in light- and heavy-duty engines can provide, on the average, an energy saving a little lower than 0.5%.

Because of their special characteristics, advanced ceramic materials provide, in principle, advantages in implementing specific parts of energy production units, such as heat exchangers, hot gas cleaners, magneto-hydro-dynamic (MHD) generator channels, solid electrolyte materials for solid oxide fuel cells (SOFC), and components and parts for boilers and gas turbines.

In the present paper the above listed items are considered to examine the status and prospects for ceramics in the field of energy production units. No consideration is given to the use of ceramics as canning materials for high temperature nuclear reactor fuel, even through the initial study of silicon carbide material, roughly 30 years ago, was primarily motivated by such an application. Consistent with the decision of the Italian administration not to install nuclear fission reactors, no activity on nuclear energy production is carried out at present in the author's Company.

Heat Exchangers

Heat exchangers provide energy transfer in the form of heat from a warm medium to a cool one. In general, they are used for heating a fluid, for instance inlet air for heaters and boilers, water for steam generators, process liquids for their evaporation and separation, etc. Sometimes they may find application in cooling a warm medium with a cooler one.

Figure 3 shows parts of corrugated and finned SiC heat exchangers obtained by extrusion in Eniricerche's ceramic laboratories.

Figure 3. Corrugated and finned SiC heat exchangers prepared by extrusion at Eniricerche.

An analysis of the different industrial areas where heat exchangers should enjoy increasing penetration indicates:

Chemical processing: Although, in general, medium to low temperatures are required, and thus ceramic heat exchangers would have no advantage over metallic ones, the corrosive nature of many chemicals makes ceramic parts attractive. However, because in this range of temperature, metal heat exchangers with protective coatings can solve the problem at lower cost, the penetration of fully ceramic heat exchangers in this sector is unlikely at present.

Primary metals industry and glass industry: To avoid contamination by combustion products the primary metals and glass industries are more and more interested in furnaces provided with radiant tubes: the flame is restrained in the tube interior and heat is radiated to the product. The high thermal conductivity and melting point of SiC with respect to metals, as shown in Table 1, makes this material very interesting for such an application. Incidentally, the market for primary metal melting and of reheating processes in the metal industry is large (in the order of many thousands of installations worldwide) and thus represents one of the largest potential markets for ceramic parts in the energy field.

Power generation heat exchange: The basic problem is to transfer heat at medium to high temperature to a working fluid, which can be under very high pressure. The use of ceramic heat exchangers would be very appropriate, even though the availability of high-pressure seals and joints, and the too-wide distribution of the fracture strength of ceramics have greatly limited the penetration of this market by ceramic parts to date.

Industrial waste heat recovery: The operative temperature for this application is typically in the medium to high range, and the environment can be corrosive. It is the type of sector to which increasing attention is directed to ceramics. There are good possibilities for the penetration of ceramic recuperators.

In conclusion, although ceramic heat exchangers can provide several advantages, metallic ones are used most by industry even though they have two major limitations: temperature and susceptibility to corrosion. However, as for temperature, some designs have been developed (parallel flow and counter flow mode of operation (7)) allowing the metal heat exchangers to handle exhaust gases with temperatures exceeding by 200-250°C those acceptable by the material (which is about 650°C for stainless steel). In practice, ceramic heat exchangers are considered cost-effective at temperatures higher than about 1000-1100°C up to about 1500°C.

However, because of corrosion under steady operating conditions by constituents such as sulphur, antimony, vanadium, magnesium or aluminium, in addition to corrosion in intermittent operation by oxygen, protective coatings are required for metal heat exchangers. Protective coatings for ceramic materials can be deposited by plasma spray or by chemical vapour deposition. Case by case, the relative opportunity for using protective coated metal or bulky ceramic heat exchangers must be decided.

Three or four major suppliers of ceramic heat exchangers are sharing a modest market at present but with potential to reach a worldwide annual value in the order of 100 M$ in five years. The materials used are mainly SiC.

Hot Gas Cleaning

Significant energy saving can be obtained by hot gas cleaning in combined-cycle systems, where the output gas from a coal combustion unit for from a coal gasification unit is supplied to a gas turbine electric energy generator. The direct supply of the hot gases to the turbine is impossible at present, because the gas cleaning devices now available, such as cyclones or electro-static filters, are not suitable for high temperature operation. Moreover, they are barely effective for the elimination of fine soot in the generated gas. The presence of soot is, on the other hand, very detrimental to the life of the turbine blades because of its erosion and corrosion effects. Preliminary results obtained in USA and Italy indicate the effectiveness of ceramic filtering, thanks to the capability of ceramics in withstanding high temperature gas stream erosion and corrosion, and to provide filtration of fine particles.

This represents a field of application of advanced ceramics where the size of the market is intrinsically limited because of the relatively small number of worldwide installed systems which can take advantage from such a technological improvement. However, the effectiveness in energy saving in each system can be, in principle, significant. On the other hand, the R&D and engineering activities are considerable for the ceramics manufacturer.

This is a typical example demonstrating the cyclical effort on the development of structural advanced ceramics, shown in Figure 1. This arises from the uncertain definition of the return on the investments due to the limited size of a market which often demands considerable attention to technological innovation. It is not by chance that recently the leading position in advanced ceramics development for energy systems has been taken by Japan where the business philosophy is different from that of Western countries. Also the return on investment is generally considered on a longer time scale, thanks partly to significant economic support by the government for advanced technologies.

MHD Generators

Channels of MHD generators are challenging items for ceramic materials because of the requirement of withstanding high-temperature, high-speed, corrosive fluids containing potassium. Moreover, possibilities exist for thermal shock arising from sudden emergency heat up and cool down operations.

Many ceramic materials (Al_2O_3, BeO, MgO, Si_3N_4, SiC, etc.) have been tested for such an application. Among them BeO exhibits a large thermal conductivity (240 W/mK, to be compared with the data of Table 1), a good resistance to thermal shock damage, high electrical resistance (10^{14} ohm cm) and bondability to metal elements, the last being essential for channel wall construction. Unfortunately, beryllium poisoning prevents the use of BeO ceramics in MHD generator channels.

Very recently, it has been shown (8) that hot-pressed SiC ceramics made from β-SiC powder with less than 1 wt.% BeO, as sintering additive, can be conveniently used as a material for coal-fired MHD generators. According to the experiments, in potassium seeded coal combustion environments no erosion of 1% BeO-SiC ceramics was detected up to 800°C; it is believed that a slag layer on the insulator provides a protective coating to the ceramic surface.

Unfortunately, predictions for market demand of MHD generators are very small in the short term; in practice these developments might find a sizeable market after year 2000.

Solid Electrolyte Material for SOFC

Solid oxide fuel cells (SOFC) promise to be the most interesting solution among alternatives in fuel cell development because of:

- an energy conversion efficiency as high as 60% (to be compared with 40% obtainable with the best phosphoric acid fuel cells);
- the generated exhaust gas at temperatures as high as 1000°C (compared with the maximum temperature of about 750°C obtainable with the molten carbonate fuel cells), allowing high yield in the energy conversion in combined-cycle systems, which do not require any gas cleaning;

- the possibility to reform the primary fuel gas internally in the cell; and last but not least
- the use of a solid-state electrolyte which reduces criticality of material corrosion and electrolyte loss.

In spite of initial forecasts indicating availability of SOFCs on a commercial basis only after year 2000, the recent concentration of effort in Europe (GEC in UK, Siemens AG in Germany and Eniricerche in Italy) increases expectations. SOFCs could be commercially available in the second half of the nineties, as anticipated for molten carbonate FCs.

Although other materials such as bismuth oxide and cerium oxide are under preliminary evaluations, the SOFCs presently under development take advantage of the ionic conduction of Y_2O_3 fully stabilized zirconia. Such material was successfully used by Weissbart and Ruka (9) since 1958 in the first operating fuel cell.

There are still problems to make SOFCs widely usable energy generation systems. One concerns the deposition at low cost of thin ($\sim$ 50 μm) non-porous layers of fully stabilized zirconia over a large area. It is firstly necessary to obtain submicron powders. Figures 4 and 5 show (10), respectively, the spherical and uniform shape of submicron powder of stabilized zirconia and the X-ray diffractometry of a sample of such a powder. The powder has been obtained from separate metallo-organic compounds of yttrium and zirconium. A colloidal suspension of uniformly

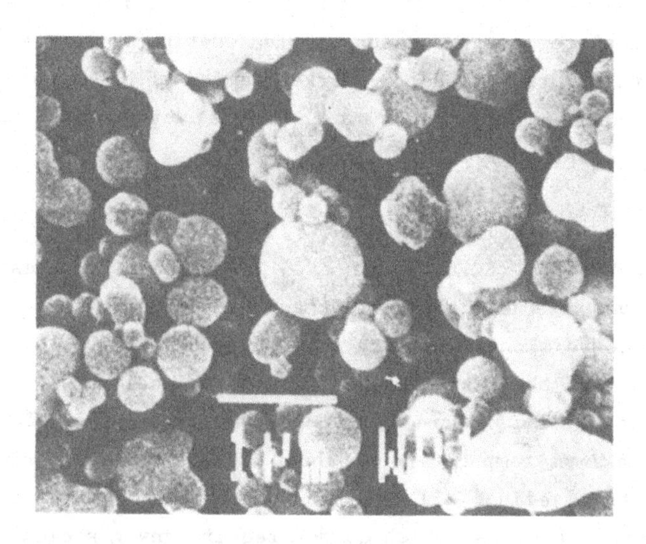

Figure 4. Submicron 8% Y_2O_3-ZrO_2 powders obtained with an original process developed at Eniricerche.

dispersed precursor particles with submicronic diameters via hydrolysis was appropriately calcinated to the solid material.

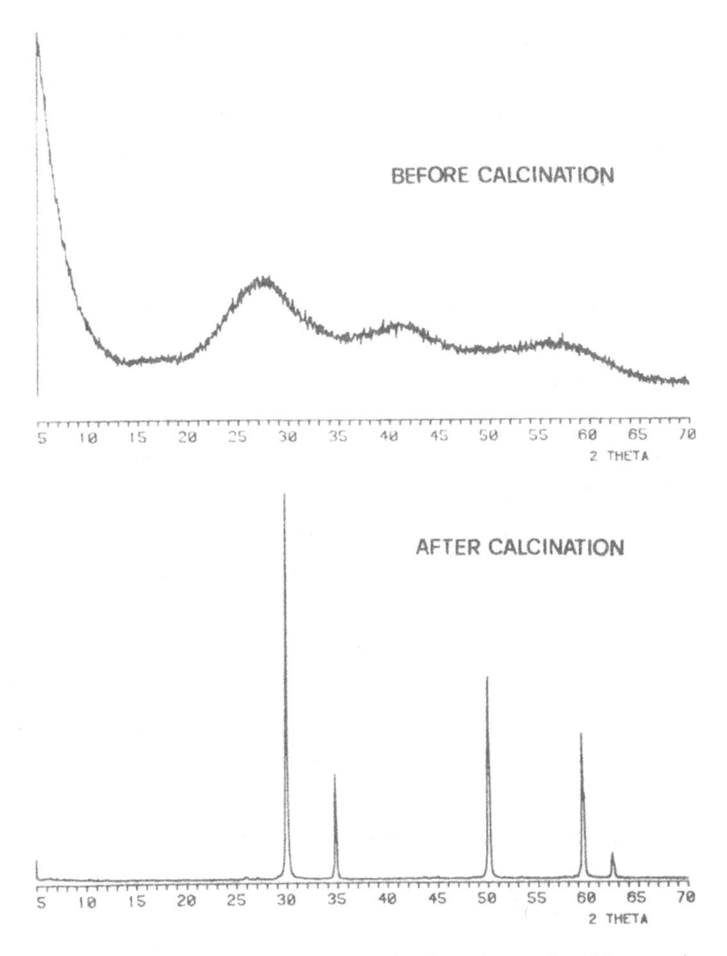

Figure 5. X-ray diffractometry of the powder shown in Figure 4.

Large area cells deposited by tape-casting appear to be prone to cracking because of thermal stresses during sintering. Blending and sintering conditions must be improved. Alternative deposition technologies are under evaluation. Oak Ridge National Laboratory, for instance, is trying to apply the 28 MHz microwave sintering procedure to Y_2O_3-ZrO_2, already successfully applied to aluminium oxide (11).

Another problem to be solved is the difference in thermal expansion coefficients between the solid electrolyte, the cathode materials and the interconnecting ones.

For the anode a porous $Ni-ZrO_2$ permits a solution, but a satisfactory replacement for strontium doped lanthanum manganite for the cathode has not yet emerged.

Also the geometry of the cell is under consideration. The two main approaches are the tubular structure developed by Westinghouse Electric Corporation and the flat corrugated design, very compact and with improved current density, proposed by Argonne National Laboratories. The most important efforts currently carried on in Europe are concerned with the flat design.

In conclusion, although solid-oxide fuel cells are promising for ceramics in energy units, the value of the current market is nil.

Components and Parts for Boilers and Gas Turbines

In boilers, supplied with pulverized coal or coal-water mixtures (12), erosion and corrosion are observed in many parts and represent the main reason for boiler out-of-service.

ENEL, the Italian organization for electric energy supply, reports (13) that in coal boilers the maintenance costs amount to 54% of the global costs to be compared with 30% in oil fed boilers.

As recently pointed out (14) combined-cycle units, where the exhaust gas of a gas turbine is recovered in a boiler feeding a steam turbine, and repowering systems, where a gas is added to an existing steam-unit in order to increase power, present almost fresh ground either with parts in ceramic materials or, more conveniently in some cases with ceramic protective layers applied to critical parts against erosion and corrosion.

In contrast to ceramic turbocharger rotors for engines, for large gas turbines it seems more convenient to use protective layers either because of the difficulty and the costs of manufacturing large dimension ceramic blades or because bursts of acceleration are certainly less important in energy units than in car engines.

A technology for the deposition of SiC coatings on high temperature alloys, such as superalloy Nimonic 80A, has been recently developed (15). It has been shown that a matching layer of TiN is necessary between the base material and SiC coating. For the deposition of both TiN and SiC layers a low-pressure chemical vapour deposition has been used. For obtaining the TiN intermediate layer a mixture of H_2, N_2 and $TiCl_4$ was introduced into the deposition chamber. Deposition was followed by a post-treatment at 1000°C with soaking in N_2/H_2 flow. For the final SiC coating a mixture of

$Ar/H_2/CH_3SiCl_3$ was used. Figures 6 and 7 show the coating in top view and cross-section. The interface is free from defects and porosity; thicknesses are ∿ 5 μm for TiN and ∿ 3 μm for SiC.

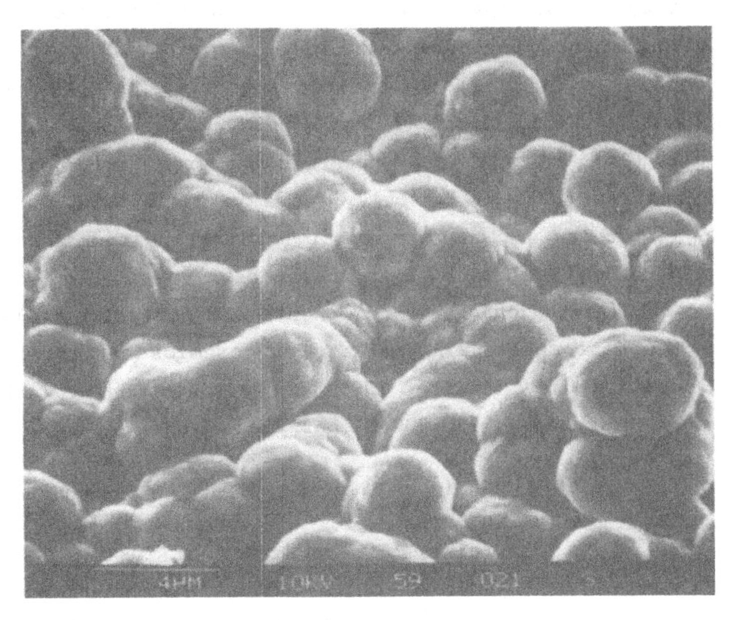

Figure 6. Top view of SiC coating layer.

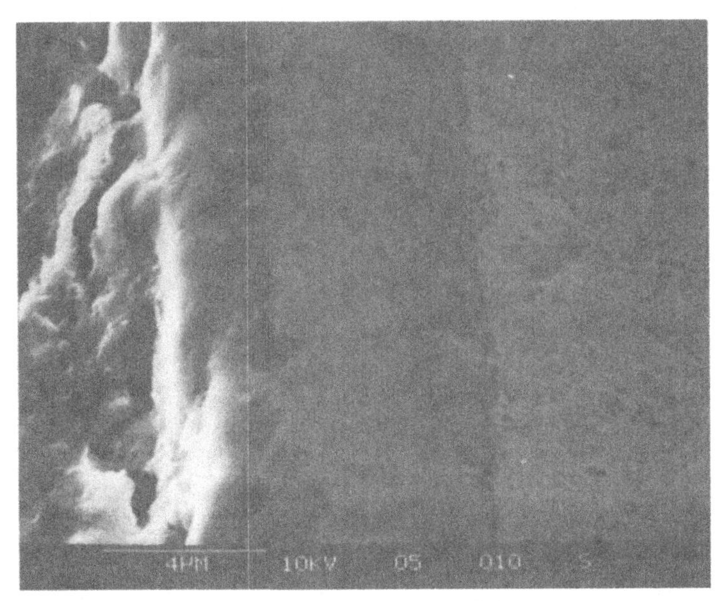

Figure 7. Cross-section of Nimonic 80A superalloy coated with a layer of TiN (5 μm) and a layer of SiC (3 μm).

X-ray diffraction analysis (see Figure 8) shows a polycrystalline cubic structure for both layers, but also indicates the presence of non-stoichiometric silicides. However, because of their limited presence in the layers (less than 3%) silicides do not affect the adhesion of the coating. Testing adhesion by scratch-testing indicates a critical load in the order of 25 N. Such a value can be considered acceptable in erosive/corrosion environments with parts not submitted to high mechanical stress.

Figure 9 (16) reports the progress progressively made in high temperature turbine blades in bulky ceramic materials, showing that ceramics, in particular silicon nitride, can withstand temperatures as high as 1220°C at 70 kg/mm^2 with a low fracture probability.

Together with heat exchangers, components and parts for boilers and gas turbines are the most challenging market for ceramics and ceramic coatings in energy production units. After 1995 the worldwide market value in such applications is estimated at 150-200 M$ per year.

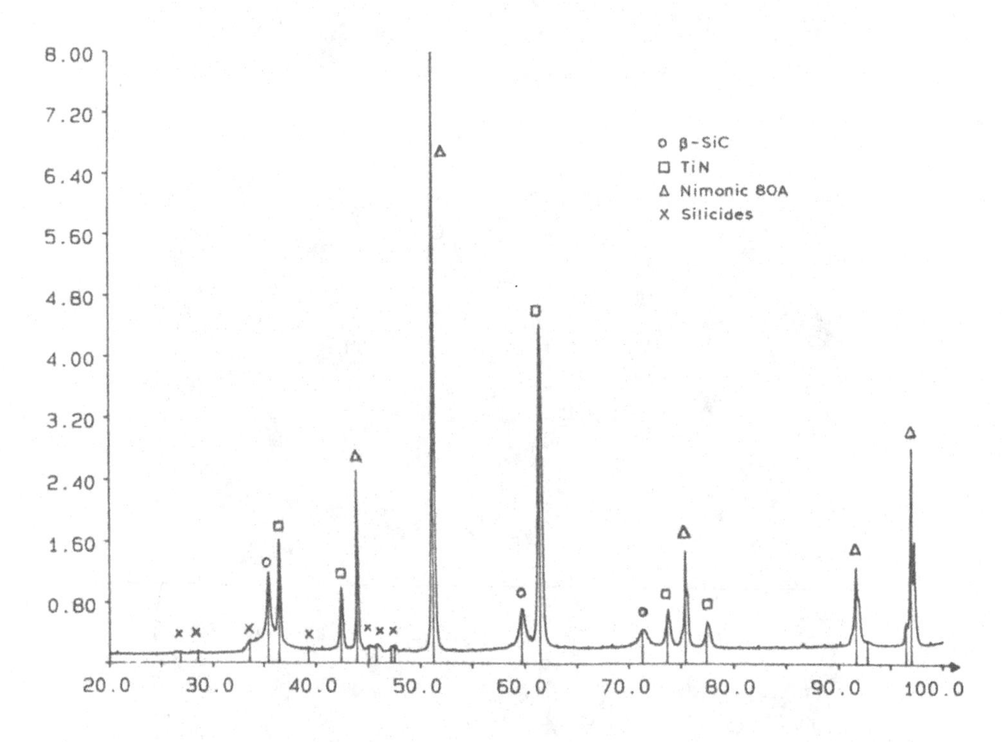

Figure 8. X-ray diffractometry of the layers, shown in Figures 6 and 7.

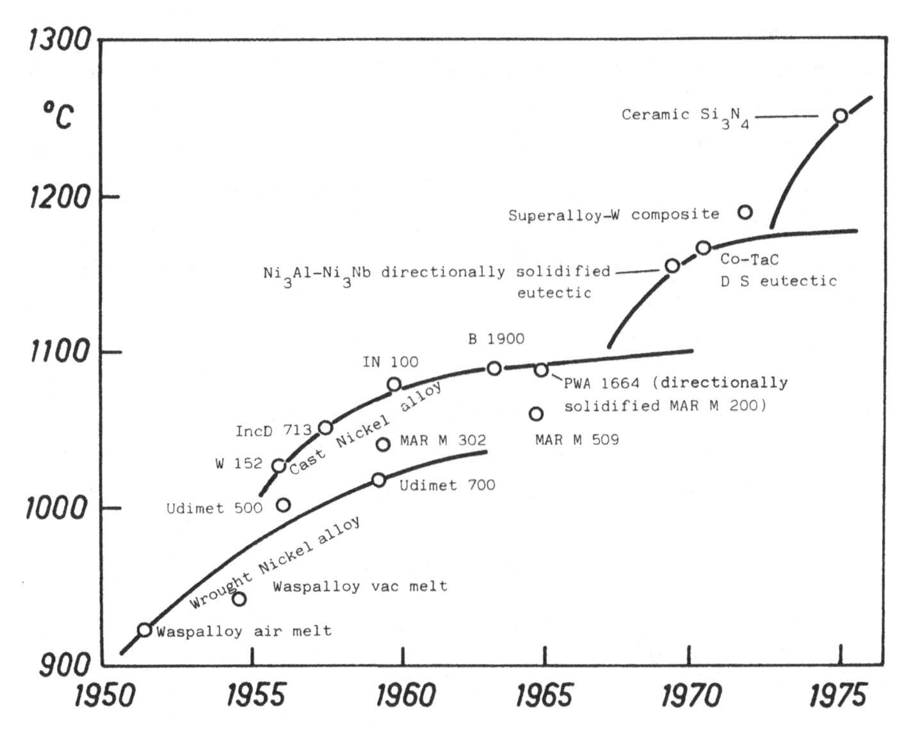

Figure 9. Evolution in the characteristics of materials used in turbines operating at high temperatures.

CONCLUSIONS

Materials and parts used in energy production units must survive in many cases in particularly severe environments, generally characterized by erosion, corrosion and sometimes by mechanical stresses. After 30 years effort on ceramics, one has to take note of the lack of a fully demonstrated capability in the field of energy, mainly because of problems connected with the reliability of performance. On the other hand, in spite of the decision of some groups to reduce significantly investments in the development of ceramics, overall worldwide effort continuously increased in the past and continues to increase today.

Nevertheless, the basic characteristics of ceramics are still very attractive for exploitation. Maybe, the pressure to attain practical exploitation quickly in the past concentrated the efforts along a too empirical approach.

In the last few years the approach to the problem has changed: larger investments have been directed to develop models for the predictability of the fracture behaviour and equipment for testing the parts in a non-destructive way; effort has been spent for proving models and for perfecting test philosophy and procedures. Today more science is directed to understanding the basic behaviour of ceramics.

If the investors remain sufficiently committed and do not loose their nerve too soon, the end of this decade or, at the latest, the dawn of the coming millenium should see ceramics introduced into many fields of technology, particularly energy production units. The determination to continue investment in Research and Development in the medium to long range prospective certainly requires the support of national and multi-national organizations, such as the EC, for the ultimate success of the enterprise.

REFERENCES

1. Lenoe, E.M., International perspectives on ceramics in heat engines, Bull. Amer. Ceram. Soc.,1984, **63**, 422; F. Omida and R. Malaman, Industria italiana ed alta tecnologia, Franco Angeli, 1989, **2**, 154.

2. Briggs, J., Ceramics as energy resistant materials, Chem. & Ind., 1986, **6**, 630.

3. SRI Inter. Tech Monitoring, Adv. Struct. Ceram., 1989.

4. Kaneno, M. and Oda, I., Sintered silicon nitride turbocharger rotor, SAE Conf., Paper N. 840013, 1984.

5. Okazaki, Y., Matsudiai, N. and Matsuda, M., Ceramic turbine wheel development for Mitsubishi turbocharger, SAE Conf., Paper N. 850312, 1985.

6. Larsen R.P., Vyas, A.D., Teotia, A.P.S. and Johnson, L.R., Outlook for ceramics in heat engines, 1990-2010: A technical and economic assessment based on a world Delphi survey, US Department of Commerce, NTIS, DE90-011262.

7. Heinrich, J., Huber, J., Forster, S. and Quell, P., Advanced ceramics as heat exchangers in domestic and industrial appliances, Indust. Ceram., 1987, **7**, 34.

8. Okuo, T. et al., Evaluation of ceramic channel performance on combustion driven MHD generator, 8th Intern. Conf. on MHD, Moscow, USSR, 1983; T. Okuo et al., Ceramics for coal-fired MHD generator channels, Indust. Ceram., 1988, **8**, 137.

9. Weissbart, J. and Ruka, R., Solid electrolyte fuel cell, J. Electrochem. Soc., 1962, **109**, 723.

10. Castellano, M., Procedure for the preparation of a zirconium oxide precursor (in Italian), Pat. Appl. No. 22590A/88, Italy, 1988.

11. Techmonitoring, Fuel Cells, 1989, 2.

12. Laganà, V., Piccinini, C., Orlandi, T., Paronuzzi, O., Procedure for the preparation of high concentration solid suspension (in Italian), Patent No. 1175943, Italy, 1984.

13. Del Pianto, G., di Maio, G., Martinella, R., Pasini, S. and Sbrama, L., Erosion by ash in pulverized coal boilers (in Italian), National Meet. on Ind. Tribology, Bologna, Italy, 1989.

14. D'Augusto, D. and Livraghi, M., Advanced materials application for power plants, AIM-ASM Intern. Conf. on Evolution of Advanced Materials, Milan, (Italy), 1989.

15. Nistico, N., Cappelli, E., Giunta, G., Minnaja, N. and Vittori, V., Optimization of metal-ceramic compatibility through substrate nitridation, Proc. 7th CIMTEC, Montecatini, Italy, 1990.

16. Van de Voorde, M., The importance of materials for advanced energy technology, High Temp Techn., 1983, 1, 195.

APPLICATIONS AND DESIGN OF POROUS CERAMIC STRUCTURES

H.J. VERINGA, R.A. TERPSTRA and A.P. PHILIPSE
National Ceramic Centre,
Netherlands Energy Research Foundation, ECN,
PO Box 1, 1755 ZG Petten, The Netherlands.

ABSTRACT

In this paper we give a number of examples of current applications of porous ceramic materials and new directions for future developments. Emphasis is given to design aspects as an important issue in determining the production method to be adopted. In this respect four categories of porous material with high potential are worth mentioning: porous ceramics for advanced catalytic processes; ceramic membranes for liquid and gas separation; porous ceramic (and non-ceramic) materials for molten carbonate fuel cells; and ceramic foams. Membranes and foams are discussed here in detail.

INTRODUCTION

For a number of years there has existed a growing interest in the development and application of porous ceramic bodies which combine well controlled microstructures with high mechanical and thermal stability. The requirements imposed on the inner pore structure of these materials are due to a number of aspects dictated by the specific design for the selected applications. In this way, the microstructural requirements are derived from factors like chemical, mechanical and thermal stability but also specific resistance to gas or liquid flow as well as their influence on diffusion of different species in the structure. A number of applications of porous ceramic structures have proved to be particularly attractive and seem to have a high market potential in the large-scale process industry. It is foreseen that further possibilities will grow quite dramatically in the next decades.

APPLICATIONS AND DESIGN REQUIREMENTS OF POROUS CERAMICS

Most of the present development of porous ceramics focuses on very accurate control of microstructure combined with thermal and mechanical stability. This is due generally to the fact that high mechanical loads in combination with high temperatures, are required by the process itself, whereas the specific properties of the porous materials in fact will limit the actual conditions for operation.

It is well known that the elasticity (E) and strength (σ) of high porosity materials have values which are considerably lower than $(\rho/\rho_{th})E$ or $(\rho/\rho_{th})\sigma$, where ρ and ρ_{th} are the bulk and theoretical densities. This is a consequence of the fact that bulk strains are to a certain extent internally accommodated as bending of ceramic filaments to reduce stored energy and internal stresses (1). For this reason, overall stresses necessary to cause bulk failure are relatively low. On the other hand, as high overall elastic strains can be accommodated, only moderate internal stress variations will occur, and may result in a material with potentially high mechanical reliability. This is particularly advantageous where temperature gradients, differential thermal expansions and combinations with other materials are envisaged. In practice this beneficial situation has been demonstrated only in a limited number of circumstances. But there are additional possibilities based on materials presently available through a better understanding of the factors which control microstructural development during processing.

Ceramic Membranes

Ceramic membranes will be applied for both liquid and gas separation where temperatures < 600°C are required (2). Generally, a ceramic membrane system consists of a number of tubes such that a maximum surface area is accommodated in as small a volume as possible. Since the present possibilities of ceramic processing give a lower limit to the radial dimensions of porous tubes of a few millimetres, the specific area per unit of volume attainable is no more than abut 500 m^2/m^3. This is less than can be achieved with the more familiar polymeric systems where fibrous materials of a diameter of less than a millimetre are used. However, some particular properties of ceramic materials change the emphasis towards ceramic membranes. Unlike a polymeric system, a ceramic membrane with inherent thermal stability and strength can be cleaned easily with aggressive (hot) chemicals and/or steam. This allows for in-service cleaning under reversing pulsed flow conditions and ex-situ cleaning by burning. Permeability and selectivity can be chosen over a very wide range by control of the ceramic processing.

A ceramic membrane system for gas separation consists of a tubular support which is coated with an intermediate and a top layer. The latter is the actual membrane layer. The different layers comprising the ceramic membrane have the purpose of separating the operating functions.

The support structure should give a high strength to the system such that no significant pressure drop will develop under operation. Therefore, a microstructure should be chosen for the support material with pores large enough not to influence the permeability of the total membrane system significantly and a porosity small enough to guarantee strength and to make application of an intermediate layer possible. The porosity of a typical support structure should not be lower than about 50% while the required minimum pore size is determined by the pore size and permeability of the layer(s) on top of the support tube. The support structure further should be such that it guarantees bonding with the intermediate layer without allowing any ceramic material to penetrate into the inner structure which would give an unwanted local decrease of permeability of the interface area.

The intermediate layer is film-coated from an α-alumina suspension, and has a porosity of 40-50%. This layer should be as thin as possible to limit pressure drop under operation at high material fluxes. The whole structure of support and intermediate layer is sintered together to give a ceramic body in which failure due to internal pressurizing occurs only at pressures > 30 bar, with no blistering of the intermediate layer below this pressure. A support structure together with the intermediate layer can already be applied as a microfiltration membrane for liquid separation, where the intermediate layer should be considered as a microfiltration layer. Ceramic processing can be controlled such that the pore radius of a microfiltration layer ranges from 0.05 to 0.25 μm. The technology is now being scaled-up for commercial production for applications in the process industry and for the solution of a number of environmental problems.

Although the most promising applications of ceramic membranes for gases lie in high temperature separation (> 300°C), where Knudsen diffusion is probably the only segregative transport mechanism, the performance of the membranes at room temperature is also of interest.

The actual γ-Al_2O_3 membrane top layer is applied on the microfiltration layer by coating with a boehmite (γ-AlOOH) sol. The top layer has a thickness of about 3 µm, a specific area of 225 m^2/g and a porosity of 55%; it has segregative properties for gases and can also be used as an ultrafiltration membrane.

Figure 1 shows the pore size distributions of the different surface layers in a complete ceramic membrane system and a micrograph is shown in Figure 2. From a design point of view, a careful choice of the permeability (defined as litre per bar, per square metre and per hour) of the microfiltration layer in combination with the permeability of the support is necessary.

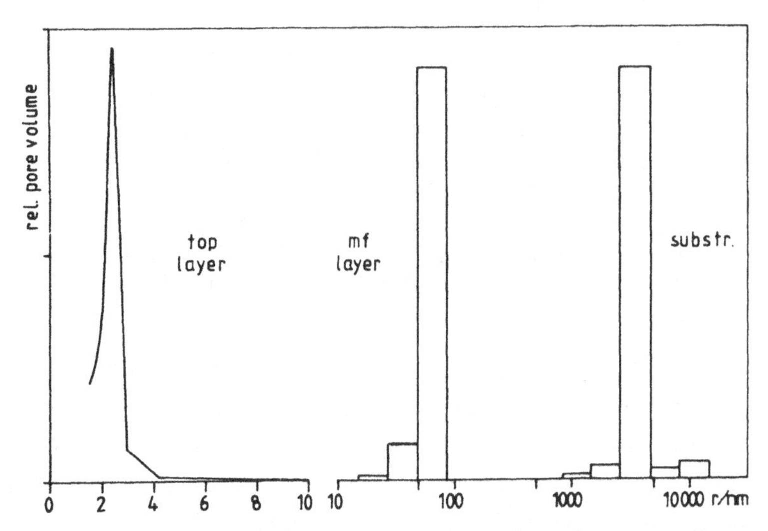

Figure 1. Pore size distribution of the three layer membrane system.

Figure 2. Microstructure of an all-ceramic three layer system.

Figure 3 shows the required permeability of the filtration layer as a function of the substrate and system permeability.

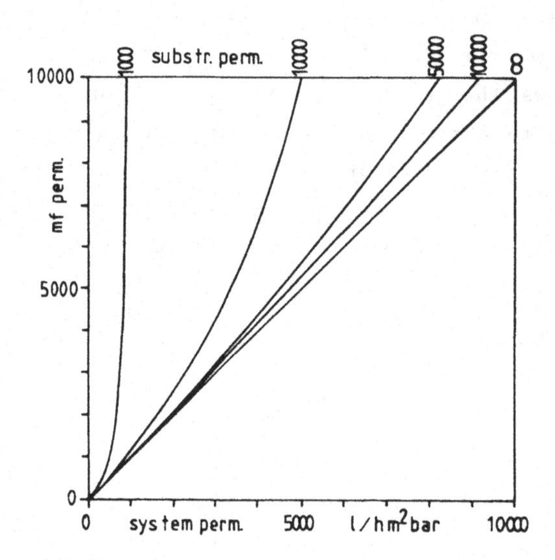

Figure 3. The permeability of the microfiltration (intermediate) layer as a function of the required system and substrate permeability.

Fabrication of membranes for liquid separation at ambient temperatures can still take advantage of metal or polymer sealing and supporting flanges. In such a case, the tubes, which have the filtration layer on the inside, should be finished at the bottom and top with a locally dense and smooth structure such that leak-tight sealing is possible. This solution has the advantage that malfunctioning filtration tubes are easily replaced. When temperatures > 150°C and corrosive atmospheres are present this solution is no longer possible and full ceramic flanges are required. This means that the tubes will have to be fitted to multihole ceramic flanges with a heat resistant ceramic bonding between the porous and dense structures; also careful control over thermal and mechanical stresses is necessary. Figure 4 shows a multitube all-ceramic membrane made by the National Ceramic Centre.

A temporary solution to the sealing problem is to seal to the metal parts at moderate temperatures, allowing an axial temperature gradient over the filtration tube to complete suppress the permeability in areas where the temperature is too low. Further development envisages ceramic sealing and separation functions combined at the process temperature to give gas separation membranes with very high selectivities for different gases.

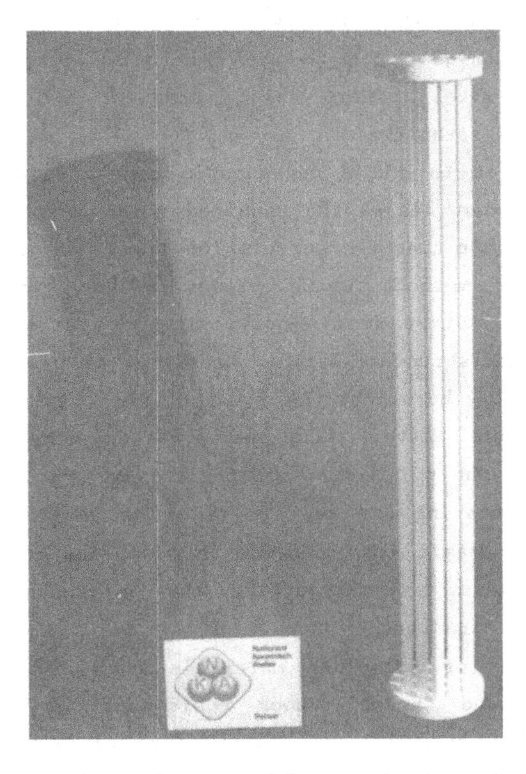

Figure 4. An all-ceramic membrane module consisting of 19 tubes sealed in alumina flanges. (Total length ∿ 70 cm.)

Replica Techniques

In several applications of porous ceramics where fluid flow through the material takes place, turbulent streaming along with a low pressure drop is desired. When used as hot gas filters or catalyst carriers, for example, ceramic bodies should have a negligible resistance to gas flow. Turbulent flow increases the encounter frequency between fluid particles and the inner surface of the material and therefore enhances the capture rate in filtration or the efficiency in catalysis. The question is which type of pore structure in the ceramics, and which pore sizes, will lead to the desired pressure flow behaviour and mutual interactions. Low pressure drops require large pores but turbulence will be stimulated also by an irregular pore structure. In such a case local flow rates of relatively large fluid volumes change rapidly and fluid inertia will be important, i.e. turbulent flow may occur. For these reasons, open structures with high tortuosity together with a very high porosity (> 80%) are of particular interest.

Porous structures with pore sizes considerably larger than the grain or agglomerate sizes of the starting powders are not easily produced by normal ceramic shaping techniques due to the imperative proportionality between pore and particle size. There exist, however, two main approaches to produce this kind of structure.

One procedure is to add a degradeable constituent to the dispersion which after burning-out leaves the required voids in the material. If the concentration is higher than the percolation limit, an open structure may be obtained. Another method, which allows better control of the pore structure (3), is to start from a polymeric material with a high porosity and to coat the inner structure with a ceramic layer by slurry infiltration. The polymer is finally burnt-out. To control the thickness and homogeneity of the coating, this latter technique requires careful control of the interaction between the polymer, the ceramic powder and the carrier fluid. Also drying and sintering are very critical and may easily give rise to large defects. A ceramic foam made by this replica technique consists of thin ceramic filaments, which outline the polyhedral cells originally present in the foam structure.

The application of foam-like materials for gas burners, where high thermal stability and thermal shock resistance is of prime interest, dictates the required material structure. Central in the design of such a structure is the gas or liquid flow through the material in relation to a given pressure drop. Clearly, the foam structure needed to produce a required pressure - flow behaviour is not readily defined in terms of pore dimensions. Nevertheless, some guidelines can be given which make the designing and testing of ceramic foams something more than trial and error. Applying Darcy's law for laminar viscous flow driven by a pressure drop ΔP across a foam of thickness L, we have:

$$\Delta P/L = \mu U/k_1 ,$$

where U is the (superficial) velocity of the fluid with viscosity μ. The Darcian permeability k_1 is an unknown function of the foam structure but is directly related to the actual foam structure. For porous media one generally finds that $\sqrt{k_1}$ roughly equals the average pore size times a geometry dependent constant, and this rule also holds for ceramic foams.

If Darcy's law applies for flow through a certain foam, the effects of fluid inertia are negligible and the flow is not supposed to be turbulent. If these effects are important, ΔP is a non-linear function of U, and the flow in relation with pressure is given by:

$$\Delta P/L = \mu U/k_1 + \rho U^2/k_2$$

where ρ is the (constant) fluid mass density, k_1 is the Darcian and k_2 the "non-Darcian" permeability. Using this equation, k_1 and k_2 can be determined from simple air flow measurements at room temperature. Since k_1 and k_2 are functions of the pore structure geometry only, pressure-flow relations can then be estimated for other fluids (water, hot gases, molten metals) provided μ and ρ are known at relevant temperatures. Further, flow rates at which turbulence may occur follow from the requirement that $U\rho k_1/\mu k_2 > 1$.

Knowledge of the permeabilities k_1 and k_2 is also useful to check the reproducibility of manufacturing procedures since they are sensitive to small, hardly visible, changes in pore structure. The quotient k_1/k_2 is of interest because it is roughly proportional to the average pore size in the foams and has a strong influence on the design aspects of systems where foam structures are incorporated.

For large-scale commercial application it is beneficial that the outer surfaces of a body with a foam structure can be sealed off with a dense ceramic structure. This will suppress liquid and gas flow in a lateral direction and it should give extra strength to the body. It is further demanded where easy replacement and rough handling should be possible without any risk of damaging the structure.

Figures 5 and 6 show the internal structure and the ceramic bodies made of foam by the replica technique. Future development work will be directed to exploring the possibilities of such materials for catalyst carriers, supporting structures for dust filtration, liquid metal filtration and many other applications.

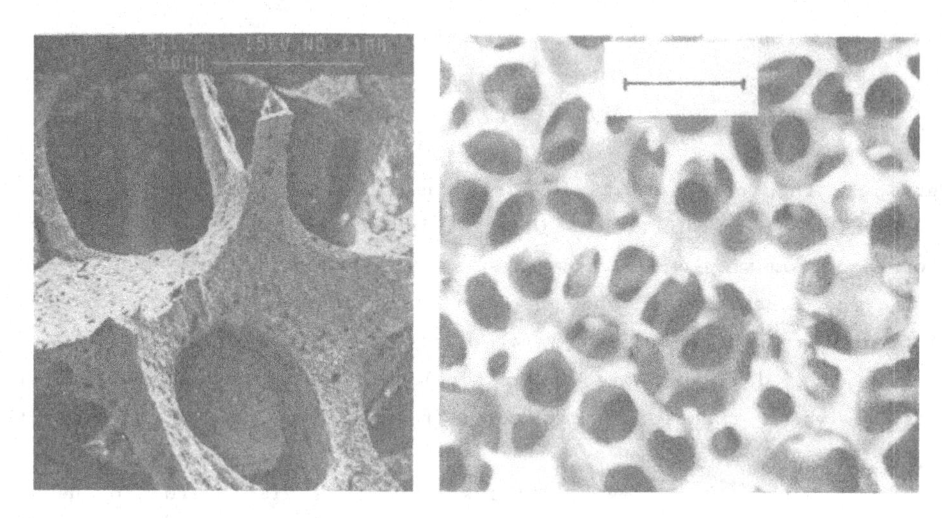

Figure 5. Structure of cordierite foams (bar is 1 mm).

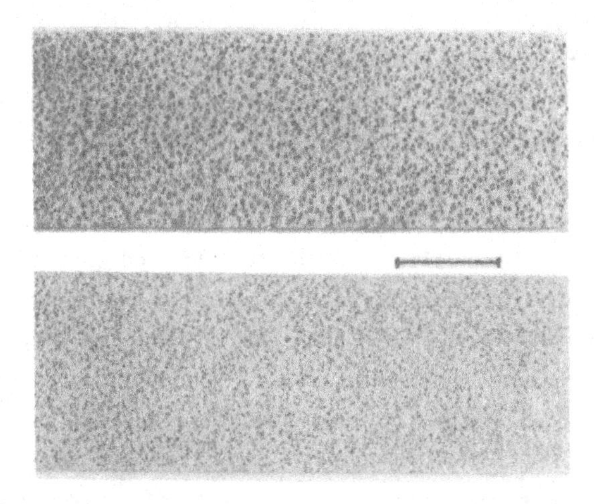

Figure 6. Alumina-silica foams which are being investigated at ECN for application as gas burner material (bar is 2.5 cm).

REFERENCES

1. Ashby, M.F., The mechanical properties of cellular solids. Metal Transaction, 1983, **14A**, 1755.

2. van Veen, H.M., Terpstra, R.A., Tol, R.P.B.M. and Verginga, H.J., Three layer alumina membranes for high temperature gas separation applications. Proc. 1st Intern. Conf. on Inorg. Membranes, Montpellier, 1989, 329.

3. Philipse, A.P., Moet, F.P., v. Tilborg, P.J. and Bootsma, D., A new application of milliporous ceramic shapes. Proc. Euro. Ceramics, 1989, **3**, Elsevier.

ALIGNED WHISKER REINFORCEMENT BY GELLED EXTRUSION

D.G. BRANDON and M. FARKASH
Department of Materials Engineering,
Technion - Israel Institute of Technology,
Haifa 32000, Israel.

ABSTRACT

Silicon carbide whisker reinforcement of a ceramic is attractive because of the excellent chemical stability of the whiskers at high temperatures. Hot-pressed, whisker-reinforced alumina is a commercial product which has been used for both wear components and cutting tool tips, while whisker-reinforced silicon nitride has been shown to have some advantages in specific wear applications. In these materials the whiskers have been partially aligned in a plane perpendicular to the direction of hot-pressing and the observed mechanical properties of the product have been markedly anisotropic.

Fibre reinforced ceramics are specifically tailored to have anisotropic properties either by fibre-winding or by the lay-up of laminate, braid or woven cloth. However, the available fibres have limited chemical stability, and fibres and fire-coatings for high temperature applications are being actively developed.

In the present project a whisker-containing ceramic slip containing a gelling agent is extruded through a fine needle into a gelling solution. The resultant thread of gelled material may be cold-compacted and hot-pressed to produce an isotropic body, but the thread also has sufficient strength to be handled as an intermediate product, for example, in fibre-winding and laminate production. The whiskers are aligned along the extrusion direction, the degree of alignment depending on the slip viscosity and the extrusion parameters. This route to the production of aligned ceramic reinforcement may have advantages for certain high temperature applications. The principles of the process are described and the resultant microstructures reported.

RAW MATERIALS

The powders and whiskers used in the present project are listed in Table 1.

TABLE 1
Raw materials

Material	Supplier	Dimensions	Impurities
α-Si_3N_4 H1	Hermann C. Starck	1.0 μm avg.	≈4% β, <1.5% O, <0.5%C
α-Al_2O_3 A16	Alcoa Chemicals	<1.0 μm avg.	<0.1% Na_2O, <0.1% SiO_2
Y_2O_3	Alpha Products		
β-SiCw SC9	ARCO Metals Co.	0.6 x 50 μm avg. <20% particles	<0.75% Ca Mn Al

SLIP PREPARATION

Matrix Powder

The Si_3N_4 powder was attrition milled with 3 wt.% Al_2O_3 and 9 wt.% Y_2O_3 in water for 4 h and then dried.

Whiskers

The whiskers were ball-milled in water for up to 4 h to reduce the average length and achieve a more uniform length distribution. The histograms in Figure 1 compare the as-received length distribution (above) with that achieved after milling (below).

Slip

The milled whiskers were ultrasonically dispersed in deionised water with small additions of PC-85 dispersant and the pH adjusted to 10 (5-10 min using a 600 watt ultrasonic probe). The milled powder was then added and the aqueous dispersion blended, using an ultrasonic probe and simultaneous magnetic stirring, for up to 1.5 h. The final slip composition is given in Table 2.

TABLE 2
The slip composition

Material	Powder	Whiskers	Dispersant	Na-Alginate	Water
Vol.%	20-21	5-7.5	≈1.5	0.2-0.25	70-75

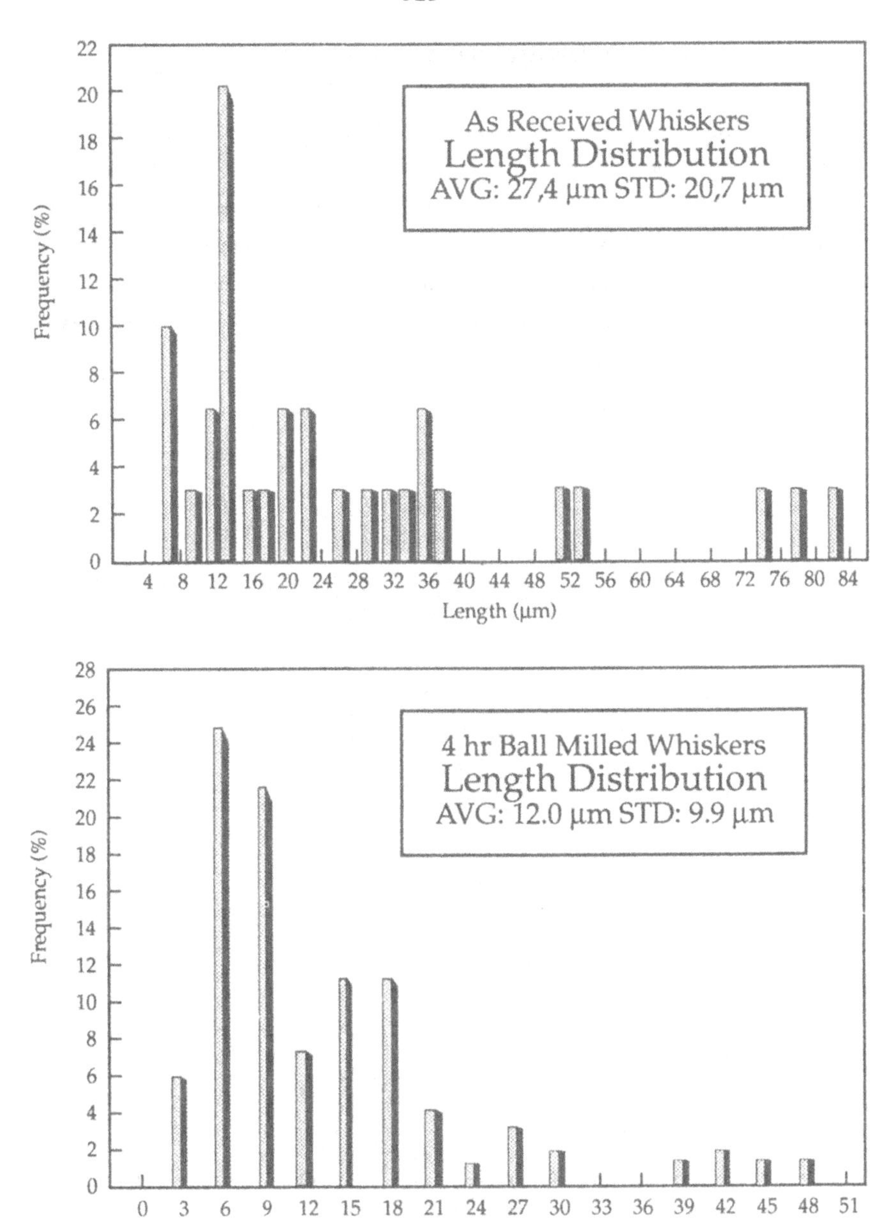

Figure 1. Histograms of fibre lengths before (above) and after (below) milling.

Viscosity

The viscosity of the slip was measured using a Brookfield viscometer and was found to range from 16,500 centipoise at 20 rpm to 110,000 at 2 rpm, indicating pseudoplastic behaviour.

EXTRUSION AND THREAD FORMATION

The blended Si_3N_4/SiC_w slip was extruded through a hypodermic syringe fitted with either a 0.5 mm ID, 80 mm long steel needle or a 0.5 mm ID. 100 mm long glass capillary. The cone entry angle into the extrusion needle was 30° in both cases. The extrusion pressure was applied via a commercial syringe pump (Orion Research Inc., USA) to achieve uniform and continuous extrusion of a thread of the blended slip (Figure 2). The extrusion rates varied from 60 to 200 mm min^{-1}.

Figure 2. Syringe pump and extrusion assembly showing thread extrusion into the gelling vessel.

Conversion of the viscous slip to a flexible thread was achieved by an ion exchange reaction, whereby the Na-alginate dissolved in the slip was converted to Ca-alginate on emerging from the polished tip of the extrusion needle into a saturated solution of $CaCl_2$. This resulted in immediate gel formation (Figure 2). The resultant gelled thread (75% water) was strong enough the withstand handling without any intermediate drying operation.

WHISKER ALIGNMENT

Assuming that the velocity profile in the cylindrical needle results in laminar flow, the whiskers in the thread are expected to align parallel to the flow direction. In order to monitor the extent of whisker alignment, random sections of thread were dried and infiltrated in vacuum with epoxy resin. The samples were subsequently mechanically polished and representative micrographs were recorded. The angle between the extrusion direction and the major axis of the sectioned whiskers θ was then recorded for some 300-600 whiskers in each sample using an image analyser (Cambridge Instruments Quantimet 970). These measurements were then used to evaluate an <u>orientation factor</u>, f_o, for the sample as follows:

$$f_o = 2\cos^2\theta - 1$$

$$\cos^2\theta = \int_0^\pi f(\theta).\cos^2\theta.\sin\theta.d\theta$$

where $f(\theta)$ is a distribution function and $\cos^2\theta$ can be calculated from the measured angular distribution by:

$$\cos^2\theta = \sum_{i=1}^{N} \cos\theta_i/N$$

For a uniform distribution $f_o = 0$, while for perfect alignment $f_o = 1$. The present results indicate moderate-to-good alignment, with values of f_o in the range 0.76 to 0.88 (Figures 3 and 4).

FILTER PRESSING

Filter pressing was used to convert the extruded thread to a high density green body of uniform microstructure before sintering. The wet thread was collected from the gelling vessel and loaded directly into the filter press. Excess liquid was then removed, initially by vacuum suction and finally by applying a static pressure to the press (Figure 5).

The green body (30 mm diameter and up to 10 mm thick) was dried and vacuum impregnated with epoxy resin for microstructural evaluation. The whiskers were found to be uniformly distributed both parallel and perpendicular to the pressing direction, but retained their preferred

Figure 3. Optical micrograph: Cross-section of dried thread. $f_o = 0.888$. Mag. x400.

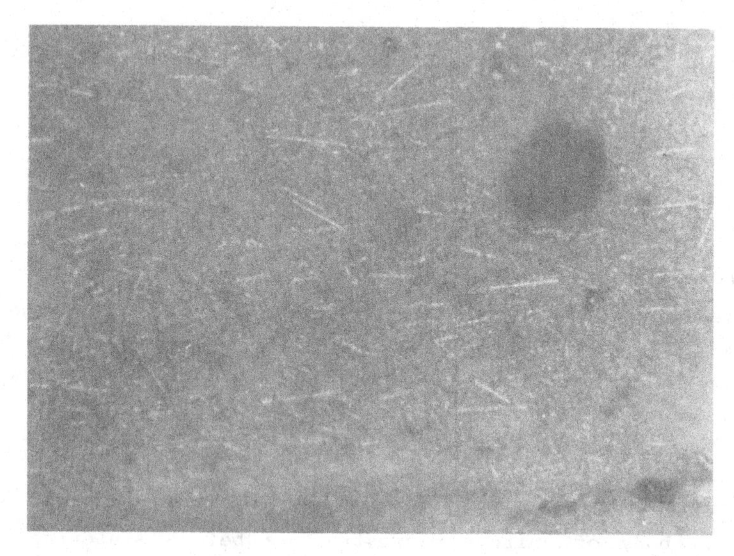

Figure 4. Optical micrograph: Cross-section of dried thread. $f_o = 0.76$. Mag. x 400.

alignment in the direction of extrusion of the original thread. This bimodal domain structure (Figure 6) is expected to impart unusual mechanical properties to the final sintered product, which is the ultimate objective of the current research programme.

Figure 5. Cold compaction by filter-pressing of thread collected from gelling vessel.

CONCLUSIONS

Extrusion followed by gel formation in solution is a very convenient route to the preparation of an extruded thread of a silicon carbide whisker reinforced green body which can be used as an intermediate product in the manufacture of a ceramic matrix composite.

Controlled flow of the slip at reasonably low viscosity is achieved by reducing the average whisker length and narrowing the length distribution, providing adequate care is taken in dispersing the solids in the aqueous media.

Thixotropic behaviour is a characteristic of the present slip rheology and gives rise to still unsolved problems or irregular flow during extrusion.

The extruded thread has sufficient green strength to be used as an intermediate product in the manufacture of green compacts or fibre-wound laminates.

Cold compaction of the randomly collected thread by filter-pressing leads to an essentially isotropic product with a bimodular microstructure which is expected to impact unusual mechanical properties after subsequent sintering.

Figure 6. Optical micrograph: Polished epoxy-filled green body. Mag. x200. Above - Perpendicular to the pressing direction. Below - Parallel to the pressing direction.

INTERMEDIATE TEMPERATURE BEND STRENGTH OF 5 WEIGHT % YTTRIA STABILIZED ZIRCONIA

G. FILACCHIONI*, E. CASAGRANDE*, U. De ANGELIS*, D. FERRARA*,
A. MORENO** and L. PILLONI*
*VEL-TECN-MATECN
**COMB-VALPRES
E.N.E.A. C.R.E., Casaccia CP 2400,
00100 Rome AD, Italy.

ABSTRACT

A ceramic fabrication, characterisation and mechanical property evaluation procedure has been established. Reliability has been demonstrated for partially stabilized zirconia: satisfactory strength and Young's modulus data have been obtained at ambient temperature and at 850°C.

INTRODUCTION

In the last years E.N.E.A. has been increasingly engaged in a R&D programme based on ceramic components for use in the energy field, produced from industrial or sol-gel powders.

A first material to be investigated has been chosen (yttria stabilized zirconium oxide) which will be used in the national programme as a structural ceramic for a wide range of applications: combustion engine components, rotor blades as well as M.H.D. retrofitting prototype electrodes ($T \leq 1200K$).

Obviously the thermo-mechanical characterization of ceramic monoliths will be obtained from different kinds of mechanical test depending on the final application (tensile, compression, bending, low and high cycle fatigue, creep, impact resistance and fracture mechanics).

E.N.E.A. has been the main promoter for the equipment of national laboratories (essentially the Casaccia and TEMAV Institutes) for new or updated experimental facilities for high temperature testing.

The preliminary feasibility step of the programme, is complete: a first batch of PSZ specimens was produced by TEMAV (ENI group) using an industrial powder, and tested at the Casaccia Centre Laboratories.

The current paper presents results of a bending test campaign carried out on the E.N.E.A. old facilities; the testing and evaluation methods are adequate for a characterization but the maximum allowable test temperature-limits utilization to "low service temperature" ceramics ($T \leq 900°C$).

MATERIAL AND EXPERIMENTAL DETAILS

Specimen Fabrication, Shape, Size and Morphology

Bend specimens have been fabricated by TEMAV Laboratories using a commercial zirconia (yttria stabilized) powder produced by the TOSOH Corporation (Grade TZ-3YB). The mean crystallite size was 251 Å and the specific surface area 13.3 m^2/g: the chemical composition is reported in Table 1. Chemical analysis has been performed by means of atomic absorption.

TABLE 1
PSZ (5% yttria) chemical composition (wt.%)

TiO_2	CeO_2	NaO_2	SiO_2	Y_2O_3	Al_2O_3	Fe_2O_3	C	S	ZrO_2
0.02	0.20	0.04	0.02	5.0	0.006	0.007	1.758	0.0117	Balance

After cold pressing (0.68 $tonne/cm^2$) a "green" density of 2.671 g/cm^3 was measured (44.14% TD). Sintering was performed in air following the TOSOH standard heating schedule (see Figure 1). Final mean bulk density was 6.04 g/cm^3 corresponding to 99.86% of theoretical density (6.05 g/cm^3). Open porosity was nil and closed porosity $\sim$ 0.5%.

Final forming was carried out by grinding and polishing to give bend specimens 48 mm length, 3 mm thickness and 4 mm width (final roughness better than 0.4 μm). No flaws or superficial defects have been detected by optical microscopy.

Hardness ranges between 1350 and 1420 kg/mm^2 (HV_{10}). The structure of samples is rather homogeneous with grain size 0.3-1 μm. Figure 2 shows the typical TEM morphology.

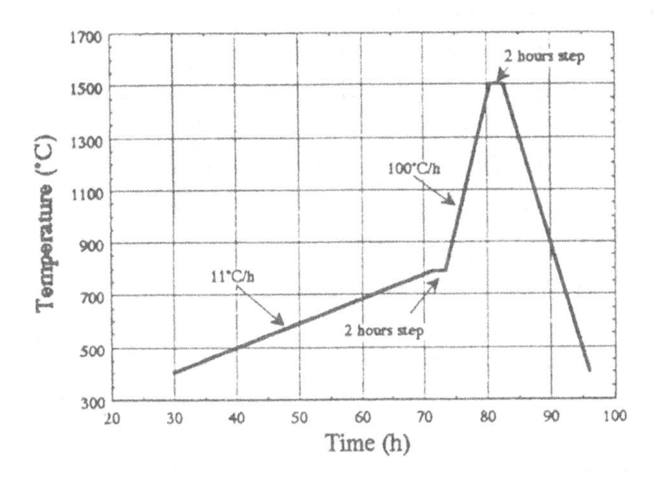

Figure 1. Standard heating schedule for PSZ grade TZ-3YB.

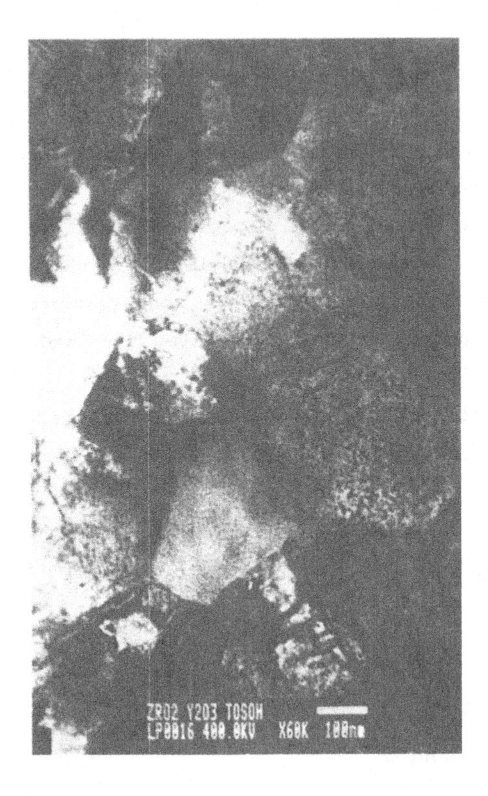

Figure 2. Typical TEM morphology of PSZ bend specimens.

Testing Device and Procedures

A three point bending device (40 mm outer span) was mounted on an electro-mechanical closed loop MAYES ESM 100 machine which was set at constant cross-head speed of 0.533 mm per min, corresponding to $1.10^{-4}\ s^{-1}$. All tests have been carried out in air. A three independent heating zone furnace was utilized for high temperature (850°C) tests. In both cases (RT and elevated temperature) we used an extensometric device which allows measurement of specimen deflection, thence the Young's modulus. Tests have been performed in accordance with MIL-STD-1942 (MR) standard (closely similar to AFNOR B41-104).

EXPERIMENTAL RESULTS

Modulus of Rupture

The MOR data are reported in Figures 3 and 4. Values of flexural strength have been calculated following the classical formula for a three point loaded beam:

$$MOR = \frac{3\ F_{max}\ L}{2\ b\ t^2}$$

where F_{max} is the breaking load, L is the support span, b is the specimen width and t is the specimen thickness.

Rupture data show a mean value of 838.5 MPa and a standard deviation of ± 97.8 (RT). At 850°C we observe a mean of 377.2 MPa ± 64.6.

A two parameter Weibull analysis has been performed and Table 2 gives the values of the population scale parameter (α) and the population shape parameter (β - the Weibull modulus).

The maximum likelihood method was used to estimate the values of β and α. The cumulative distribution function, given by the formula

$$F(x) = 1 - \exp\left[-\left(\frac{x}{\alpha}\right)^{\beta}\right]$$

and the density function

$$f(x) = \frac{\beta}{\alpha}\left(\frac{x}{\alpha}\right)^{\beta-1} \exp\left[-\left(\frac{x}{\alpha}\right)^{\beta}\right]$$

are shown in Figures 5 and 6.

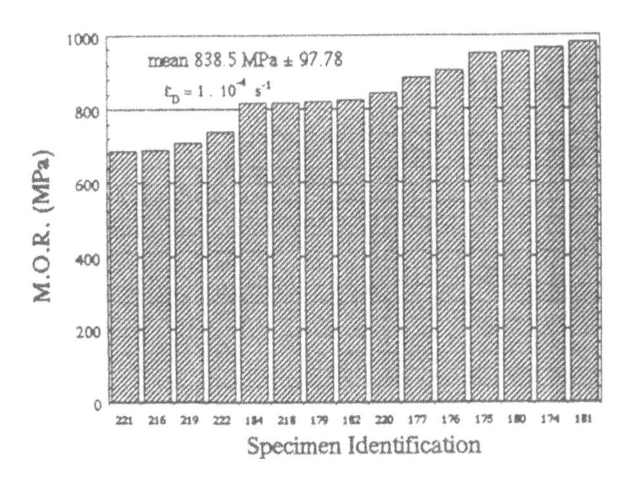

Figure 3. PSZ flexural strength at 23°C.

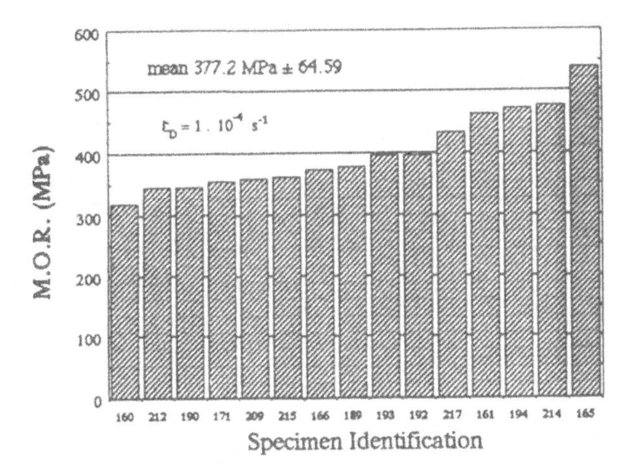

Figure 4. PSZ flexural strength at 850°C.

TABLE 2
Two parameter Weibull analysis of PSZ MOR

T (°C)	α	β	Mean of Population	Standard Deviation	R^2
23	881.9	10.02	922.9	69.76	0.991
850	426.9	6.64	458.2	52.8	0.989

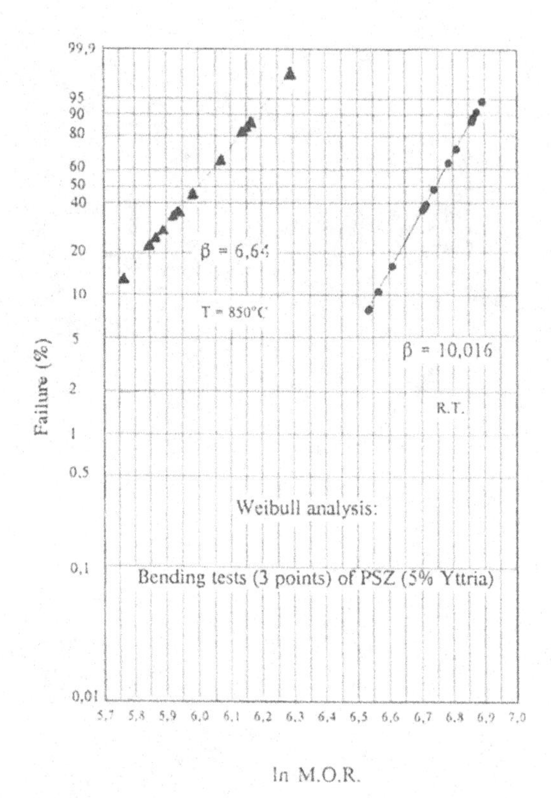

Figure 5. Cumulative distribution function (Weibull analysis) for PSZ.

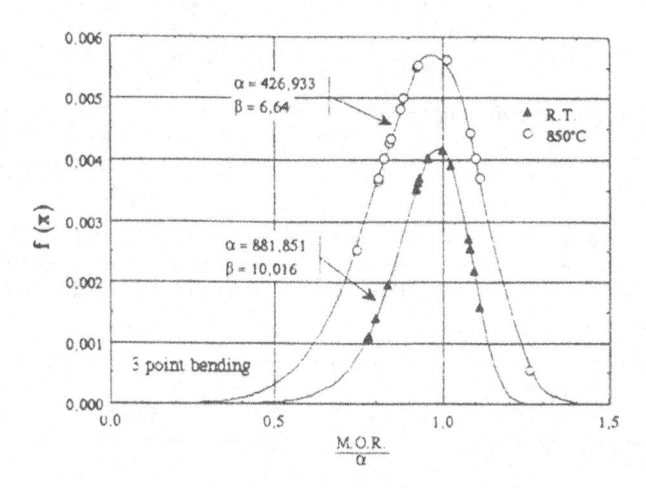

Figure 6. Density function of Weibull analysis for PSZ.

Young's Modulus

An analogic-digitalized recording of load and deflection allows computation of Young's modulus by means of the formula

$$E = \frac{F\,L^3}{f\;4\;b\;t^3}$$

where F is the applied load and f is the specimen deflection at loading axis. The values obtained agree very well with values in the technical literature for dense PSZ: at room temperature a Young's modulus of 206.0 GPa (± 9.6) was found and at 850°C, 175.2 GPa (± 8.3), see Figure 7.

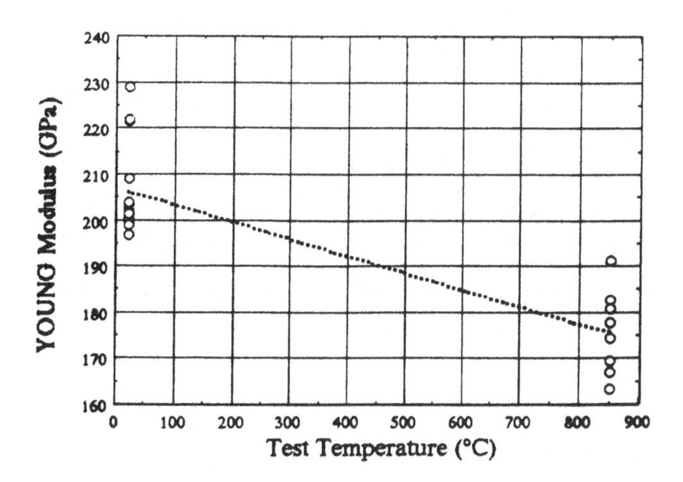

Figure 7. Young's Modulus of PSZ.

CONCLUSION

The results of this first integrated campaign (fabrication and evaluation) indicate that the current procedures are substantially reliable. In fact

(i) the experimental instrument is accurate

(ii) the data logging and handling system are able to provide reliable MOR and Young's modulus values

(iii) the fabrication route is well established for a commercial powder and data are consistent with those in the technical literature.

A key limiting aspect is the low (about 900°C) maximum allowable test temperature. This serious shortcoming will be overcome by new equipment being acquired ($T \leq 1400$°C).

CERAMIC COMPOSITE MATERIALS

A Contribution to an Overall Technology Assessment

F. CASTEELS, J. SLEURS and R. LECOCQ
SCK/CEN,
Boeretang 200, B-2400 Mol,
Belgium.

ABSTRACT

Technology Assessments have been performed to study the influence of various aspects on the introduction of new material development technology. Ceramic composite materials are chosen as example. The methodology and results are described. An R&D summary for the development of these materials is included.

INTRODUCTION

Within the framework of SCK/CEN activities in the field of new technologies development, a study has commenced on the influence of technological, economical, social, health, safety and environmental aspects influencing the introduction of new material development technologies. For ceramic composite materials Technology Assessment (TA) analyses are being carried out. The analysis of different alternatives of producing ceramic composite materials has been chosen. The TA study has been formulated on the basis of existing knowledge; certain basic assumptions are analysed with regard to prior conditions and potential consequences, and evaluated against a wide range of criteria. Thus, the results of this TA study do not consist in forecasts but in the clarification or even the quantification of the positive as well as negative consequences potentially connected with the options selected. This type of analysis aims at supporting decision-making processes in technology, economics, politics and business.

METHODOLOGY

For each of the above mentioned aspects, the following analyses have been or will be carried out:

- critical analyses of data in the open literature, data banks and selected patent literature;
- completion of data after consultation of experts;
- identification of missing data.

Field and telephone interviews are conducted with the aim of obtaining complementary information:

- consultation of key executives of leading corporations, trade organisations, technology development organisations, financial institutions;
- attention will be paid to technological, economical, safety and health aspects of introducing ceramic composite materials.

The data are analysed using different mathematical tools such as correlation calculations, multiple regression analyses, cross tabulation. Multicriterion decision methods are used to implement statistical data in relation to the probability of economical, technological, commercial and socio-economic success (1-4).

RESULTS

The study will result in the identification of (5):

- Names and addresses of companies and research centres active in the field of ceramic composite materials.
- Industrial and academic research and development programmes.
- Manufacturers of raw materials: powders, particulates, fibres and whiskers: properties of continuous SiC fibres; commercial alumina fibres; characteristics of commercially available SiC whiskers.
- A critical analysis of the major fabrication routes and properties of the products.
- Identification of opportunities (market oriented) for powder suppliers, equipment manufacturers, manufacturers of end products and manufacturers of fibres, whiskers and other starting materials (7,8).
- Identification and future of competitive technologies, e.g. classical ceramics, metal matrix composite materials.

- Health and safety aspects (6).
- Special requirements for manipulating ceramic composite materials in the workshop (6).
- Development of monitoring techniques enabling manufacture of ceramic composite materials in the workshop (10).

DETAILED ANALYSES OF MARKET POTENTIAL AND MAJOR FABRICATION ROUTES OF CERAMIC COMPOSITE MATERIALS

Market Potential (8)

Rapid developments are underway for a limited number of high temperature composite materials for structural applications. The most important market in ceramic composite materials is related to whisker reinforced cutting tools and silicon carbide matrix base components for application in missiles, rockets and modern defence systems.

However, a significant growth in the market is anticipated in heavily loaded wear parts and different aero-engines. The total market volume is presently about 13 MECU and will increase up to about 100 MECU before the end of this century. Carbon-carbon composites have been accepted as standard materials for brakes in military and large civil aircraft. The growth in the demand is proceeding rapidly in a number of market segments. The total market will be above 800 MECU before the end of the 1990s.

Major Fabrication Routes (9)

Fibres, whiskers and platelets are used to overcome limiting properties of technical ceramics, namely their limited resistance against thermal shock and their low ductility. Ceramic composites can be manufactured by the following technologies:

- Impregnation
  - CVI (chemical vapour infiltration). At this moment, components with limited dimensions have been produced. This technology needs further development. C-C and SiC-SiC based components have been produced. The price is between 300 and 5000 ECU/kg, and is mainly applied as brake material.
  - Infiltration by a molten metal.

- Classical shaping technologies in combination with a sintering step
  Al_2O_3-SiC and Si_3N_4-SiC ceramic composite materials are made with a homogeneous structure using different technologies: slurry processes for continuous fibre CMCs, sol-gel and polymer precursor processes, sintering, hot pressing and other routes.

- Reaction-bonded silicon nitride (RBSN)
  The technology for manufacturing reaction-bonded silicon nitride components can be used for different ceramic composite materials by combination of slip-cast silicon-silicon carbide fibres combined with nitridation and hipping. Technical applications have not yet been identified.

- Reinforcement by particles
  - Hexolog: SiC reinforced by TiB_2 particles.
  - Rockwell International: silicon nitride reinforced by ZrO_2 particles.
  - Different developments are in progress in various research and development laboratories.

- Lanxide process

CONCLUSIONS

The results of this study provide the following information:

- The potential market segments for different types of ceramic matrix composite materials.
- Ceramic matrix composite materials are applied in cutting tools, brakes and different parts of aeroplanes.
- A number of shaping technologies are only available on a laboratory scale.
- Costs of shaping technologies have to be reduced to enable application in other market segments, e.g. wear parts.
- Industrial development of technology is hampered due to health and safety risks.
- Development of monitoring techniques for the environment in workshops is necessary.

From these analyses a description of the R&D activity has been formulated (see Annex 1) (3).

ANNEX I - DEVELOPMENT OF CERAMIC MATRIX MATERIALS

This is a generic programme aimed at the development and characterisation of ceramic-ceramic composite materials.

Objectives

- To develop high performance components by increasing the knowledge of fundamental, technological and service aspects determining operation of components.
- To develop quality assurance procedures based on non-destructive testing and performance prediction techniques from property data.
- To generate data banks of relevant properties of new materials.
- To study the mechanisms of materials degradation in service conditions.

The programme will contribute to three aspects which are of importance for this type of material: i.e. development of materials, characterisation and use of components.

<u>Development of materials</u>:

- synthesis of starting materials necessary for the preparation of conventional, zirconia toughened, fibre-reinforced and carbon-carbon composites;
- development of manufacturing processes (hot pressing, cold pressing and sintering, reaction bonding, slip casting, plasma spraying and "in situ growth";
- continued studies on fabrication of composites with minimum cost and reliable properties;
- development of NDT quality assurance procedures;
- optimisation of product: improvement of properties mentioned below by introducing new fabrication schemes and adaptation of chemical compositions of composite material.

Characterisation

The behaviour of ceramic-ceramic composites has to be studied in the temperature range and environment of interest. Particular attention has to be paid to:

- creep and fatigue studies at high temperature;
- toughness at high and low temperature;
- synergetic effects on the properties of fibre-reinforced composites due to the effect of prolonged exposures to high temperatures in corrosive and inert environments;
- thermal stability including optimization of the interface bonding between components of the composite;
- increase of structural stability against thermal cycling delamination.

Response to requirements for application in components

- optimization of mechanical, thermal, tribological properties in view of construction of components with tailored properties;
- development of an adequate design methodology;
- development of fabrication techniques for large scale components (unidirectional solidification, in situ growth);
- development of NDT technology applicable to large size components;
- development of joining and "in service" repairing techniques for large size components.

REFERENCES

1. Roubens, M., Preference Relations on Actions and Criteria in Multi-criteria decision making, European Operational Research, 1982, **10**, 51-55.

2. Dujmovic, J.J., Extended Continuous Logic and the Theory of Complex Criteria, Publikaije Electrotechnikog Fakultita, Beograd, Serija Matematika i Fizika, 1975, 489-541.

3. Nihoul, J., Casteels, F. and Van Asbroeck, Ph., Materials for Energy Systems, An approach to a long term EC Programme, 1984. Confidential.

4. Roy, B., Classement et choix en présence de points de vue multiples, RIRO, 1968, **2**, 57-75.

5. Sleurs, J., et al., Personal communication, Technological aspects of injection moulding techniques, (Document to be prepared, 1990).

6. Lecocq, R., Effetts cancérigènes des fibres minérales, Intermediate Internal Report CER/90/N/106.

7. Gorham International Inc., Injection Moulded Ceramic, Cermet and Cemented Carbide Powders: Market Forecasts and Applications, Technology Assessment and Business Opportunities to 1996, (Volumes I and IV), 1989.

8. Briggs, J., Advanced Ceramic Matrix, Metal Matrix and Carbon-Carbon Composites, The Current and Potential Markets Materials Technology Publications, 1990.

9. Lecocq, R., Fabrication et propriétés des composites à matrices céramiques: un aperçu général, Intermediate Internal Report RL/mvds/89/249.

10. The European Ceramic Fibres Industry Association Guidance Note, Sampling and Gravimetric Determination of Airborne Total Dust in Ceramic Fibre Workplaces: A Method Recommended by ECFIA, 1989.

INDEX OF CONTRIBUTORS

Printed in the United States
By Bookmasters